Dinosaur Tracks

Dinosaur Tracks

Tony Thulborn

Reader, Department of Zoology,
University of Queensland, Australia

CHAPMAN AND HALL

LONDON • NEW YORK • TOKYO • MELBOURNE • MADRAS

UK	Chapman and Hall, 11 New Fetter Lane, London EC4P 4EE
USA	Chapman and Hall, 29 West 35th Street, New York NY10001
Japan	Chapman and Hall Japan, Thomson Publishing Japan, Hirakawacho Nemoto Building, 7F, 1-7-11 Hirakawa-cho, Chiyoda-ku, Tokyo 102
Australia	Chapman and Hall Australia, Thomas Nelson Australia, 480 La Trobe Street, PO Box 4725, Melbourne 3000
India	Chapman and Hall India, R. Sheshadri, 32 Second Main Road, CIT East, Madras 600 035

First edition 1990

© 1990 Tony Thulborn

Typeset in 11/12½pt Goudy Old Style by Mayhew Typesetting, Bristol
Printed in Great Britain by St Edmundsbury Press Ltd
Bury St Edmunds, Suffolk

ISBN 0 412 32890 9

British Library Cataloguing in Publication Data
Thulborn, Tony
    Dinosaur tracks.
    1. Dinosaur.   Extinction.   Theories
    I. Title
    567.91
        ISBN 0-412-32890-9

Library of Congress Cataloging-in-Publication Data
available

Contents

Contents

List of Plates

Acknowledgements

It would have been impossible for me to write this book without the generous assistance of colleagues around the world. Among the many who cheerfully responded to my requests for information I wish to thank in particular Don Baird, Stanley Colliver, Walter Coombs, Justin B. Delair, Paul Ensom, Jim Farlow, Doris Fredholm, Gene Gaffney, Hartmut Haubold, Natascha Heintz, Karl Hirsch, Giuseppe Leonardi, Ralph Molnar, Christian Montenat, Mike Raath, Bill Sarjeant, Frank Stacey, William Lee Stokes, Masahiro Tanimoto, Maggie Taylor, Cyril Walker, Derek Waterfield, Ken Woodhams and Jack T. Woods. Jiro Kikkawa helped by translating Japanese and Korean literature, and I owe a special debt of gratitude to Mary Wade, who shared many hours of toil on the dinosaur tracks at Lark Quarry. Terry Dyer and Peter Gofton offered the benefit of technical advice and practical assistance on several occasions. Julie Blackwood, Tom Bliesner, Robyn Bright, Natalie Gorman, Amy Jansen, Jonathan Marshall and Kerry Watson provided tracings for Figure 4.15. Some of my fieldwork and basic research was supported by funding from the Australian Research Grants Scheme and by a Special Studies Programme from the University of Queensland. The resources of the Dorothy Hill Library (Geology Department, University of Queensland) were of inestimable value; not least among those invaluable resources were the library's staff, Margaret Eva and Judith Henderson. Michael Benton and an anonymous reviewer offered some heartening comments on various drafts of the manuscript, while Tim Hardwick, Bob Carling, Peter Martin and Jeremy Macdonald provided expert help with the technicalities of publication. I must also thank John Lazarus, who (if he recalls!) suggested that I should write this book.

Most illustrations of dinosaurs in Chapter 9 are the work of Laurie Beirne, who undertook a great deal of research before setting pen to paper. The other illustrations are my own or are adopted from sources acknowledged either in the captions or in the following list. Finally, my wife Susan deserves more than customary thanks for her patience and unfailing encouragement.

I express sincere thanks to the following individuals and organizations for their permission to reproduce copyright material:

American Museum of Natural History, New York. Plate 2 (centre), *Psittacosaurus* ['*Protiguanodon*'] skeleton (neg. no. 311488; photo Julius Kirschner); Plate 4 (top left), dinosaur track used as a bath (neg. no. 319835; photo R.T. Bird); Plate 10 (bottom left), excavation at Paluxy River (neg. no. 125128; photo R.T. Bird). All photographs courtesy Department of Library Services, American Museum of Natural History.

Asociación Paleontológica Argentina, Buenos Aires. Plate 11 (centre), *Delatorrichnus* trackway; reproduced from R.M. Casamiquela 1966.

Mr Laurie Beirne. Figures 9.1, 9.7, 9.10, 9.11, 9.13 and 9.15.

Cambridge University Press, Cambridge. Sections of text in Chapters 8 and 9, and accompanying illustrations (Figures 9.2, 9.3, 9.8 and 9.9; reproduced from R.A. Thulborn 1989).

Centre National de la Recherche Scientifique, Paris. Plate 12 (bottom right), sauropod track in Niger; photo © CNRS, reproduced from P. Taquet 1972.

Professor Dong Zhiming. Plate 7 (bottom right), theropod track from the Jurassic of Sichuan.

Dover Publications Inc., New York. Figure 3.2; reproduced from 1972 reprint of G. Heilmann 1927.

Elsevier Science Publishers B.V., Amsterdam. Plate 6 (centre and bottom centre), 'squelch marks' and *Anchisauripus*-type footprint with coarse pebbly filling; both reproduced from M.E. Tucker and T.P. Burchette 1977.

Dr James O. Farlow. Plate 4 (centre), dinosaur track mounted in bandstand at Glen Rose; Plate 11 (top right), manus-pes couple of *Caririchnium*; Plate 12 (top left, centre left and bottom left), manus and pes of *Brontopodus birdi*, and sauropod tracks at the Davenport Ranch site.

Gustav Fischer Verlag, Stuttgart. Plate 2 (top right and bottom right), bite-marked vertebrae and Lower Cretaceous coprolite; both reproduced from O. Abel 1935.

Geological Survey of Canada, Ottawa. Plate 9 (centre left), *Gypsichnites* track; Plate 10 (top left), *Columbosauripus* track; Plate 11 (top left), manus-pes couple of *Tetrapodosaurus*; all reproduced from C.M. Sternberg 1932.

Mr D.P. Grant. Plate 4 (bottom left), dinosaur tracks in Balgowan colliery, Queensland.

Dr Natascha Heintz. Plate 4 (bottom left), casting *Iguanodon* tracks at Spitsbergen; Plate 9 (bottom right), *Iguanodon* track at Spitsbergen;

Plate 10 (centre left), *Megalosaurus* track at Spitsbergen (photo by Leif Koch).

Institute of Vertebrate Paleontology and Paleoanthropology, Academia Sinica, Beijing. Plate 2 (bottom left), nest with dinosaur eggs; reproduced from C-C. Young 1965.

Professor Dr M. Kaever. Plate 13 (top left), Barkhausen tracks; reproduced from M. Kaever and A.F. de Lapparent 1974.

Dr Giuseppe Leonardi. Plate 11 (bottom left), *Sousaichnium* trackway.

Dr Christian Montenat. Plate 7 (top centre and top right), cast and mould of *Grallator variabilis* track; Plate 8 (top left), *Talmontopus* track; Plate 10 (top right), slab with tracks of *Grallator variabilis*; all reproduced from A.F. de Lapparent and C. Montenat 1967.

Museum of Northern Arizona Press, Flagstaff. Plate 13 (top right), *Navahopus* trackway; reproduced from D. Baird 1980.

National Geographic Society, Washington DC Plate 4 (bottom right), dinosaur tracks on sale near Glen Rose; reproduced from R.T. Bird 1954.

The National Museums and Monuments of Zimbabwe, Bulawayo. Plate 2 (top left), gastroliths from a prosauropod; reproduced from M.A. Raath 1974.

Palaeovertebrata, Montpellier. Plate 3 (bottom), Lesotho dinosaur tracks by moonlight; Plate 6 (left), *Moyenisauropus* trackway with tail-drag; Plate 11 (bottom right), manus-pes couple of *Moyenisauropus*; all reproduced from P. Ellenberger 1974.

Queensland Department of Mines, Brisbane. Plate 4 (bottom left), dinosaur tracks in Balgowan colliery, Queensland; reproduced from Anonymous 1952.

Queensland Museum, Brisbane. Plate 6 (top centre and top right), two photographs of *Skartopus* track under different lighting conditions; Plate 7 (top left, centre right and bottom left), three photographs of *Wintonopus* tracks, showing drag-marks and slide-marks; Plate 9 (top right and centre right), two photographs of *Wintonopus* tracks; Plate 14 (bottom), photomosaic of Lark Quarry site; all reproduced from R.A. Thulborn and M. Wade 1984.

Professor M.A. Raath. Plate 2 (top left), gastroliths from a prosauropod.

E. Schweizerbart'sche Verlagsbuchhandlung, Stuttgart. Plate 3 (centre), *Iguanodon* tracks on display in Munich Museum; reproduced from E. Stechow 1909.

Smithsonian Institution, Washington, DC Plate 1 (centre left and bottom right), portraits of Edward Hitchcock and James Deane. Reprinted by permission of the Smithsonian Institution Press, from 'Contributions to a history of American state geological and natural history surveys', *United States National Museum Bulletin* No. 109, by George Perkins Merrill, 1920.

Société Géologique de France, Paris. Plate 7 (top centre and top right), cast and mould of *Grallator variabilis* track; Plate 8 (top left), *Talmontopus* track; Plate 10 (top right), slab with tracks of *Grallator variabilis*; all reproduced from A.F. de Lapparent and C. Montenat 1967. Plate 13 (top left), Barkhausen tracks; reproduced from M. Kaever and A.F. de Lapparent 1974.

Dr Philippe Taquet. Plate 12 (bottom right), sauropod trackway in Niger.

Texas Christian University Press, Fort Worth. Plate 12 (top right and bottom left), R.T. Bird measuring a sauropod track, and sauropod tracks at the Davenport Ranch site; both reproduced from R.T. Bird 1985.

Dr Maurice E. Tucker. Plate 6 (centre and bottom centre), 'squelch marks' and *Anchisauripus*-type footprint with coarse pebbly filling; both reproduced from M.E. Tucker and T.P. Burchette 1977.

United States Geological Survey, Denver. Plate 3 (top), dinosaur tracks in Chile; reproduced from R.J. Dingman and C.O. Galli 1965.

Preface

Ten years ago a book about dinosaur tracks would have appealed to a handful of specialists. Today, it is likely to attract wider interest, in the wake of some major controversies about the natural history of dinosaurs. Few readers will be completely unaware of those spirited debates about 'cold-blooded' versus 'warm-blooded' dinosaurs, and eyebrows are no longer raised at the suggestion that dinosaurs may still be alive and kicking – in the guise of birds. Issues such as these have prompted many biologists and palaeontologists to take a serious second look at the everyday lives and habits of dinosaurs, and in doing so they have begun to turn their attention to the long-neglected study of fossil tracks – the direct testimony of dinosaur behaviour.

This resurgence of interest in dinosaur tracks might legitimately be described as a renaissance, and its extent may be gauged from the success of the First International Symposium on Dinosaur Tracks and Traces, held in May 1986 at the New Mexico Museum of Natural History, Albuquerque. The proceedings of that symposium, which attracted no fewer than 60 contributions from researchers in 14 countries, were published recently by Cambridge University Press under the title *Dinosaur Tracks and Traces*, edited by D.D. Gillette and M.G. Lockley. In the space of a decade the study of dinosaur tracks has escalated from a minor and neglected pursuit into a formidable scientific enterprise, with exponents around the globe. The past few years have witnessed scores of new discoveries, many of them providing remarkable glimpses into the behaviour of dinosaurs; and perplexing slabs of rock, marked with footprints that once attracted the scrutiny of eminent naturalists, are nowadays being resurrected from museum basements, dusted off and studied afresh. In these circumstances it seems appropriate to take stock of existing knowledge and to provide a general review of the fossil tracks and other traces left by dinosaurs.

Originally, this book was intended to be an extensive survey of all dinosaur trace fossils, including stomach stones (gastroliths), eggs, nests, droppings (coprolites) and feeding traces such as bite-marks. Unfortunately, the limitations of time and space have precluded all but the briefest mention of anything aside from dinosaur footprints. Yet, even with this narrowed focus, I fear that my text will barely scratch the surface of a surprisingly big and labyrinthine subject. Despite its inadequacies this book is, so far as I am aware, the first comprehensive review of dinosaur tracks to be published in the English language. I am certain that it is very far indeed from perfection, but I am confident that it is better than nothing.

The first chapter is a brief survey of dinosaur fossils in general, and serves merely as scaffolding for the chapters that follow. Those subsequent chapters deal almost exclusively with fossil footprints, and they represent my best attempt to find a path through a morass of scientific literature and a jungle of terminology. Wherever possible I have tried to explain my subject in plain English, using the simplest and most appropriate terms, and I have made a determined effort to avoid the tangles of nomenclature that surround the names of many fossil footprints: in general, I have adopted the names that are most widely used, on the assumption that readers would find these easiest to locate in the literature. It is assumed that all readers will be familiar with the principal types of dinosaurs, and that they will have some grasp of basic concepts in biology and geology. Even so, I have deliberately opened my discussion of fossil footprints at an elementary level. This low-key approach seems justified by the fact that fossil footprints are rarely considered in modern textbooks of palaeontology.

The factual basis for this book resides, of course, in the published works of countless other scientists. Unfortunately, that literature is vast and scattered, as will be explained in Chapter 4. There are a few major treatises, mostly in languages other than English, but nonetheless it has been necessary for me to ferret out information from numerous original sources. This fact alone accounts for the extensive bibliography. I have tried to make the bibliography as useful as possible by citing all titles and periodicals in full. Earlier works on dinosaur tracks rarely provided such details, so that it is sometimes difficult or impossible for today's researchers to locate the original literature. In some instances, I was unable to discover any published reference to what seemed to be a noteworthy fact; in such cases I have ventured to express my own opinions, though in suitably guarded fashion. There are still some enormous gaps in scientific understanding of dinosaur tracks, even concerning simple matters of preservation and morphological variation, and I have not hesitated to point out some of those gaps that seem to deserve immediate attention.

My overriding concern is that this book should be of practical value. It does not merely review the findings and techniques of other scientists, but also tries to explain *how* to study dinosaur tracks – what to look for and how to evaluate it. Consequently, I have devoted considerable space to documenting specific examples, common errors and misinterpretations. Chapter 7, for example, is given over entirely to the subject of problematical and anomalous tracks. Some palaeontologists have remarked, in all seriousness, that the greatest difficulty in dealing with trace fossils is simply recognizing that they are trace fossils in the first place, and this is certainly true in the case of dinosaur tracks. In the past 3 years I have investigated nine reports of dinosaur tracks in Queensland, mostly from experienced naturalists or professional geologists; four transpired to be erosion features or inorganic sedimentary structures, two were aboriginal carvings of kangaroo tracks, and only three were genuine dinosaur tracks. Such experience convinces me of the need to point out all those potential pitfalls that await the unwary.

Tony Thulborn
Brisbane, March 1989

Note. All illustrations of foot skeletons and footprints are of the *right* side, unless specified otherwise.

1

Dinosaur fossils

'Hold!' cried Media, 'yonder is a curious rock. It looks black as a
whale's hump in blue water, when the sun shines.'
'That must be the Isle of Fossils,' said Mohi. 'Ay, my lord, it is.'
'Let us land, then,' said Babbalanja.

Herman Melville, *Mardi* (1849)

The fossil record contains two sources of information about dinosaurs.
First there are **body fossils** which, as their name implies, are remnants
of dinosaur carcasses. Usually these are the hardest and most durable
parts of the body – namely, the teeth, bones and skeletons that figure
so prominently as museum exhibits. Second, and less familiar, are
trace fossils, which may be defined as traces of the life activities of
dinosaurs. These include footprints and trackways, along with more
unusual items such as teeth-marks, nests, eggs and droppings.

The term 'trace fossil' was introduced into the English language in
1956 by Professor Scott Simpson of the University of Exeter, who
translated the terminology established by K. Krejci-Graf in Germany
in 1932 (Simpson 1956, 1957). Among the many other terms used to
signify trace fossils in general, and fossil tracks in particular, are
'ichnites', 'ichnofossils', 'bioglyphs' and *Lebensspuren* (German, 'life
traces').

Conditions that suited the preservation of trace fossils were not
always favourable for the preservation of body fossils, and vice versa.
These two sorts of fossils tend to be discovered separately, and, as a
result, the scientific study of trace fossils, or **palaeoichnology**, evolved
as a discipline that was only loosely connected to the rest of palaeontology. The name palaeoichnology, which is derived from the Greek
palaios (ancient) and *ichnos* (footprint or track), is sometimes shortened
to palichnology or simply ichnology – though, strictly speaking, the
latter encompasses the traces of all animals, living and extinct. Today,
palaeoichnology has its own special methods and terminologies, its
own unusual systems of classification and its own body of literature.
These facts might seem to imply that palaeoichnology is an esoteric
and narrowly specialized field of science. Yet, in reality, it is no more

complicated or mysterious than any of the related branches of biology and geology: it just happens to be less widely understood.

Body fossils and trace fossils provide complementary insights into the biology of dinosaurs: the body fossils represent hard factual evidence about dinosaur anatomy whereas the trace fossils are explicit clues to the behaviour and life habits of the animals. The corollary is equally obvious: to gain the most complete understanding of dinosaurs it is essential to study their body fossils *and* their trace fossils. Nevertheless, many books about dinosaurs pay remarkably little attention to trace fossils, which are often relegated to footnotes or mentioned as curiosities (if they are mentioned at all). While this book is intended to redress the balance, by drawing attention to the tracks and traces of dinosaurs, these fossils should not, and indeed cannot, be treated as isolated objects. To grasp their full meaning it will be necessary to draw on information from a variety of sources – from anatomy, biomechanics, animal behaviour, sedimentology, geophysics, and so on. Consequently, it should become apparent that palaeo-ichnology is very far indeed from a narrow specialization.

BODY FOSSILS

These remnants of dinosaur carcasses include not only bones and teeth, but also a variety of softer tissues, such as cartilage and keratin, along with impressions of the skin.

The bones comprise those of the internal skeleton together with various spikes, plates and studs of bone that were embedded in the skin. Also to be included are the bony horn-cores on the skulls of ceratopsians, and the gastralia (or so-called ventral ribs) that reinforced the belly region in theropods and prosauropods. The petrified bones and teeth sometimes reveal fascinating details of histological structure (e.g. de Ricqlès 1976, 1980; Reid 1984, 1987), and they may even retain organic components such as amino acids, protein residues, collagen fibrils and osteocytes (e.g. Pawlicki *et al.* 1966; Wyckoff and Davidson 1979; Armstrong *et al.* 1983).

Many ornithischians had a regular lattice or trelliswork of ossified tendons alongside the neural spines of the vertebrae. This complex system of tendons probably functioned as a form of scaffolding, helping to maintain the arch of the backbone over the pillar-like hind-limbs (Dollo 1887; Lull and Wright 1942; Galton 1970). In some agile-looking theropods a dense sheath of these tendons served to stiffen the tail, thereby enhancing its function as an organ of balance (e.g. Ostrom 1969a,b; Hamley in press). An even more elaborate system of ossified tendons reinforced the backbone of some ankylosaurs, perhaps endowing these animals with exceptional endurance (Molnar and Frey 1987). Calcified ligaments have been reported in a few dinosaurs, though some are more likely to be displaced fragments of ossified tendons (Moodie 1929a; Norman 1980).

Dinosaurs resembled living reptiles in having life-long replacement of their teeth (Edmund 1960). As each tooth came to the end of its functional life it dropped out from the jaw while a new tooth erupted to replace it. Tooth loss was not so dramatic as it might sound, because in many cases the root was resorbed and only the abraded tooth crown was shed from the jaw. Not surprisingly, discarded teeth are among the commonest of dinosaur body fossils.

Dinosaur tissues softer than bone, tooth enamel or dentine were rarely preserved as fossils. In a freshly excavated dinosaur skeleton the individual bones are sometimes separated by narrow spaces that were once occupied by cartilage and other soft tissues. Evidently, the durable bones resisted decay and survived the processes of fossilization whereas the intervening soft tissues did not. At the ends of well-preserved dinosaur bones there is frequently a porous zone that marks the gradual transition of calcified cartilage into uncalcified cartilage (Reid 1984).

In a few instances the keratin claws and beaks of certain dinosaurs left imprints in the sediment, though the keratin itself is hardly ever preserved as a petrifaction. Rare exceptions include remnants of claws in the small theropod dinosaur *Compsognathus* (Ostrom 1978) and vestiges of a horny beak in some of the hadrosaurs (Cope 1883; Sternberg 1935; W.J. Morris 1970). Remnants of a keratin horn-sheath are known to have survived in at least one example of a ceratopsian dinosaur (Hatcher *et al.* 1907: 145).

Some ancient fossil fishes were buried in sediments so fine-grained and impermeable that there persisted traces of muscle fibres, kidney tubules and blood corpuscles. And well-preserved tadpoles of Palaeozoic amphibians may even show evidence of feathery gills, internal glands and eye pigments. Among the many other examples of fossil vertebrates that retain clear indications of the soft tissues are frogs, lizards, crocodiles, ichthyosaurs, pterosaurs, birds and mammals (Whittington and Conway Morris 1985). Unfortunately, no dinosaurs seem to have been preserved in such magnificent detail. Faint grooves in the sediment enclosing a skeleton of the coelurosaur *Compsognathus* were once thought to be imprints of muscle fibres (Nopcsa 1903) but seem more likely to be erosion features (Ostrom 1978). And fossil droppings attributed to a carnosaur have been found to contain un-digested scraps of muscle and other soft tissues derived from the animal's prey – presumably, another dinosaur (Bertrand 1903). A few patches of petrified skin have been reported in sauropods (Brown 1941: 293).

As dinosaur carcasses rotted and decayed their body cavities were sometimes filled with sediment. Those natural fillings, called endo-casts, internal moulds or *steinkerns* (German, 'stone kernels'), are also classified as body fossils. In addition, it is possible to obtain artificial endocasts by filling the vacant body cavities of fossils with materials such as plaster, latex, resin or plastic. Some of the most intriguing endocasts are fillings of the braincase, or endocranial casts. These display the general configuration of the dinosaur brain, the arrange-ment of the cranial nerves, and the delicate tubes and sacs forming the membranous labyrinth of the inner ear (Jerison 1969, 1973; Hopson 1977, 1979). Fillings of the expanded neural canal in the sacral region (endosacral casts) represent the so-called 'second brain' of some dinosaurs.

Occasionally, dinosaur carcasses were buried before the skin had rotted away, and in those circumstances the enclosing sediments might retain faithful imprints of the skin (e.g. Brown 1916, 1917). Such body imprints were most likely to be formed where the skin was reinforced with studs and plates of bone, or where the carcass had been dehydrated or baked hard in the sun, as in the case of the famous 'mummified' hadrosaurs (Osborn 1912; Lull and Wright 1942). Skin imprints that originated from living dinosaurs, rather than corpses, should technically be classified as trace fossils.

TRACE FOSSILS

These indications of the life activities of dinosaurs include footprints, stomach stones, droppings, nests, eggs, and feeding traces such as bite-marks (Figure 1.1 and Plate 2).

As might be expected, the fossil **tracks** made by dinosaurs reveal that these animals normally went about their everyday business in an unhurried fashion. Tracks of running dinosaurs are less common, as

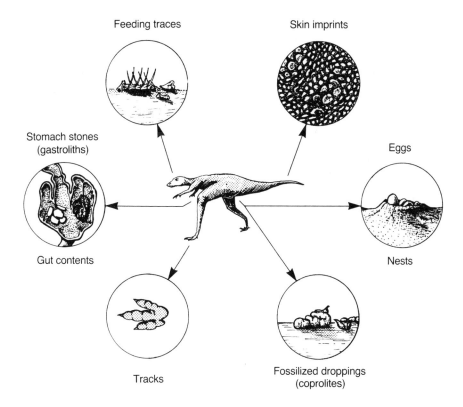

Figure 1.1 Principal types of dinosaur trace fossils.

are the traces produced by dinosaurs wading or swimming through shallow waters. Other fossil traces, made by dinosaurs squatting or lying on the ground, show impressions from the undersurface of the tail and the belly. Footprints and trackways are among the most abundant and widespread of dinosaur trace fossils. They have been the subject of intensive scientific studies for well over 150 years and will be examined in detail in the following chapters.

Feeding traces include markings made by dinosaurs rooting through the mud in search of food, and teeth-marks left by predaceous dinosaurs on the bones of their victims (Plate 2, p. 48, top right; see also W.L. Beasley 1907; Brown 1908; Jensen 1988). Gut contents comprise fossilized remnants of food preserved in the mouth, gullet, stomach or intestinal region of various dinosaurs, ranging from chicken-sized predators (Figure 1.2) to gigantic plant-eating brontosaurs (Stokes 1964). Such occurrences sometimes provide unexpected

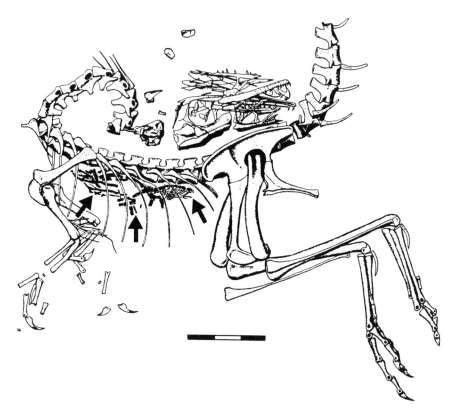

Figure 1.2 Fossil gut contents inside the skeleton of a small predaceous dinosaur, *Compsognathus*, from the Upper Jurassic of Germany; scale bar indicates 3 cm. Undigested bones of this dinosaur's last victim, the small lizard *Bavarisaurus*, are visible inside the ribcage (arrowed). Several ribs are omitted for the sake of clarity, along with much of the tail (at top right). The dinosaur's neck is thrown back, with the disarticulated skull lying upside-down above the hips. (Adapted from Ostrom 1978.)

glimpses into the behaviour of dinosaurs. For instance, the skeleton of one small theropod, *Coelophysis*, was found to contain the remains of an even smaller *Coelophysis* – mute testimony of cannibalism (Colbert 1983: 170–1). Reports of dinosaur feeding traces and gut contents are scattered through the scientific literature, but have never been compiled into an exhaustive review. However, the hunting and feeding techniques of predatory dinosaurs were investigated by J.O. Farlow (1976) and G.S. Paul (1987, 1988), while the diets of various plant-eating dinosaurs have been pondered by many researchers (e.g.

Weaver 1983; Weishampel 1984a; Farlow 1985b; Wing and Tiffney 1987).

Stomach stones or **gastroliths** are similar to those that occur in crocodiles. They have been found in a variety of dinosaurs, all of which seem to have been herbivores or omnivores, but their functional significance is still a matter of debate. These pebbles might have been swallowed to relieve the pangs of hunger, to assist in crushing and grinding the food, or to serve as ballast when an animal ventured into water. This last idea is consistent with the fact that gastroliths are quite common in water-dwellers, including ancient amphibians (Warren and Hutchinson 1987), crocodiles (Cott 1961) and plesiosaurs (Darby and Ojakangas 1980). By contrast, most dinosaurs seem to have been terrestrial animals, with little need for such ballast, and it is more likely that they swallowed pebbles to assist in grinding up their food, in much the way that some birds swallow particles of grit. Gastroliths may be regarded as authentic if they are discovered inside, or at least near, the skeleton of a dinosaur (see Plate 2, p. 48, centre). The polished pebbles that are sometimes offered in souvenir shops as dinosaur gastroliths have been described more accurately as 'gastro-myths' (Jepsen 1964). The subject of dinosaur gastroliths has recently been reviewed by J.O. Farlow (1985a: 209-11).

Fossilized droppings, or **coprolites**, are relatively common but are not easy to interpret (Plate 2, p. 48, bottom right). In the first place, it is difficult to distinguish the coprolites of dinosaurs from those of other animals; and, in the second place, it is difficult to determine which sorts of dinosaurs produced which sorts of coprolites. Despite these problems a fair number of coprolites have been attributed to dinosaurs with reasonable confidence (e.g. Majer 1923; Matley 1941; C.R. Hill 1976). Least questionable are those examples that were associated with dinosaur skeletons (e.g. Bertrand 1903; Jaekel 1914; Larsonneur and de Lapparent 1966) or dinosaur footprints (e.g. E. Hitchcock 1844a; P. Ellenberger 1974). Coprolites often contain undigested remnants of food, thus furnishing some clues to the dietary preferences of dinosaurs.

It is also possible that some dinosaurs regurgitated the indigestible residues of their food, in the way that owls produce their pellets, rather than expelling them in the faeces. However, such pellets would also have the potential to survive as fossils and might, indeed, be mistaken for coprolites. Occasionally, the surface layer of a coprolite has been identified as urolite - urinary waste which invests or is mixed with the faeces, as in living birds and reptiles (e.g. Duvernoy 1844). This suggestion is supported by chemical analysis of coprolites, which sometimes reveals the presence of uric acid (e.g. Dana and Dana 1845). The literature on coprolites (including those of animals other than dinosaurs) was thoroughly reviewed by G.C. Amstutz in 1958 and by W. Häntzschel and his colleagues in 1968.

Dinosaur **eggs** are not particularly rare, though they are abundant only in certain late Cretaceous deposits and in definite geographic areas – namely, southern France, the Spanish pre-Pyrenees, Mongolia, China, India and central North America (Plate 2, p. 48, bottom left). The rarity of dinosaur eggs in Triassic and Jurassic sediments has prompted speculation that early dinosaurs might have laid soft-shelled eggs, which rarely survived as fossils, or that some dinosaurs might have given birth to live young. By comparison, the profusion of hard-shelled eggs in the late Cretaceous has been regarded as a dinosaurian response to counter dehydration of the eggs in a period of increasing aridity (Sochava 1969; Erben *et al.* 1979). Dinosaur egg shells are structurally similar to those of birds (Erben 1970) and they sometimes retain organic components, including proteins, cuticle and shell membrane (e.g. Voss-Foucart 1968; Kolesnikov and Sochava 1972). Progressive thinning of the egg shells has been implicated in the extinction of dinosaurs at the end of the Cretaceous (Erben 1972; Erben *et al.* 1979), though not without some criticism (Dughi and Sirugue 1976; Penner 1985).

Dinosaur eggs varied considerably in their shape and size. Some were ellipsoidal or ovoid, but many were spherical, conical or almost cylindrical. The biggest eggs, which were presumably laid by bronto-saurs, were spheroidal in shape and attained a maximum length of slightly more than 30 cm. Some dinosaurs are known to have deposited their eggs in crater-like excavations or in mounds (e.g. Granger 1936; Young 1965; Kerourio 1981; Horner 1984b) whereas others seem to have left their eggs on the surface, sometimes in clusters or strung out in lines (Dughi and Sirugue 1958, 1976). In a few instances the eggs have been found to contain embryos (e.g. Sochava 1972; Kitching 1979; Horner and Weishampel 1988) or have been discovered in **nests** that also contained hatchlings (e.g. Bonaparte and Vince 1979; Horner and Makela 1979; Mohabey 1987). At some sites dinosaurs nested in groups or colonies, where the parent dinosaurs may have tended their young until they were big enough to fend for themselves (Horner 1982). Such favoured nesting grounds may have been visited by dinosaurs on a regular basis for periods extending over many years (Horner 1984b; Breton *et al.* 1985; Srivastava *et al.* 1986).

The literature on dinosaur eggs is large but scattered. Most earlier discoveries were listed by G. Borgomanero and G. Leonardi (1981), while the numerous finds in China were reviewed by S. Zhen and his colleagues in 1985. In addition there have recently been important discoveries of dinosaur eggs in India (e.g. Mohabey 1982; Sahni *et al.* 1984; Jain and Sahni 1985; Vianey-Liaud *et al.* 1987). There are also reports of dinosaur eggs from Korea (Yang 1986) and Portugal (e.g. van Erve and Mohr 1988), though these await detailed study. The structure of dinosaurian egg shell has been investigated in great detail by H.K. Erben (1970; see also Sochava 1970, 1971), and its physiological implications have been examined by R.S. Seymour (1979) and by D.L.G. Williams and his colleagues (1984).

Various fossil burrows are known to have been excavated by creatures as diverse as lungfishes (Dubiel *et al.* 1987), beavers (Martin and Bennett 1977) and mammal-like reptiles (Smith 1987). No burrows have yet been attributed to dinosaurs, though some were certainly capable of digging (Coombs 1978b).

THE DISTINCTION BETWEEN TRACE FOSSILS AND BODY FOSSILS

Trace fossils were produced by living animals whereas most body fossils originated from corpses. The only common exceptions to this general rule are the worn-out teeth that were discarded by living dinosaurs. Despite this difference there are both practical and theoretical difficulties in distinguishing trace fossils from body fossils.

For example, it may be impossible to decide whether a skin imprint was made by a drifting carcass or by a living dinosaur; in the first case it would qualify as a body fossil, in the second as a trace fossil. Fossil eggs present another difficulty because some palaeontologists consider them to be body fossils (e.g. Frey 1973, 1978) while others regard them as trace fossils (e.g. Vialov 1972). There is also some uncertainty about the status of coprolites (Simpson 1975: 40), though most palaeontologists would probably classify them as trace fossils.

Then there are some curious ambiguities. Consider the coelurosaur *Compsognathus*, shown in Figure 1.2: within its ribcage are the skeletal remains of its last victim, a small lizard identified as *Bavarisaurus*. Here, then, is an example of a trace fossil, in the form of gut contents, inside a body fossil. However, the lizard skeleton is also a body fossil in its own right – so that it is simultaneously a body fossil and a trace fossil. Indeed, it might be argued that every body fossil is also a trace fossil, because it is an indication of the reproductive activities of its parents and ancestors.

In short, the term 'trace fossil' is a subjective label. Trace fossils are sometimes defined in a rather narrow sense, as sedimentary structures generated by the life activities of ancient organisms – that is, as **biogenic sedimentary structures**. Definitions of this nature, though much more precisely and objectively formulated, are given by A. Seilacher (1953) and R.W. Frey (1973), and they would certainly encompass traces such as dinosaur tracks. Other descriptions are so much broader and looser that they would embrace all fossils that are not obviously parts of organisms (e.g. Ager 1963). There are even definitions that would accommodate biological processes such as excretion and egg-laying (e.g. Vialov 1972). And, from the completely different viewpoint of molecular biology, T.J.M. Schopf (1981: 169) has argued that all fossils might be dubbed trace fossils because they express so little of the genetic code in ancient organisms.

In view of these conflicting opinions it is necessary to make an arbitrary decision. For present purposes, dinosaur trace fossils will be considered to comprise the following:

1. footprints, trackways and body imprints made by living dinosaurs;
2. feeding traces and gut contents of dinosaurs;
3. dinosaur gastroliths;
4. waste products of dinosaurs, including coprolites and urolites;
5. nests and eggs of dinosaurs.

All these fossils resulted from the life activities of dinosaurs and none of them could be labelled unequivocally as a body fossil. This list excludes indications of injury and disease. R.L. Moodie managed to compile a gruesome catalogue of dinosaurian ailments, including bacterial infections (1928), dental abscesses (1930a), fractures (1926), arthritis and tumors (1921, 1923), but few of these pathological features would seem to qualify as trace fossils. More debatable examples include ceratopsian dinosaurs with head wounds that were probably sustained during combat (e.g. Sternberg 1927, 1940), pathological egg shells (e.g. Erben 1970), and sauropods that may have regenerated their damaged tails (Brown 1941; Tanimoto 1988). Abnormal dinosaur tracks, left by animals with deformed or injured feet, certainly do qualify as trace fossils and will be discussed at an appropriate point.

TRACE FOSSILS AND DINOSAUR BIOLOGY

Phylogeny and systematics of dinosaurs

Trace fossils can contribute very little to the study of dinosaur phylogeny. They are not, by definition, parts of dinosaur bodies, and so they are unlikely to furnish any new facts about dinosaur anatomy. Nevertheless, trace fossils do have subsidiary roles in the study of dinosaur phylogeny.

In the first place they sometimes provide supplementary information about dinosaur anatomy. So, for instance, fossil footprints have confirmed that the theropod dinosaurs had a grasping type of foot structure that was inherited by birds (Thulborn and Hamley 1982).

Secondly, trace fossils may betray the existence of dinosaurs that are unknown from body fossils. Those otherwise unknown animals may then be used in reconstructing hypothetical evolutionary lineages (e.g. Haubold 1969) and in pinpointing major evolutionary events such as faunal replacements (e.g. F. Ellenberger *et al.* 1969; Haubold 1986).

Thirdly, and most importantly, trace fossils will allow the transformation of a phylogenetic tree into a phylogenetic scenario – defined by N. Eldredge (1979: 192) as 'a phylogenetic tree with an overlay of adaptational narrative'. Trace fossils reveal the natural history of ancient organisms, thus providing a good deal of that adaptational narrative. By superimposing the evidence of natural history on a straightforward evolutionary tree it may be possible to identify inconsistencies in the tree, such as apparent reversals in habits or habitats of the organisms. Any serious inconsistencies might warrant the abandonment of an unsatisfactory tree and the search for a better one. In this sense the natural history of the organisms, which is ascertained primarily from trace fossils, can be used to test a phylogenetic tree, which is based primarily on the evidence of body fossils.

Trace fossils can make no contribution to the task of defining and classifying dinosaurs. Definitions and classifications are better based on the hard evidence of anatomy, in the form of body fossils, rather than on the more subjective behavioural evidence of trace fossils.

Natural history of dinosaurs

Trace fossils constitute the material basis for **palaeoethology**, or the study of behaviour in extinct animals (sometimes described as the study of 'fossil behaviour'). Fossil footprints, for example, are vital clues to the life habits of dinosaurs, often providing information that would be unobtainable from the study of bones and teeth. While such clues are not always easy to decipher, they do not, in general, have the same limitations as dinosaur body fossils.

First of all trace fossils are relatively common. A dinosaur had only one skeleton to leave as a potentially valuable body fossil, but during its lifetime it could roam far and wide, through a range of environments, leaving behind it thousands of traces – footprints, droppings, teeth-marks, nests and eggs.

Secondly, traces such as footprints could not have been transported from one environment into another: they indicate exactly which sorts of dinosaurs inhabited which environments. This is important evidence when one considers that dinosaur skeletons were often preserved out of context – in environments quite different from those inhabited by living dinosaurs.

Next, trace fossils are richly informative about dinosaurian life habits, and they can sometimes be deciphered in a straightforward fashion. The size and shape of a footprint often allow accurate identification of a track-maker, in just the way that a hunter will identify an unseen mammal from its spoor. And in some circumstances it is possible to decide if a dinosaur was behaving normally, in dinosaurian terms, or whether it was engaged in some more unusual pursuit, such as hunting or swimming. Footprints also provide clues to the general appearance, body dimensions and weight of dinosaurs, while measurements of trackways can be used to determine their gaits and speeds. Average and maximum speeds have been calculated for each of the major types of dinosaurs and may also be predicted for any dinosaur that is represented by a skeleton or a trackway. It has also proved possible to quantify the locomotor abilities of dinosaurs, and then to measure their abilities against those of living animals. Findings such as these may provide some small clues to that most vexed question of dinosaur physiology – warm-blooded or cold-blooded?

In summary, the study of trace fossils allows us to construct a remarkably detailed picture of dinosaurs as living animals. It provides glimpses of dinosaurs going about their everyday business, sleeping and eating, visiting the local water-hole to drink or to forage. There is evidence of dinosaur nesting grounds, of careful nest-building, and of parent dinosaurs tending their youngsters. There are traces of plant-eating dinosaurs moving in herds through their feeding grounds, of dinosaurs browsing through thickets of tree-ferns. And there is evidence of predation, of solitary hunters stalking their prey, and of opportunists and scavengers roaming in packs.

Much of this evidence may be gleaned from a careful reading of the fossil tracks left by dinosaurs. The aims of the following chapters are to describe those tracks, to explain their scientific value and their limitations, and to show how they may be used in reconstructing the lives and habits of dinosaurs.

2

The preservation of dinosaur tracks

The tides rise very high, and when they are lowest, large areas
recently overspread with red mud are laid dry, and are often
baked in the sun for many days, so that the mud becomes
consolidated and retains permanently the impressions of rain-drops,
and the tracks of birds and animals which walk over it.

C. Lyell, 'Notes on some Recent foot-prints on
red mud in Nova Scotia' (1849)

SIMPLIFIED MODEL

The sequence of events leading to the preservation of dinosaur tracks
is explained diagrammatically in Figure 2.1. First of all a dinosaur
traversed an area of soft sediment, leaving its footprints as it did so
(diagrams a and b). It might, for example, have walked across the
mudflats of an ancient estuary, leaving its tracks in the wet sediment
exposed by the falling tide. The next rising tide might deposit more
sand and mud over the newly formed footprints (diagram c), and once
they were buried in this way they would be largely protected from the
destructive effects of sun, wind and water. Continued accumulation of
sediments would result in deeper burial of the footprints, and the
consequent changes in pressure, temperature and water chemistry
would bring about the complex process of lithification, or the
transformation of soft wet sediment into harder and drier rock. The
layers of sediment would be compressed and reduced in thickness; the
water would be squeezed out from between the grains of sand and
mud, which would be packed more tightly and, in many instances,
cemented by mineral deposits. Ultimately, the lithified sediments
would be raised by earth-movements and exposed by erosion. The
layers of rock containing the footprints might then be exposed in
hillsides, river beds, cliffs, quarries or roadside cuttings. Finally, those

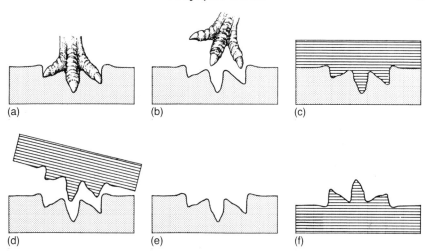

Figure 2.1 Simplified model to explain the formation and preservation of dinosaur tracks: (a) a dinosaur's foot impressed in the substrate; (b) the foot is withdrawn, leaving its imprint; (c) the footprint is filled and buried by accumulating sediment; (d) after lithification, the sediments are split open to reveal the original footprint and its filling; (e) the original imprint or natural mould; (f) the footprint filling or natural cast, shown inverted.

layers might be split open, by natural weathering or by an inquisitive fossil-hunter, to reveal the ancient footprints (diagram d).

Notice that each footprint will be represented by two fossils – an imprint, or **mould**, in the upper surface of the substrate (diagram e), and its infilling, or **cast** (diagram f). The terms 'cast' and 'mould' are used throughout this book because they are simple, easy to remember and widely used. Equivalent terms such as **concave epirelief** (= mould) and **convex hyporelief** (= cast) abound in the literature dealing with invertebrate traces but are decidedly cumbersome and are not so commonly applied to the tracks of vertebrates. The terms 'positive footprint' and 'negative footprint' are ambiguous and best avoided. Most people have no trouble in identifying moulds as foot-prints, but they may experience genuine difficulty in recognizing that casts are also fossil footprints. This conceptual difficulty arises because casts are usually studied by turning them upside down (Figure 2.1f). Consequently, the footprints appear in the form of raised reliefs, rather than cavities, and the left and right directions are reversed. The easiest way to overcome this conceptual problem is to remember that casts are the *fillings* of the original footprints (see Plate 5, p. 68, bottom).

FACTORS AFFECTING PRESERVATION

The process of footprint preservation is a long and sometimes complicated story that has been greatly simplified in the foregoing account.
In reality, the successful preservation of dinosaur footprints would
have depended on many and variable factors including the geography
of the environment, the physical properties of the substrate and the
behaviour of the track-maker.

Depositional environment

Footprints were most commonly preserved in environments that
experienced periodic or cyclic accumulation of sediments. Their formation and preservation would have entailed the following sequence of
events:

1. influx and deposition of sediment;
2. halt or slackening-off in the deposition of sediment;
3. newly laid sediments are trodden by dinosaurs;
4. footprint moulds are consolidated;
5. influx and deposition of sediment fills the footprint moulds, buries
   them and restarts the cycle.

Note that the sediments did not accumulate at a constant rate. There
would have been periods when little or no sediment was being
deposited (stage 2) and when the recently laid sediments could have
been traversed by terrestrial animals including dinosaurs (stage 3).
Sediments that were trodden in this way must, obviously, have been
exposed to the open air or covered by shallow water. So, naturally
enough, dinosaur tracks are often associated with organic and inorganic features that indicate shallow-water conditions or subaerial
exposure. These features include ripple-marks, sun-cracks, pits made
by raindrops, evaporites (or their pseudomorphs), plant roots, insect
trails, algal mats and markings made by wind-blown vegetation (e.g.
Moodie 1929b; Prentice 1962; Tucker and Burchette 1977; Courel *et
al.* 1979). In addition, the rocks containing fossil footprints sometimes
have a distinctive red colour, which may have resulted from the
oxidation of iron minerals while the wet sediments were exposed to
the air.

Suitable cycles of sediment accumulation occur in relatively few
environments. First, there are beaches, shore-lines, estuaries and tidal

lagoons, where freshly laid sediments may be exposed to the air on falling tides. Next, there are floodplains, deltas, lakes and water-holes. In these settings extensive blankets of sediment may be deposited in seasonal floods and exposed during the intervening dry periods. Then there are swamp environments, where carpets of plant debris may be interbedded with seams of mud and silt. Finally, there are deserts, where footprints impressed on sandflats and dunes may be buried by drifting sand. In addition, deserts may experience occasional sheet-floods that deposit wet sediments. These are all continental or 'marginal' environments, and they are all, with the possible exception of deserts, lowland settings. Some types of dinosaurs probably flourished in uplands environments, where erosion predominated over sedimentation, but their traces were less likely to be preserved as fossils.

Even within the appropriate environments footprints tend to be preserved only in particularly favourable spots. For instance, there is only a narrow zone around the shores of Lake Turkana, in Kenya, where vertebrate tracks are at all well preserved (Laporte and Behrensmeyer 1980). This zone extends landwards for a few tens of metres from the water's edge and is blanketed by sediments that are sufficiently moist and plastic to retain the impressions of an animal's feet. Offshore from this zone the constant disturbance of the bottom sediment by wave action tends to obliterate any tracks, and in the other direction, farther inland, the substrate is usually so hard and dry that animals fail to leave adequate impressions of their feet. Ichnologists studying marine environments have identified a series of distinct environmental zones, each with its own characteristic suite of trace fossils (predominantly those of invertebrates). There is, as yet, no comparable zonation for the terrestrial and 'marginal' environments in which dinosaurs left their footprints, though preliminary studies indicate that such zonation is certainly feasible (Frey and Pemberton 1986, 1987).

Sediments that contain abundant footprints rarely produce fossil bones, while sediments that contain dinosaur skeletons tend to have few footprints. It is not known with certainty why bones and footprints should occur separately, though it is likely that the two sorts of fossils were preserved under somewhat different circumstances (Lessertisseur 1955: 13–14). For example, rapid supply of sediment was probably necessary to bury dinosaur carcasses, but this requires fast-running water that would probably erode fresh footprints. On the other hand, footprints can survive lengthy exposure in the open air, but carcasses might not because they would attract the attention of scavengers. Then, after burial, acidic groundwaters can easily dissolve the minerals of bones and teeth (e.g. Carpenter 1982a) but would have no appreciable effect on footprints. It is also pertinent to recall that body fossils can be transported from one environment to another whereas footprints cannot. These are just some of the many factors that are likely to have favoured the preferential survival of footprints in some circumstances and the preservation of bones and teeth in others.

Whatever the reason, skeletons and footprints are so rarely found together that it might seem difficult to decide which sorts of dinosaurs made which sorts of tracks. Fortunately, there are many reliable criteria for identifying dinosaurian track-makers (discussed in Chapters 5, 6 and 7), and there are also some instances where the bones and footprints of dinosaurs do occur at different levels within a single succession of rocks. A good example is furnished by W. Langston's (1960) report of two hadrosaur fossils at a single site in Canada: here, a hadrosaur footprint was found in a sandstone speckled with plant debris whereas a hadrosaur skeleton was discovered in a different sandstone unit that was much richer in mica. The compositional differences of the two sandstones doubtless indicate that the footprint and the skeleton were preserved under slightly different conditions.

Consolidation and burial

It is sometimes implied that rapid burial was important for the successful preservation of all fossils. This is not true in the case of footprints, where the style of burial was as important as the rate of burial.

Imagine some footprints in the mud surrounding a water-hole. If the mud were still soft, an influx of water and sediment might well scour and erode the footprints rather than bury them intact (e.g. Tucker and Burchette 1977: 205, fig. 4). But if the same prints had been exposed to the wind and sun for a few days, or longer, they might have dried out and hardened to such an extent that they would survive inundation (e.g. Hunt 1975, fig. 158). In other words, some interval should elapse between the imprinting of tracks and their burial. Ideally, that interval should be adequate to allow consolidation of the footprints, though its exact duration would depend on many variables including the texture and wetness of the substrate, the depth of the footprints and climatic conditions. Brief exposure may be sufficient to consolidate footprints, but protracted exposure is almost certain to destroy them, for eventually the prints would be eroded by rain and wind-blown sand, or shattered by the development of sun-cracks (e.g. Wuest 1934). Even so, there are exceptional cases where footprints are known to have survived very lengthy exposure. Tracks of mules that crossed the floor of Death Valley, in California, before the year 1900 were still plainly visible in the 1970s (Hunt 1975, fig. 123).

Consolidation did not always involve the drying out of wet sediment. Footprints formed under shallow water were sometimes reinforced by the growth of an algal mat, and those impressed in dry wind-blown sand must also have been consolidated, most probably by a film of mist or dew that evaporated to leave a thin crust (McKee 1945) or even by an organic covering such as a spider's web (Ekdale and Picard 1985).

Consistency of substrate

If a substrate were too hard and dry, a dinosaur would leave no footprints at all, or only very faint ones. At the other extreme the substrate should not have been excessively soft or wet: in waterlogged sediments footprints tend to slump and collapse as an animal withdraws its feet (Plate 6, p. 94, centre). Ideally, the substrate should have been of medium consistency, neither too soft nor too hard. In fact, its physical properties should have resembled those of wet plaster, fresh Plasticine (modelling clay), uncooked pastry or potter's staple – the very materials that are used in experiments to obtain footprints from living animals.

The finest dinosaur tracks were probably impressed in sediments of about the ideal consistency, and they sometimes show remarkable details of foot structure, such as tubercles, scales, claws and skin creases (Plate 7, p. 118, bottom right). Most dinosaur footprints are somewhat less than perfect; they may show clear outlines of the toes, but little in the way of finer details. The least informative prints are faint or irregular hollows formed in sediments that were somewhat too hard or too soft to retain faithful impressions of the feet.

Footprints are often partly or completely encircled by a raised rim

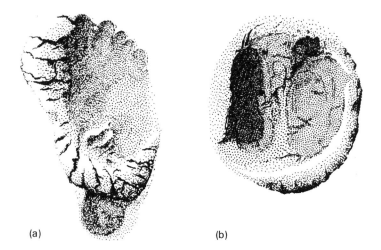

(a) (b)

Figure 2.2 Examples of footprints bordered by a raised rim of displaced sediment. (a) Human footprint on a sandy beach; about 26 cm long. Note the compression-cracks traversing the friable sandy sediment. (b) Footprint of an artiodactyl, in Oligocene limestone, France; about 9 cm long. Here the cohesive calcareous mud has bulged up into a smooth rim, without fissures. (Adapted from photographs by the author (a) and in Leonardi 1987 (b).)

of sediment that was displaced by the impact of the track-maker's foot (Figure 2.2; Plate 3, p. 58, top). This raised rim, or **bourrelet**, is best developed in the tracks of bigger and heavier animals, and its structure depended in large measure on the consistency of the substrate. A cohesive sediment, like moist mud, tended to bulge up between the track-maker's toes, and around the margins of the foot, to form a swollen or inflated rim in which the substrate surface is largely unbroken. Friable or crumbly sediments, like moist sand, responded differently to the impact of the track-maker's foot: here, the displaced sediment tended to spill over the edges of the footprint and to accumulate on the original substrate surface.

Dinosaur footprints are usually found in water-laid clastic sediments such as sandstones, siltstones and mudstones, though they sometimes occur in more unusual substrates, providing that these were once of an appropriate consistency. For example, dinosaur tracks were occasionally preserved in the soft and incoherent surfaces of ancient sand dunes. Here the footprints would have been buried by slumping or drifting sand. Other footprints were formed in carpets of plant debris, when dinosaurs walked across ancient peat bogs and swamps. Those footprints are sometimes encountered as casts that are left hanging from the roof of a coal mine, once the lithified plant debris, now in the form of coal, has been removed. Such footprint casts have been described as 'a constant danger to miners because they sometimes break along the bedding and fall to the floor' (Shrock 1948: 179; see also Parker and Rowley 1989; Plate 4, p. 62, bottom left). Still other footprints were impressed on ancient tidal flats and beaches composed of shell debris, as well as in limestones and muds that accumulated in coastal lagoons. Dinosaur footprints are known from sediments inter-bedded with lava flows (Ferrusquia-Villafranca *et al.* 1978) and might also be expected to occur in deposits of volcanic ash. Mammal tracks are reported from ancient ash falls (e.g. Leakey 1979; Renders 1984) and there seems no reason why those of dinosaurs should not have been preserved in similar fashion. Perhaps the most unusual substrate known to have been trodden by a dinosaur was the interior of a nest containing a clutch of dinosaur eggs (Zhao 1979). In short, footprints may have been left in virtually any circumstances where dinosaurs traversed suitable substrates. Those substrates might have been clastic sediments, organic debris or volcanic ash, but all that really mattered was their consistency.

Tenacity of substrate

If the substrate was adhesive, a dinosaur would tend to disfigure its own footprints as it tried to pull its feet clear. Alternatively, the feet might become encased in sticky sediment, so that the animal would produce large and shapeless prints. In some circumstances the sediment clinging to the foot fell off in lumps as a dinosaur swung its leg forward at each step. Dinosaurs that did kick up and fling aside clods of sticky sediment are known to have left trackways that can only be described as 'messy' (e.g. the sauropod tracks reported by Bird 1944: 65).

In the worst possible circumstances a dinosaur might misjudge the firmness of the ground and become mired. Then it might perish, stuck fast. Some of the African sauropods *Brachiosaurus* and *Tornieria* may have died in this way (Russell *et al.* 1980: 172), and a similar fate is suspected to have befallen some examples of the Canadian horned dinosaur *Leptoceratops* (Sternberg 1970: 4). Even the lightweight bipedal dinosaurs might not have been completely immune to such risk, to judge from the accidents that befall ground-dwelling birds (e.g. Mantell 1850: 336; de Deckker 1988, pl. 5A).

To some extent all dinosaurs would have distorted their own footprints as they withdrew their feet. Such disfigurement would be most obvious in sticky and clay-like substrates, but it also affected footprints in less-tenacious sediments. Withdrawal of the foot from wet sediment often creates a suction effect, so that the walls of the footprint tend to be drawn inwards. In most dinosaur footprints such effects were negligible, but in a few cases there were major distortions of footprint shape (e.g. Figure 5.18; Plate 8, p. 134, bottom left).

At the other extreme there would have been little or no tendency for dry wind-blown sand to adhere to a track-maker's foot. Here, the loose sand would slump inwards as the dinosaur withdrew its foot, thus obscuring details of footprint structure. On sloping surfaces, such as the sides of sand dunes, the impact of the foot might also cause sand to slump downslope and away from the foot. This produced a characteristic sand crescent bordering the footprint on its downslope side (Leonardi 1980a: 566).

Texture of substrate

The best-preserved footprints occur in fine-grained sediments such as siltstones and mudstones. Reasonably good, and sometimes superb, footprints may be found in sandstones, whereas prints of moderate quality may be collected from coarse-grained or irregular substrates including accumulations of plant debris and shell fragments. Dinosaur footprints have even been retrieved from materials as coarse as conglomerate and breccia (e.g. Sollas 1879). However, the coarser-grained clastic sediments, especially sandstones, tend to be more durable and resistant to weathering than the finer-grained ones such as mudstones. As a result, footprints tend to be discovered more commonly in sandstones than in mudstones, though there are some fortunate exceptions to this general rule.

Homogeneity of substrate

The substrates traversed by dinosaurs were rarely homogeneous, so that the preservation of footprints tends to vary from place to place, even within the space of a few metres. A series of footprints from a single dinosaur will often include some shallow prints or discontinuities, formed wherever the animal chanced to tread on an

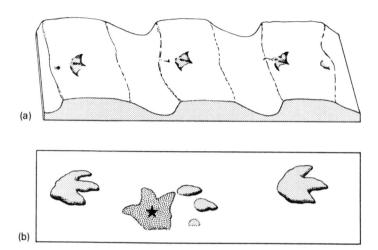

(a)

(b)

Figure 2.3 Examples of discontinuous trackways. (a) Trackway of a seagull traversing a ripple-marked beach; about 75 cm long. Firm sediment on the crests of ripples retained clear imprints of the left foot, but slushy sediment in intervening troughs collapsed on withdrawal of the right foot.
(b) Sequence of three footprints (casts, shown inverted) made by a large ornithopod dinosaur, from the Upper Cretaceous of Colorado, USA; total length about 5.1 m. Two right footprints are well preserved, but the intervening left print is so faintly impressed that it was initially overlooked. The other footprint (indicated by star) was made by a different animal. (Adapted from photographs by the author (a) and in Brown 1938 (b).)

unfavourable patch of sediment (Figure 2.3). Variations in footprint depth are particularly obvious where dinosaurs traversed ancient shore-lines: on the relatively firm sediments that were exposed to the air the animals produced rather shallow footprints, but in the softer and wetter sediments below the water-line their tracks tend to be noticeably deeper and better-defined (e.g. Figure 2.7).

Sometimes the surface of the substrate dried out into a firm crust while the underlying sediment was still saturated and more fluid. Even if this crust had been very thin it could have exerted some important controls on footprint preservation. For instance, it is known that lightweight dinosaurs might run across such a crust without leaving perceptible traces whereas heavier animals would break through to produce deep footprints (Thulborn and Wade 1984: 443). If the crust had dried out to such an extent that it became brittle or friable the impact of a dinosaur's foot sometimes generated a series of fractures radiating from the footprint. Similar compression-cracks may be seen around footprints made on sandy beaches (Figure 2.2a).

Substrates comprising interbedded seams of sand and mud were sometimes exceptionally well suited to the preservation of fossil footprints. Here, an animal's foot might punch out a plug of mud and press it firmly into the underlying sand, thus producing a deep and sharply defined footprint (Laporte and Behrensmeyer 1980). In other cases, the track-maker's foot might plunge right through a surface layer of mud and come to rest on the firmer sandy sediment beneath.

The cyclic or periodic accumulation of sediment commonly produces a distinctive pattern of rock-layering that has an important bearing on the preservation of footprints. Often, each influx of water introduces a mixture of rock particles, ranging from large sand grains down to fine flakes of mud. These particles tend to fall out of suspension in sequence: the coarsest and heaviest grains of sand settle down first, to be followed by finer sand grains and silt and, ultimately, by the finest dust-like particles of mud. As a result, each layer of sediment reveals a natural grading from bottom to top: the base is a coarse-grained sediment, such as sandstone, whereas the upper surface is often a mudstone or shale. This natural grading has important consequences for the preservation of dinosaur footprints. First of all, the exposed surface of newly laid sediment will have a fine-grained texture that is well suited to retaining detailed imprints of an animal's feet. By comparison, the sediment filling those imprints will tend to be relatively coarse-grained. In other words, coarse-grained casts fit tightly into fine-grained moulds (Figure 2.4).

The sharp textural difference between cast and mould usually ensures that there will be a clean parting between the two, either when the rocks are weathered naturally or when they are deliberately split open. The cast may show details of footprint structure that are not apparent on the mould, and vice versa (e.g. Sarjeant 1971: 344–5). There are also some less-fortunate consequences. Sandstone casts tend to be more durable than moulds in mudstone or shale, and in natural settings, such as a cliff or the wall of an abandoned quarry, weathering destroys the softer rock layers while the harder sandstone layers project as ledges. The only surviving indications of footprints may be casts on the undersides of the sandstone ledges and fallen slabs. A casual observer might walk right over the sandstone slabs without being aware that there are footprints preserved beneath them. It is for

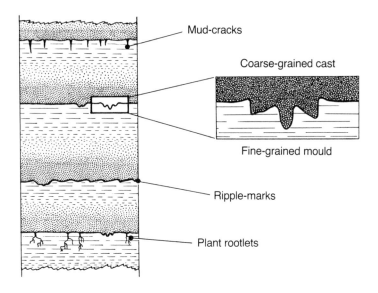

Figure 2.4 Diagrammatic succession of graded sediments, showing characteristic preservation of footprints and associated sedimentary structures at several horizons. Note that the footprints are impressed in the finest-grained sediment and are filled by the coarsest-grained sediment.

this simple reason that so many fossil footprints go unnoticed, even by experienced palaeontologists. To search successfully for footprints it is often necessary to look underneath overhanging ledges and to turn over fallen slabs.

Sandstone casts, which are most resistant to weathering, tend to retain less detail than the moulds in finer-grained sediments. Fortunately, there are many exceptions, where the moulds prove to be more durable than casts. At sites along the Paluxy River, in Texas, USA, footprint moulds in beds of Cretaceous limestone are filled by a much softer clay. The limestones originally accumulated as calcareous muds in lagoons and tidal flats, and the dinosaur footprints impressed into them were buried under sheets of silt and clay (Cole *et al.* 1985: 37). The calcareous muds were subsequently lithified to become durable limestones whereas the footprint fillings of silt and clay remained soft and incoherent. At the Lark Quarry site in Queensland, Australia, footprint moulds occur in a claystone overlain by sandstone. Here, the dense and fine-grained claystone proved to be

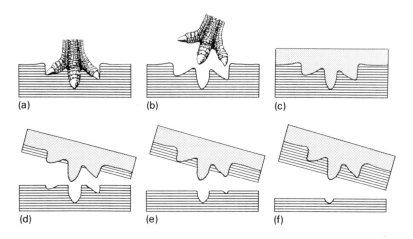

Figure 2.5 Underprints. The foot penetrates a laminated substrate (a) and is cleanly withdrawn (b). After burial and lithification (c) the fissile substrate may be split open at successively deeper levels (d–f) to reveal correspondingly less-complete sections through the footprint.

slightly harder than the sandstone and was naturally reinforced by a surface film of iron oxide and sand grains (see Figure 2.9, layer 2). In the Wealden (Lower Cretaceous) sediments along the coast of south-eastern England the preservation of dinosaur tracks varies from site to site, depending on the relative hardness of the sediments and the manner in which they are exposed by erosion. Thus, some footprints occur as natural casts on the undersides of sandstone ledges along the cliffs, whereas others appears as moulds in bedding planes exposed on the foreshore (Delair and Sarjeant 1985: 144–7).

In many instances the substrate is laminated, comprising successive layers of sediment that differ slightly in their composition. Such laminated rocks are often fissile – that is, their component layers of sediment may be split apart quite easily. Several styles of footprint preservation may be encountered in laminated and fissile sediments. First, a laminated substrate was sometimes firm enough for an animal's foot to penetrate several layers of sediment. If the substrate is split open at successively deeper layers the footprint will be found to be less and less complete. These deeper-lying sections through footprints have been referred to as **underprints** or **subtraces** (Figure 2.5d–f). When the laminated substrate was more plastic it sometimes

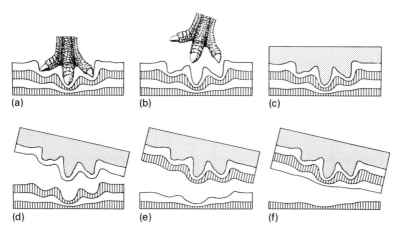

Figure 2.6 Transmitted prints. Impact of the foot (a) buckles the underlying layers of sediment, and the foot is cleanly withdrawn (b). After burial and lithification (c) the fissile substrate may be split open at successively deeper levels (d–f) to reveal correspondingly shallower versions of the whole footprint. In effect, a single foot-impact has generated an entire stack of footprints.

responded to the impact of a dinosaur's foot by buckling (Figure 2.6). Here the impact of the foot was transmitted through a succession of sediment layers to form a stack of casts and moulds. These **transmitted prints** or **ghost prints** usually become shallower and more vaguely-defined at successively lower levels, though in rare instances they may be better-defined than the primary footprints that overlie them (e.g. van Dijk 1978).

Footprints can be transmitted through considerable thicknesses of sediment. A horse traversing a beach has been known to cause disturbances to a depth of 25 cm (van der Lingen and Andrews 1969), and a dinosaur footprint from the Jurassic of Yorkshire was transmitted through at least 5 cm of sandy sediment (Whyte and Romano 1981). In some cases prints are known to have been transmitted obliquely through the substrate, rather than straight downwards (E. Hitchcock 1858, pl. 6, fig. 2). Transmitted prints were rarely produced by small and lightweight dinosaurs but may be surprisingly common in the tracks of bigger and heavier ones. In fact, it is likely that many of the dinosaur footprints decribed in the scientific literature are transmitted prints rather than the primary casts and moulds (see Plate 9, p. 140, bottom right). Finally, if laminated sediments were unusually fluid, or rested on a foundation of much firmer sediment, the impact of dinosaur's foot might simply deform them rather than create transmitted prints.

The sediments filling natural moulds sometimes have laminations that mirror the contours of the underlying imprints. These overlying indications of the buried footprints are usually so vague that they are unlikely to be mistaken for the primary moulds and casts. They have sometimes been termed 'over-tracks' (Langston 1986) or 'overprints' (Sarjeant 1988).

Slope of substrate

Most dinosaur tracks were impressed in water-laid clastic sediments, normally deposited in nearly horizontal sheets. These substrates would have been virtually flat, aside from minor features of relief such as ripple-marks and run-off channels. However, the same is not true for wind-blown sands, which may accumulate in steep-sided dunes. It has sometimes been claimed that the tracks of animals heading *up* sand dunes are the only ones likely to be preserved. When moving down a slope of unconsolidated sand an animal tends to slide as it attempts to maintain balance and to control the speed of its descent. In that slithering descent the loose sand usually slumps as the feet are lifted, thus obscuring the footprints (Reiche 1938; McKee 1944, 1947). Thus, in ancient dune deposits, such as the Coconino Sandstone of Arizona, USA, it has been supposed that the fossil tracks are almost exclusively those of animals that walked uphill (Gilmore 1926). However, these assumptions about the prevalence of uphill tracks have been challenged by L. Brand (1979), whose observations and experiments produced some contrary findings. Moreover, fossil tracks in the dune sands of the Botucatu Formation, Brazil, have been found to bear no consistent relationship to the original slope of the substrate (Leonardi and Godoy 1980). Nevertheless, it is noticeable that animal tracks on the sloping faces of modern sand dunes do sometimes have a preferred orientation – either along the contours or straight up and down the line of maximum slope (e.g. McKee 1982, fig. 11) – and there is some evidence that mammals may select optimal paths in relation to their body weight and the slope of the ground (Reichman and Aitchison 1981). Clearly, the relationship of footprint preservation to slope of the substrate stands in need of more research.

Similar assumptions have been extended to footprints that occur in

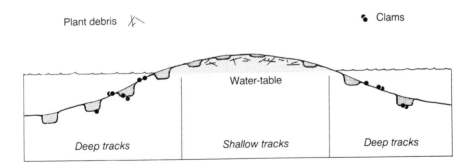

Figure 2.7 Schematic cross-section through a trackway site in the Upper Jurassic of Colorado, USA, showing the relationship between the depth of sauropod tracks and the original topography of the substrate. Variations in the depth of the footprints reveal the location of the shore-line and the approximate level of the ancient water-table. The location of the shore-line is confirmed by the distribution of fossil plant remains and clams. Note that some clams were trampled deep into the substrate by the feet of dinosaurs. (Adapted from Lockley 1986a.)

water-laid cross-bedded sediments. Here, a prevalence of 'uphill' tracks might indicate that track-makers tended to swim when travelling down-current but dropped down and walked along the bottom when moving upstream (Brand 1979: 37).

Water cover

Some footprints were impressed in sediments exposed to the air whereas others were formed in sediments covered by shallow water. Where a single dinosaur crossed a shore-line, travelling from firm sediments exposed to the air into softer sediments under water, the differences in footprint preservation may be quite obvious (e.g. the brontosaur track mentioned by Sarjeant 1988: 129). Comparable differences are seen where groups of dinosaurs travelled along shore-lines: the footprints of the animals wading through the water tended to be deeply impressed in waterlogged substrates whereas those walking on the firmer sediments above the water-line left somewhat shallower prints (Currie 1983; Figure 2.7). Preservational differences related to the depth of the water-cover have also been noted (Frey 1975: 34) in the Lower Cretaceous limestones of the Paluxy River in Texas, USA, where

Some of the [dinosaur] tracks seem to have been made on mudflats (they are associated with mudcracks), others in extremely shallow water (they are deeply impressed in the nondesiccated substrate), and still others in water just deep enough to buoy the animal's body slightly (the tracks are slurred and lightly impressed).

At some sites dinosaur footprints are known to have accumulated over a matter of days or weeks while the substrate was gradually draining free of water and becoming firmer. In these situations the earliest-formed footprints tend to be more deeply impressed than the later ones (e.g. Figure 11.4).

The depth of water sometimes exerted a control on the minimum size of the footprints that were preserved. This effect was noted by P.J. Currie (1983) in a series of hadrosaur tracks from the Lower Cretaceous of Canada, where it seems that only the bigger animals had legs sufficiently long to touch bottom and leave recognizable footprints. Currie suspected that these big hadrosaurs were accompanied by juveniles, but that these latter were too small to have planted their feet on the bottom and were obliged to progress by swimming rather than wading.

Finally, the depth of water sometimes affected a dinosaur's method of locomotion and, hence, the morphology of its tracks. Thus the tracks left by wading dinosaurs are noticeably different from those made by dinosaurs that swam along by periodically thrusting their feet against the bottom. The swimming animals sometimes took longer strides, because they tended to float between one footfall and the next, and their footprints may show only the tips of the toes (Coombs 1980b).

UNUSUAL TYPES OF PRESERVATION

In exceptional cases the topography of casts and moulds may be reversed, so that the moulds appear as raised reliefs on the upper surface of the substrate (= **convex epireliefs**) while the casts are complementary cavities in the overlying sediment (= **concave hyporeliefs**). L. Courel and G. Demathieu (1984) listed five mechanisms that might account for such cases of inverted relief.

1. Sediment adhering to the underside of an animal's foot was drawn up into a mound.

2. The foot's impact compressed a column of sediment that subsequently resisted further compression during lithification.
3. The foot's impact compressed a column of sediment that subsequently resisted erosion while the surrounding sediments were winnowed away.
4. Wind-blown sand accumulated to form miniature dunes in an otherwise normal footprint.
5. A footprint impressed in fine-grained sediment acted as a trap for waterborne sand grains; subsequently, the sand-filled footprint resisted compaction whereas the surrounding sediments were greatly reduced in thickness.

These effects are sometimes seen in the tracks of living animals (Hughes 1884: 184):

> In some cases in the footprint of a frog the line of the toes is marked in relief . . . not depressed, as one would naturally expect, on the upper surface of the mud. This appears to happen where the mud has just that consistency, which causes it to stick to that part of the foot which is well pressed into it, and which, on being withdrawn, lifts it up, as a spoon or finger draws up after it a column of treacle or honey.

A second example is furnished by a series of footprints in a desert area of New Mexico, USA (Anonymous 1982: 37). When the prints were first observed, in 1932, they were impressed to a depth of about 6 cm below the surrounding surface. The same prints 42 years later were found to be represented by soil pedestals about 3 cm high. Evidently, the compacted soil beneath the footprints resisted erosion by wind and rain while the surrounding soil was winnowed away. Footprints capping pedestals of compacted substrate are sometimes encountered on modern beaches and dunes (e.g. Lewis and Titheridge 1978, fig. 1D), and in the snow on ski-slopes (Teichert 1934).

Such cases of inverted relief are uncommon in the fossil record. Courel and Demathieu (1984) mentioned examples from the Triassic of France and the Cretaceous of the United States and introduced the term **counter-relief** (*contre-relief*) to describe them. Some examples, such as those in the Cretaceous limestones of the Paluxy River, Texas, are probably overlooked because they are far from conspicuous, often being less than 1 mm high.

Occasionally, the relief of a footprint is inverted by present-day erosion of a lithified substrate. M.E. Tucker and T.P. Burchette (1977, fig. 7) showed a fine example of this effect in a Triassic dinosaur footprint from South Wales. In this instance the footprint mould had been impressed quite normally in a sandy substrate but had been filled with coarse pebbly sediment. On exposure to the elements this pebbly filling proved to be exceptionally resistant to weathering whereas the surrounding sandstone was eroded more rapidly to a lower level (Plate 6, p. 94, bottom centre).

In some instances the depressed interior of a footprint mould differs in colour from the surrounding rock. However, there are also reports of fossil footprints that have no relief – either normal or inverted – but appear merely as superficial colour stains. Examples from the Cretaceous of Texas have been discussed by G. Kuban (1989b), who was unable to provide a satisfactory explanation for their origin. It is possible, though unlikely, that the impact of a dinosaur's foot could generate local chemical changes in the substrate, and that these might be emphasized through mineral alteration or weathering at a much later date. A more likely explanation is that shallow footprints acted as traps for flakes and particles of plant debris, and that this finely disseminated organic matter was responsible for chemical changes leading to discoloration. Perhaps the simplest explanation is that these footprints do show relief, but it is so slight (or so weathered) as to be undetectable.

Finally, weathering sometimes induces a pronounced change in the colour of fossil footprints. For instance, one small footprint from the Triassic coal measures of Queensland, Australia, was indistinguishable in colour from the surrounding white sediment when it was first discovered. Yet, after a year's exposure to the elements, it had turned dark reddish-brown in colour, presumably through the oxidation of finely disseminated iron minerals in the floor of the footprint.

DESTRUCTIVE FACTORS

Even footprints that were formed under ideal circumstances might subsequently be damaged, distorted and ultimately destroyed by natural agencies. Those that were not sufficiently well consolidated might be scoured by wind or water, while those that were exposed for unduly long periods might be shattered by sun-cracks, overgrown by plant roots or trampled by other animals. In other instances dinosaur footprints were disturbed by invertebrates ploughing through the sediment or were obscured by the growth of algal mats. It is not uncommon to find two or more generations of prints on a single bedding plane, each showing different characteristics of preservation.

The parting between cast and mould sometimes acted as a channel for the seepage of fluids and the accumulation of mineral deposits. Such mineral growths tend to obscure the surface details of footprints but may otherwise serve to reinforce moulds in soft sediments. Footprints may also be affected by chemical and physical alteration of the enclosing sediments and by the development of joints or stress-fractures (Plate 8, p. 134, top right).

Finally, the preservation of footprints will be affected by weathering. Where the bedding planes intersect erosion surfaces at a high angle, as in cliffs, quarry faces or hillsides, the more durable layers of rock may stand out as ledges bearing footprint casts on their undersurfaces. But if the same rocks are exposed in river beds or foreshores they tend to be eroded and stripped away bed by bed, so that the footprints are exposed as moulds on bedding planes. In the long term, weathering is always destructive to fossil footprints, which may be obliterated very rapidly in exposed situations. In the short term, however, it may enhance the quality of prints by revealing details that were previously obscured by adherent rock. For this reason footprint slabs have sometimes been propped up out of doors to 'improve' through natural weathering (e.g. Sarjeant 1974, fig. 11).

GEOLOGICAL SIGNIFICANCE OF FOOTPRINT PRESERVATION

It is hardly surprising that dinosaur tracks should provide information about dinosaurs and their life habits, but it is not so widely appreciated that tracks and other traces are of practical value to geologists, particularly in revealing the nature of ancient sedimentary environments (e.g. West and El-Shahat 1984; Curran 1985; Alonso 1987). In reality, dinosaur tracks may tell us as much about palaeo-environments as they do about dinosaurs (Lockley 1986a). Most of this book is concerned with footprints as sources of information about dinosaurs, but at this point it may be worthwhile to add a few remarks about the geological significance of footprint preservation.

First, footprints are useful as geopetal indicators (Shrock 1948: 177):

the arrangement of casts and moulds betrays the original orientation of sediments that may have been steeply tilted or overturned. Other types of dinosaur trace fossils, including nests and coprolites, are of similar value. Deformed footprints are also useful in revealing the direction and intensity of stresses that have affected the enclosing sediments (Plate 8, p. 134, top right).

Next, fossil footprints are often an indication that sediments originated in particular settings – more specifically those 'marginal' environments that experienced periodic or cyclic accumulation of sediments. Further, as M.G. Lockley (1986a) has demonstrated, dinosaur tracks may be especially useful for pinpointing and tracing ancient shore-lines (see Figure 2.7).

Fossil footprints may also give valuable clues to the original physical properties of sedimentary deposits, such as their plasticity and their water content, though as yet there seems to have been little in the way of systematic research on this subject. However, it has proved possible to ascertain the degree to which some Jurassic dinosaur footprints, and hence the enclosing sediments, were compacted since their formation (Lockley 1986a: 43), and to infer from that the approximate level of the water-table when the footprints were formed.

In addition, the tracks of swimming or wading dinosaurs may give fairly precise indications to the depth of water covering ancient sedimentary environments (e.g. Perkins 1974: 145). From the size of the footprints it is possible to judge the length of the track-maker's legs, which reveals the maximum depth of water that could have covered the substrate. For example, the track of a swimming sauropod, described by R.T. Bird (1944), indicated that water depth was about 3.5 m. In another instance, the tracks of swimming theropods, described by W.P. Coombs (1980b), indicated a water depth of about 2 m. (Figure 2.8). Such information is sometimes valuable for palaeogeographic reconstructions: the distribution of dinosaur tracks assisted P. Allen (1959) in charting the extent of terrestrial and shallow-water environments in the Anglo-Paris basin during the early Cretaceous. In addition, deflections in the tracks of swimming dinosaurs, and other animals, are sometimes useful in revealing the direction of water currents at the time the substrate sediments were deposited.

Occasionally, footprints were responsible for the preservation of fossils that might otherwise have been destroyed. Deep prints may act as traps for bones, teeth and the remains of small organisms (Laporte and Behrensmeyer 1980), and in one case some freshwater clams were

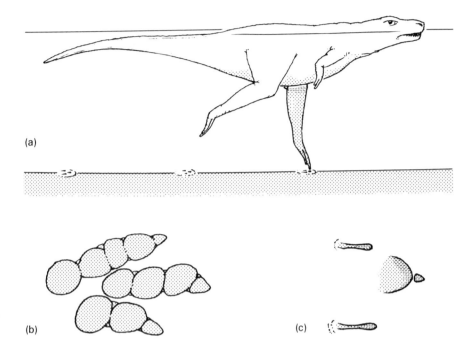

Figure 2.8 A swimming dinosaur and its tracks. (a) Outline restoration of an early Jurassic theropod, about 2 m high at the hip, which swam along by thrusting off the bottom with its feet. (b) Left footprint made by a similar dinosaur in its normal walking gait on land. (c) Left footprint made by an animal swimming as shown. (Adapted from Coombs 1980b.)

preserved by being trampled into the mud by dinosaurs (Lockley 1986a: 45). In other instances footprints are known to have served as microhabitats for communities of invertebrates (Frey 1975: 34).

Fossil footprints are potentially valuable for purposes of stratigraphic zonation and correlation. They are abundant in continental sediments where body fossils may be rare or lacking, and there is little likelihood that they could ever have been transported or re-worked. Moreover, distinctive types or assemblages of footprints are sometimes confined to relatively narrow stratigraphic intervals. For example, P.E. Olsen and D. Baird (1986) demonstrated that one distinctive type of dinosaur track, which they named *Atreipus*, was characteristic of late Triassic sediments whereas another, *Anomoepus*, was confined to the early Jurassic. Beyond that, the wide geographic distribution of

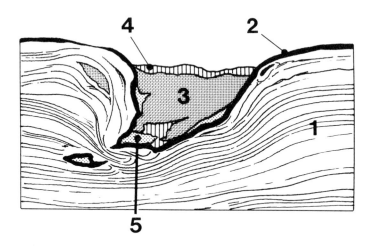

Figure 2.9 Colour variation in a fossil footprint. Diagrammatic vertical section through a small dinosaur footprint, about 3.5 cm deep, from the mid-Cretaceous of Queensland, Australia. 1, Substrate of pink claystone with purple bands; 2, greenish-black layer of sand grains cemented by ironstone; 3, bright-red sandstone filling the footprint mould; 4, reddish-buff sandstone; 5, bright yellow-green sandstone. (Adapted from a photograph by the author.)

Atreipus allowed Olsen and Baird to correlate the Upper Triassic and Lower Jurassic sequences of the eastern United States with those of Germany. The biostratigraphic potential of amphibian and reptile tracks has been exploited with great success in the Permian and Triassic rocks of Europe (e.g. Demathieu and Haubold 1972; Haubold 1973; Haubold and Sarjeant 1974). By comparison, the significance of dinosaur tracks in Jurassic and Cretaceous biostratigraphy remains largely unexplored.

Finally, the footprints of dinosaurs have proved useful in revealing long-term climatic cycles in the remote past. A thick series of Triassic lake sediments studied by P.E. Olsen and his colleagues (1978) could be divided into a sequence of definite cycles, each finishing with shallow-water conditions and subaerial exposure, as indicated by the occurrence of dinosaur tracks. On the basis of varves (annual banding patterns in the sediments) it was determined that the duration of each major cycle was in the order of 1000 to 10 000 years.

3

The search for
dinosaur tracks

We left the woodlot, climbed a fence, and started for the bend
in the river. Ryals told a lengthy tale of his experience in
quarrying tracks.

'I've had a heap o' fun at it,' he said. 'Don't put much food on
the table, but then . . . what does? Hereabouts, 'bout the only
money-makin' jobs is cuttin' cedar posts, bootleggin', and quarryin'
dinosaur footprints. And the other two is hot, hard work.'

R.T. Bird, *Bones for Barnum Brown* (1985)

EARLY DISCOVERIES

Fragmentary dinosaur bones roused the curiosity of naturalists as
early as 1677 (Halstead 1970; Delair and Sarjeant 1975; Buffetaut
1980). By contrast, the footprints and other traces of dinosaurs were
not so conspicuous, for they did not attract the scrutiny of scientists
until the nineteenth century, and even then they were misunderstood
for many years.

The first authenticated discovery of dinosaur tracks was made about
the year 1802, when a boy by the name of Pliny Moody, working on
his father's farm at South Hadley, Massachusetts, USA, ploughed up
a slab of stone bearing five small footprints (Steinbock 1989, fig. 3.3).
This unusual object must have generated some interesting discussions
in the Moody household, for they decided to use it as an ornamental
feature in the doorway of their farm. The footprint slab was destined
to serve its decorative role for about 7 years, until Pliny Moody left
home to attend college. It was then purchased by a certain Dr Elihu
Dwight, who seems to have regarded it as a conversation-piece: it was
reported that 'Dr. Dwight used pleasantly to remark to his visitors,
that these were probably the tracks of Noah's raven' (E. Hitchcock
1844a: 297). Some 30 years later, about 1839, the specimen was

acquired by Professor Edward Hitchcock, of Amherst College, Massachusetts, who had already embarked on an intensive study of fossil tracks in the red sandstones of the Connecticut Valley (Plate 1, p. 42, centre left). Eventually, the footprints discovered by Pliny Moody were described by Hitchcock (1841) under the name *Ornithoi-dichnites fulicoides*, so named because of their resemblance to the tracks of the American coot, *Fulica americana*.

Hitchcock had been prompted to investigate fossil tracks by another chance discovery in 1835. At that date the citizens of Greenfield, Massachusetts, resolved to obtain paving stones for one of their streets, and when these were delivered from a local quarry it was noticed that some of them bore markings rather like the footprints of a turkey. Writing nearly a century after the event, Hitchcock's son (C.H. Hitchcock 1927: 163) described this fortunate discovery as follows:

> In March 1835, W.W. Draper in returning home from church passed the house of Wm. Wilson in Greenfield, Massachusetts. Some slabs about to be placed on the sidewalk leaned against the fence. A light snow upon them happened to slide off suddenly as he passed and he saw these same impressions – and remarked to his wife – 'there are some turkey tracks made 3,000 years ago'.

News of these curious markings spread through the neighbourhood and came to the ears of the local physician, Dr James Deane, who immediately wrote to Hitchcock to inform him of the discovery (Plate 1, p. 42, bottom right). At first, Hitchcock was somewhat scep-tical. From his practical experience as the State Geologist of Massachusetts he was well aware that common sedimentary structures are readily mistaken for footprints. But eventually, at Deane's insistence, Hitchcock examined the markings for himself and found that they were indeed footprints of great antiquity. Hitchcock was enthralled and inspired by the discovery. He began to scour the local quarries for new and better examples and sought diligently for comparative structures in the feet and tracks of living animals. In 1836, he published his first scientific account of the Connecticut Valley tracks, endorsing the popular opinion that they were the fossil footprints of antediluvian birds. Appropriately, he termed these foot-prints 'Ornithichnites', meaning 'stony bird tracks'. Soon, Hitchcock's research had advanced to the point where he could distinguish the tracks of birds (*Ornithoidichnites*) from those of reptiles (*Sauroidichnites*) and four-footed creatures that he suspected to be mammals (*Tetrapodichnites*). Subsequently, he revised and amplified his classification of the footprints, adding many new types (Hitchcock 1841, 1843, 1844a, 1858). Eventually, Hitchcock's classification was elaborated to such a degree that the various types of footprints were designated by a combination of two names, **ichnogenus** and **ichnospecies**, and were allotted to zoological groupings.

Meanwhile, James Deane had kept up more than a passing interest in these events and had pursued his own researches on fossil tracks (e.g. 1844a, 1861). However, Hitchcock seized on the subject and developed it into what he regarded as an entirely new field of science called 'ichnolithology' – happily shortened to 'ichnology'. Hitchcock determined to take the leading role in this science. He collected fossil footprints so assiduously that they filled the entire ground floor of a specially designed museum, the so-called Appleton Cabinet (see Plate 1, p. 42). He also erected a system of nomenclature for fossil tracks, and published his findings in elaborate monographs. Almost inevitably there were clashes of opinion and personality, for both Deane and Hitchcock claimed priority for the discovery of the Connecticut Valley tracks (Deane 1844b; E. Hitchcock 1844b; Bouvé 1859). Deane argued, with some justification, that it was he who had brought these tracks to the attention of the scientific world. Hitchcock argued, with equal justification, that the tracks were meaningful to science only because of his personal endeavours.

In a style unfamiliar to modern scientists Hitchcock even ventured to promote his beliefs in verse. The following extracts are from a lengthy work that was published posthumously (C.H. Hitchcock 1927: 174–81).

The Sandstone Bird
By Edward Hitchcock
Scene – Banks of the Connecticut River.
Geologist alone examining the footmarks of a bird.
(*Ornithichnites giganteus*)
Foot-marks on stone! how plain and yet how strange!
A bird track truly though of giant bulk,
Yet of the monster every vestige else
Has vanished. Bird, a problem thou hast solved
Man never has: to leave his trace on earth

Too deep for time and fate to wear away.
A thousand pyramids had mouldered down
Since on this rock thy footprints were impressed;
Yet here it stands unaltered though since then,
Earth's crust has been upheaved and fractured oft.
And deluge after deluge o'er her driven,
Has swept organic life from off her face.
Bird of a former world, would that thy form
Might reappear in these thy ancient haunts. . .

. . .Bird of mighty foot (Oh vain)
Ornithichnites called by name;
Science thus her ignorance shows,
On a footmark to impose
Name uncouth; while by my arts
Into life the biped starts.
Bird of sandstone era, wake!
From thy deep dark prison break.
Spread thy wings upon our air,
Show thy huge strong talons here:
Let them print the muddy shore
As they did in days of yore.
Pre-adamic bird, whose sway
Ruled creation in thy day,
Come obedient to my word,
Stand before Creation's Lord. . .

Hitchcock never accepted that his Ornithoidichnites could be the tracks of dinosaurs. At the time he undertook his pioneering studies on fossil tracks, dinosaurs were not well known to science (Figure 3.1). In fact, the very word 'dinosaur' was not coined until 1841 (and published by Owen 1842), even though nine genera had already been established on fragmentary remains from Britain and Europe (Colbert 1968: 31). And it was not until 1856 that the bones of a dinosaur were first reported from North America (Leidy 1856). Yet these early discoveries were tantalisingly incomplete. No one had any real inkling of what a *whole* dinosaur had looked like, let alone how it might have moved about or what sort of footprints it might have produced. As there was no certain evidence to the contrary, palaeontologists assumed that dinosaurs must have resembled existing reptiles, and duly portrayed them as grotesquely overinflated lizards or crocodiles.

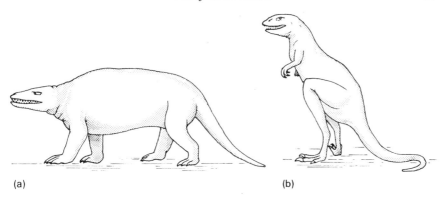

(a)   (b)

Figure 3.1 Two early conceptions of dinosaurs. (a) The European carnosaur *Megalosaurus*, as envisaged by Richard Owen in 1854. A restoration based on fragmentary fossils, and in default of evidence to the contrary, *Megalosaurus* was assumed to have been a quadruped, like existing reptiles. (b) The North American carnosaur *Dryptosaurus* (*Laelaps*), a close relative of *Megalosaurus*, as envisaged by Edward Cope about 1869. By this date the discovery of nearly complete dinosaur skeletons had revealed that dinosaurs such as the carnosaurs were bipedal animals. In this restoration, *Dryptosaurus* is modelled along the lines of a reptilian kangaroo, resting in tripod-fashion on hindlimbs and tail. (Adapted from Colbert 1968; Desmond 1975.)

Hitchcock had little reason to imagine that his Ornithoidichnites could be the tracks of such monstrous quadrupeds. Instead, he continued to envisage Triassic landscapes teeming with birds of every shape and size (1848: 250–1):

> I have experienced all the excitement of romance, as I have gone back into those immensely remote ages, and watched those shores along which these enormous and heteroclitic beings walked. Now I have seen, in scientific vision, an apterous bird, some twelve or fifteen feet high, – nay, large flocks of them, – walking over the muddy surface, followed by many others of analogous character, but of smaller size. Next comes a biped animal, a bird, perhaps, with a foot and heel nearly two feet long. Then a host of lesser bipeds, formed on the same general type. . . Strange, indeed, is this menagerie of remote sandstone days.

Hitchcock's menagerie proved far too strange for some of his

contemporaries, who found it difficult to imagine an ancient world populated by swarms of bird-like creatures. And his critics were only too ready to point out that the largest Ornithoidichnites were far bigger than the feet of any known birds, living or extinct. Nevertheless, Hitchcock stoutly defended his views. In 1840, he received an unexpected fillip when the eminent English anatomist Richard Owen demonstrated that New Zealand had once been the home of gigantic flightless birds known as the moas (Buick 1931, 1936). Despite their great size the moas and their kin were far too recent to have produced the Connecticut Valley tracks, which had been formed about 190 million years earlier (late Triassic and early Jurassic). Even so, Hitchcock could, and did, argue that truly enormous birds had existed in the past, and he noted with satisfaction that the moas were certainly big enough to have produced his cherished Ornithoidichnites. The distinguished geologist Charles Lyell also lent the weight of his authority to Hitchcock's cause: in his presidential address to the Geological Society of London, Lyell maintained that the fossil evidence 'should, I think, remove all scepticism in regard to the ornithic nature of most of these [Connecticut Valley] bipeds' (1851: lxi).

In the latter half of the nineteenth century Hitchcock's arguments began to lose their persuasiveness. A new and different idea was gaining currency – that the footprints in the Connecticut Valley had been made by *dinosaurs*. This inexorable shift in scientific opinion was generated by a series of unprecedented discoveries in the fossil record. For a start, dinosaurs were becoming much better known. Many and more complete skeletons were being unearthed in Europe and North America, and it soon became apparent that the Mesozoic era had been an age dominated by dinosaurs, and not by birds (Colbert 1968;

Plate 1 Top: The so-called Appleton Cabinet, a natural history museum constructed at Amherst College, Massachusetts, in 1863, by bequest of Samuel Appleton of Boston. Edward Hitchcock's collection of fossil tracks filled the entire ground floor. Centre left: Professor Edward Hitchcock, founder of the science of ichnolithology. Centre right: The interior of the Appleton Cabinet. Hitchcock's collection comprised more than 20 000 fossil footprints, though not all of these originated from dinosaurs. Bottom left: Footprint slabs on display in the Appleton Cabinet, about the year 1920. Bottom right: Dr James Deane, who directed Hitchcock's attention to the existence of fossil footprints in the Connecticut Valley, USA.

Figure 3.2 The track-makers of the Connecticut Valley, USA, as
visualized by Gerhard Heilmann in 1927. In the foreground, a coelurosaur
scampers along the shore and two ornithopods feast on water-plants. Two
more ornithopods are wading in the shallows, while a long-necked
prosauropod peers over the rocky headland.

Lanham 1973; Buffetaut 1987). Then, on examining these newly
found skeletons, palaeontologists began to realize that some dinosaurs
had been bipedal animals, like birds. This much seemed obvious from
the fact that certain dinosaurs had hindlimbs that were very much
bigger and stronger than their forelimbs. At first there was merely a
suspicion that such dinosaurs might have stood at rest by supporting

themselves on hindlegs and tail, rather like reptilian versions of kangaroos (Leidy 1858). But later, and almost predictably, this idea was extended to the suggestion that some dinosaurs might have progressed by hopping and leaping in kangaroo-fashion (Cope 1866). While it is now known that bipedal dinosaurs moved their hindlegs alternately, like ostriches, these imaginative comparisons with kangaroos did at least contain one tiny seed of truth – that some dinosaurs moved around on two legs, and not on all fours (see Figure 3.1). Hitchcock's theories receded even further into the background when detailed anatomical studies began to reveal that certain dinosaurs had feet virtually identical to those of birds (Cope 1867; Huxley 1868, 1870). Then, in 1861, the discovery of the coelurosaur *Compsognathus* shattered the popular image of dinosaurs as lumbering giants, for here was a delicate little dinosaur, scarcely bigger than a chicken (Figure 3.3). And, more significantly, it bore an 'extraordinary resemblance' to a bird (Huxley 1868). Finally, there came the discovery of *Archaeopteryx*. Hailed as an extremely primitive bird, barely distinguishable from its reptilian ancestors, *Archaeopteryx* delivered a crushing blow to Hitchcock's theories. If a bird so primitive had existed in the late Jurassic it was unlikely that many and diverse birds could have flourished at the much earlier date when the Connecticut Valley tracks had been formed. Undaunted, Hitchcock used the evidence of *Archaeopteryx* to argue his opinions (1865: 33) 'with more confidence than ever'! But the mounting evidence was leading to one inescapable conclusion: Hitchcock's Ornithoidichnites were far more likely to be the tracks of dinosaurs than those of birds.

On the 31 December 1867 the leading American naturalist E.D. Cope expressed his firm conviction that the most bird-like tracks in the Connecticut sandstones had been made by theropod dinosaurs. Thirty-eight days later T.H. Huxley, the renowned and influential defender of Darwinian evolution, argued that some large three-toed footprints found in the Wealden (Lower Cretaceous) rocks of southeastern England were 'of such a size and at such a distance apart that it is difficult to believe they can have been made by anything but an *Iguanodon*' (Huxley 1868: 364). Turning his attention to the tracks in the Connecticut Valley, Huxley conceded that some might be traces of birds but maintained that others were definitely those of reptiles, including bipedal dinosaurs.

Thus, two-thirds of a century after Pliny Moody's chance discovery, there came wide agreement that Hitchcock's Ornithoidichnites were

Figure 3.3 Not all dinosaurs were giants, despite popular misconceptions. Here, two small dinosaurs are shown alongside a common pigeon. At the rear is *Compsognathus*, a coelurosaur from the Upper Jurassic of Europe, and the tiny creature at the right is a juvenile specimen of *Psittacosaurus*, a primitive ceratopsian dinosaur from the Lower Cretaceous of Mongolia. (Adapted from Coombs 1980a.)

in many instances the tracks of bipedal dinosaurs (Figure 3.2). Edward Hitchcock had been the first person to publish a detailed account of dinosaur fossils from North America, and he had founded the scientific study of dinosaur tracks. Yet, ironically, he went to his grave (in 1864) convinced that he had been studying the tracks of antediluvian birds.

Hitchcock's views, however mistaken they may seem today, exerted profound influence on his contemporaries. Reports of his discoveries were incorporated in many classic works of geology, including William Buckland's *Bridgewater Treatise*, Charles Lyell's *Principles of Geology*, Gideon Mantell's *Medals of Creation*, Hugh Miller's *Testimony of the Rocks*, and the *Geological Sketches* of Louis Agassiz. More unexpectedly, allusions to the tracks of the Connecticut Valley are to be found in literary works, including those of Herman Melville, Oliver Wendell Holmes and Henry David Thoreau (Dean 1969, 1979).

Indeed, it is very likely that Longfellow's poetic image of 'footprints on the sands of time' is a direct reference to Hitchcock's discoveries. D.R. Dean (1969: 644) accorded Edward Hitchcock the accolade of being 'the last American geologist to leave a personal mark upon our creative literature'.

LATER DISCOVERIES

Since Hitchcock's day dinosaur tracks have been discovered throughout the world. A full history of those discoveries has yet to be written and would, undoubtedly, fill a whole book. What follows here is merely a selection of notable discoveries and interesting interpretations, many of which will be discussed more fully in the following chapters.

North America

Hitchcock's studies on the tracks of the Connecticut Valley were taken up and extended by his son, C.H. Hitchcock (1866, 1889), and were subsequently revised and amplified by R.S. Lull (1904, 1915, 1953). The red sandstones of Connecticut, Pennsylvania and New Jersey have since continued to yield an apparently inexhaustible supply of dinosaur footprints, documented in works such as those by M.R. Thorpe (1929), W. Bock (1952), D. Baird (1957) and P.E. Olsen (1980). It has been estimated that museums and private collections contain some 40 000 footprint slabs obtained from 40 sites in Massachusetts and Connecticut alone (Baird 1957). And, somewhat surprisingly, there are persistent echoes of Edward Hitchcock, for when some tracks were unearthed in New Jersey in 1966 reporters found that local farmers were still describing them as 'ossified turkey tracks' (Dean 1969: 641). At some localities the footprints are so abundant that they feature as tourist attractions or are sold as curiosities. The discovery of a new site with more than 1000 footprints was announced by J.H. Ostrom in 1967 (see also Anonymous 1967) but this has yet to be described in detail. Elsewhere, Ostrom (1972) has examined directional trends in assemblages of dinosaur trackways, thereby providing some evidence of dinosaurs as gregarious animals. In 1980, some unusual tracks from the sandstones of the Connecticut Valley were identified by W.P. Coombs as the traces of swimming dinosaurs (see Figure 2.8). Similar tracks have since been discovered

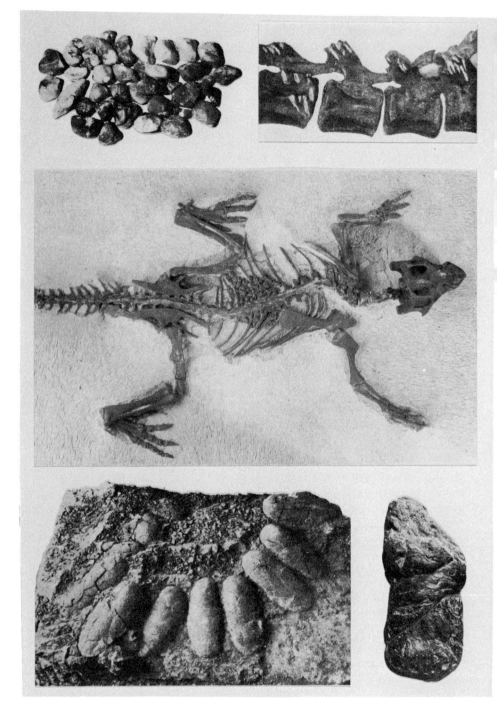

in the Cretaceous rocks of Canada (Currie 1983) and Brazil (Leonardi 1984a). In addition, Coombs has identified the tracks of tiny juvenile dinosaurs, some no bigger than pigeons (1982), and has reviewed the locomotor adaptations of dinosaurs in general (1978a). Footprints comparable in age to those of the Connecticut Valley have also been reported in North Carolina and Virginia (Olsen *et al.* 1978; Weems 1987), as well as in Canada (Olsen and Baird 1982).

The southwestern region of the United States has provided a wealth of dinosaur tracks, many of which were reviewed and discussed by M.G. Lockley (1986b). Some examples from the Cretaceous of Colorado were originally ascribed to a bipedal dinosaur even bigger than *Tyrannosaurus* (Brown 1938), though their true significance is still a matter of debate (Russell and Béland 1976; Russell 1981; Thulborn 1981). A rich assemblage of dinosaur tracks in the Lower Jurassic of Arizona, originally described by S.P. Welles (1971), is currently being studied by M. Morales and E.H. Colbert (1986). The Arizona footprints include those of theropods (*Dilophosauripus*, *Kayentapus*), along with perplexing traces of a bipedal dinosaur that seems to have taken immensely long strides (*Hopiichnus*; see Thulborn and Wade 1984: 453–4). Theropod footprints reported from Colorado and Utah (Peterson 1924; Mehl 1931) included some enormous examples, up to 80 cm long, which were subsequently named *Tyrannosauropus* by H. Haubold (1971). One of the most intriguing of recent discoveries is a rich assortment of footprint casts on the roof of a coal mine in central Utah. These tracks were apparently made by plant-eating dinosaurs

Plate 2 A variety of dinosaur trace fossils. Top left: A collection of gastroliths or 'stomach stones', weighing nearly 0.5 kg, retrieved from the ribcage of a prosauropod dinosaur, from the Lower Jurassic of Zimbabwe; greatly reduced. Top right: Tail bones of a brontosaur with teeth-marks inflicted by a predatory dinosaur, possibly *Allosaurus*, from the Upper Jurassic of Wyoming, USA; greatly reduced. Centre: Skeleton of *Psittacosaurus*, a primitive ceratopsian dinosaur from the Upper Cretaceous of Mongolia (much of the tail omitted). A compact mass of gastroliths is visible within the ribcage; *Psittocosaurus* was about 2 m long. Bottom left: A nest of dinosaur eggs from the Upper Cretaceous of southern China. The nest contained at least 18 eggs, each about 18 cm long, which were probably laid by an ornithopod dinosaur. Bottom right: A coprolite or fossil dropping, perhaps derived from a dinosaur, from the Lower Cretaceous of southern England; about 8 cm long.

threading their way though a forest and stopping to browse on the vegetation (Balsley 1980; Hickey 1980). Some of Hitchcock's ichnogenera, such as *Anchisauripus* and *Anomoepus*, have been identified as far afield as New Mexico (Baird 1964), and in 1980 the first convincing example of a prosauropod trackway (*Navahopus*) was reported by D. Baird from the Lower Jurassic of Arizona (Plate 13, p. 306, top right). There have also been discoveries of dinosaur tracks in many other parts of the USA, including Alaska, Arkansas, California, Kansas, Oklahoma, South Dakota and Wyoming.[1]

The Cretaceous rocks of Texas are particularly rich in dinosaur tracks. Some of the most important finds were made by R.T. Bird, who was the first to recognize the enormous basin-like footprints of sauropod dinosaurs (Plate 12, p. 278, top right). Those footprints contributed a great deal to current understanding of sauropod biology: they revealed that these giant dinosaurs could walk on dry land, that they could also swim, and that they sometimes moved around in herds. One famous set of tracks collected by Bird is on permanent display in the American Museum of Natural History, New York, and appears to show evidence of a sauropod being followed or stalked by a large theropod dinosaur (Plate 10, p. 160, bottom left). Bird's autobiography (1985) contains a wealth of important information about his discoveries, and the several articles that he published in *Natural History* magazine did much to popularize the study of dinosaur tracks, not least because of their eye-catching illustrations and their evocative titles – such as 'Thunder in his footsteps' and 'A dinosaur walks into the museum' (see also Farlow *et al.* 1989). Texas is also famous, or notorious, as the source of supposedly 'human' footprints preserved alongside those of dinosaurs (J.D. Morris 1980). Careful investigations have revealed that these so-called 'man-tracks' are actually incomplete or weathered dinosaur tracks, erosion features or carvings. Recent and continuing studies on the dinosaur tracks of Texas include those by W. Langston (1974, 1979, 1986), J.O. Farlow (1981, 1987), R.H. Sams (1982), G. Kuban (1989a,b) and J.G. Pittman (1989).

Dinosaur tracks are abundant in the Cretaceous rocks of Canada. They were first reported from British Columbia by F.H. McLearn (1923) and from Alberta by C.M. Sternberg (1926). Subsequently, Sternberg made more detailed study of those in British Columbia (1932, 1933a,b), identifying the tracks of theropods, ornithopods and a quadrupedal ornithischian (originally thought to be a ceratopsian, but later regarded as an ankylosaur; Plate 11, p. 208, top left). Other finds include the footprints of hadrosaurs (Langston 1960; Currie and Sarjeant 1979) and coelurosaurs (Storer 1975). A great resurgence of interest in Canadian dinosaur tracks is evident in articles such as those by R. Kool (1981), W.A.S. Sarjeant (1981; also Mossman and Sarjeant 1983) and P.J. Currie (1989). Among the more recent discoveries are late Triassic tracks in New Brunswick (Sarjeant and Stringer 1978) and early Jurassic examples in Nova Scotia (Olsen and Baird 1982; Grantham 1989). These last are thought to include some of the smallest dinosaur footprints found anywhere in the world (Anonymous 1986a; though see Chapter 7).

Central and South America

In 1978, I. Ferrusquia-Villafranca and his colleagues announced the first discovery of dinosaur tracks in Mexico. These abundant footprints are probably middle Jurassic in age and represent both theropods and ornithopods. The much earlier mention of a large 'bird' track in Mexico (see Degenhardt 1840) might conceivably refer to one of the earliest discoveries of dinosaur tracks anywhere in the world – though it has so far proved impossible to locate the original report.

There have been numerous reports of dinosaur tracks in South America, in Argentina, Bolivia, Brazil and Chile.[2] Some notable discoveries include the probable trackway of an ankylosaur in the Upper Cretaceous of Brazil (von Huene 1931a), a rich assemblage of tracks, including those of sauropods and stegosaurs, in the Jurassic of Chile (Dingman and Galli 1965; Plate 3, p. 56, top), and numerous tracks of ornithopods in Brazil (Leonardi 1979b, 1980b; Plate 11, p. 208, bottom left). Also worthy of mention are the suspected tracks of juvenile ornithopods (Leonardi 1981), hadrosaur footprints (Alonso 1980) and coelurosaur tracks that include imprints of the hands as well as of the feet (Casamiquela 1966; Plate 11, p. 208, centre). The South American discoveries have been reviewed by G. Leonardi (1984a, 1989), who has also compiled multilingual glossaries of ichnological terms (1979a, 1987).

Britain

In Britain, dinosaur tracks were first discovered in the Lower Cretaceous rocks of southern England (e.g. Tagart 1846; Mantell

1847; Beckles 1851, 1852). Understandably, these large three-toed footprints were thought to be those of the ornithopod *Iguanodon*, which is represented by skeletons from the same sediments. Unfortunately, it then became common practice to identify *Iguanodon* as the source of virtually *any* large footprints, so that genuine ornithopod tracks became thoroughly confused with those of theropods. The resulting tangle of identifications and misidentifications has yet to be unravelled.

Following the earliest discoveries there have been sporadic reports of early Cretaceous ornithopod and theropod footprints up to the present day.[3] Most dinosaur footprints from the Cretaceous rocks of southern England are quite large (30 cm or more), and while smaller footprints do exist, these have not, on the whole, attracted so much attention. In 1846, S.M. Saxby noted the occurrence of small (5–7 cm) tridactyl footprints, some with traces of webbing, and some years later S.H. Beckles (1862) mentioned the discovery of slightly larger tracks (7–8 cm). Small five-toed prints, originally suspected to be handprints of *Iguanodon*, were discovered at the Isle of Purbeck, Dorset, in 1939 and were duly cemented into a garden path! Eventually, they were examined by J.B. Delair (1963), who identified them as a new ichnogenus (*Purbeckopus*) of uncertain affinities. At the same time Delair also described an unusual T-shaped footprint (ichnogenus *Taupezia*), possibly from a small theropod. Several subparallel trackways were uncovered in the Lower Cretaceous of Dorset in 1962 (see Delair and Lander 1973, fig. 1), and were subsequently the focus of some controversy about the number, identity and behaviour of the track-makers (see Figure 11.7b,c). Two long and intersecting trackways of the *Iguanodon* type were noticed on the foreshore at Cooden, East Sussex, in 1980. These, together with other tracks exposed along the southeast coast of England, have been carefully photographed and mapped by K. Woodhams and J. Hines (1989; see also Delair and Sarjeant 1985, figs 7–10; Delair 1989). In 1977, an important series of trackways, including those of theropods, was discovered by W.T. Blows in foreshore exposures on the Isle of Wight (Blows 1978). And, more recently, an extensive series of dinosaur tracks, both large and small, was revealed during the excavation of a building site at Swanage, Dorset, in 1981. A full account of this discovery is now in preparation (Ensom 1982).

Comparatively few dinosaur tracks have come to light in the Jurassic rocks of Britain. Several discoveries were made along the coast of Yorkshire in the early years of the twentieth century (Sarjeant 1974: 343-4) and sporadically thereafter (Black *et al.* 1934; Wilson *et al.* 1934; Sarjeant 1970). Most of these finds seem to have been tridactyl footprints from bipedal dinosaurs, though it was rarely made clear if these were ornithopods or theropods. However, the single footprint described by Sarjeant (1970) was specifically attributed to an ornithopod. Some well-preserved dinosaur tracks were noticed on the foreshore at Scarborough, Yorkshire, in 1967, but seem never to have been described. Photographs of the site (Delair and Sarjeant 1985, fig. 3) show several trackways, including at least one from an ornithopod. In 1981, M.A. Whyte and M. Romano described a small tridactyl footprint from the Middle Jurassic of Yorkshire; they also reviewed the earlier discoveries in that county, tentatively identifying one footprint (Black *et al.* 1934) as that of a carnosaur and some others (Brodrick 1909) as those of an ornithopod and a quadrupedal dinosaur. Theropod footprints have been identified in the Jurassic of Buckinghamshire (Delair and Sarjeant 1985, fig. 5), and, in 1984, J.E. Andrews and J.D. Hudson reported the first definite discovery of a dinosaur footprint in Scotland. This latter footprint, a large tridactyl cast from the Isle of Skye, is of Jurassic age and was probably made by an ornithopod. (Previously, Sarjeant (1974, fig. 4) had illustrated a small three-toed print on a block of sandstone that was found in a peat bog in Caithness - but the original provenance of this specimen is a mystery.) Recently, some tracks of large quadrupedal dinosaurs have been found in the Upper Jurassic of Dorset (Anonymous 1987). They are probably sauropod or ankylosaur tracks, in either case the first to be discovered in Britain.

There are relatively few reports of Triassic dinosaur tracks from Britain, though footprints of coelurosaurs, some with indications of webbing between the toes, have been found in the English Midlands (Sarjeant 1967; Wills and Sarjeant 1970). Another footprint from the Midlands was tentatively assigned to Hitchcock's ichnogenus *Otozoum* (Sarjeant 1970). The Triassic rocks of Cheshire formerly provided a rich supply of reptilian tracks, discussed in a series of perceptive articles by H.C. Beasley (see Sarjeant 1974: 295-309; 1985). Included among the Cheshire tracks were various three-toed forms: some may have been incomplete examples of the mysterious five-toed footprints known as *Chirotherium* (e.g. Beasley 1896) whereas others, attributed to birds, were possibly dinosaurian (e.g. Cunningham 1846; Harkness 1850). There are no definite reports of dinosaur tracks from the

Triassic rocks of Scotland or Ireland. A series of five theropod footprints from the Triassic of Glamorgan, South Wales, was described in detail by W.J. Sollas (1879), who attributed them to the ichnogenus *Brontozoum*. The fact that these footprints were impressed in a slab of tough breccia probably ensured their survival: previously, the footprint slab had served as the front doorstep of the village inn! This remarkably resilient specimen has been well illustrated by M.G. Bassett and M.R. Owens (1974: 17). Nearly a century later an assemblage of more than 400 Triassic footprints was described from South Glamorgan by M.E. Tucker and T.B. Burchette (1977). The footprints were of two types, large and small, but were both classified in another of Hitchcock's ichnogenera - *Anchisauripus* (Plate 6, p. 94, bottom centre). Tucker and Burchette considered that the footprints had been made by prosauropod dinosaurs, but Delair and Sarjeant (1985) pointed out that they were more likely to be the tracks of theropods and transferred the larger footprints to still another of Hitchcock's ichnogenera - *Gigandipus*.

The fossil footprints of Britain have been comprehensively reviewed by W.A.S. Sarjeant and J.B. Delair (Sarjeant 1974; Delair and Sarjeant 1985). In addition, Sarjeant has provided the fullest account of vertebrate trace fossils that is available in the English language (1975).

Continental Europe

Europe, too, has a long tradition of scientific research on fossil tracks and traces, including those of dinosaurs. Some of the earliest studies were undertaken in Germany, in the wake of interest generated by the discovery of mysterious hand-shaped imprints known as *Chirotherium* (see Chapter 4). G. Demathieu and H. Haubold (1972) listed a variety of dinosaur footprints from the Middle and Upper Triassic rocks of Germany. Most of these were regarded as the tracks of small theropods and were referred to the ichnogenus *Coelurosaurichnus*; others may represent Hitchcock's ichnogenus *Otozoum*. Some parallel trackways, described by F. von Huene (1941) from the Middle Triassic of Germany, were originally attributed to prosauropods but are more likely to be the tracks of animals other than dinosaurs. About the year 1921 a group of nearly 50 large footprints was discovered in an outcrop of Upper Jurassic sandstone at the village of Barkhausen, near Osnabrück (Plate 13, p. 306, top left). Many years later the first detailed study of the site was undertaken by M. Kaever and A.F. de Lapparent (1974), who concluded that the footprints were made by a herd of sauropods pursued by a carnosaur. This interpretation is now open to question, since it transpires that the carnosaur travelled in the *opposite* direction to the sauropods (Prof. M. Kaever, *pers. comm.*).

There have been several reports of small trackways from the lithographic limestones (Upper Jurassic) of Bavaria – famous as the source of the primitive bird *Archaeopteryx*. From time to time these curious trackways have been attributed to pterodactyls, to small dinosaurs such as *Compsognathus* and even to *Archaeopteryx* itself (e.g. Jaekel 1929; Wilfarth 1937). However, some detailed detective work by K.E. Caster (1939, 1940, 1941) demonstrated that these perplexing little tracks were certainly the work of horseshoe crabs!

The Lower Cretaceous rocks of Germany, like those of England, have yielded numerous footprints ascribed to *Iguanodon* (e.g. Struckmann 1880; Grabbe 1881a,b; Ballerstedt 1905, 1914). Most of them are normal three-toed forms (Plate 3, p. 56, centre), but some show imprints of only two toes (Ballerstedt 1922) whereas others have four (Dietrich 1927). A more recent discovery, described by U. Lehmann (1978), is of 23 footprints representing several parallel trackways of the *Iguanodon* type. The Cretaceous rocks of Germany have also provided tracks of carnosaurs (e.g. Abel 1935), sauropods (e.g. Sommerkamp 1960; Hendricks 1981) and an unidentified quadruped that was possibly an ankylosaur (Ballerstedt 1922).

Current understanding of dinosaur tracks has been shaped, to a very large extent, by definitive or encyclopaedic studies that emerged from Germany. Some of the most influential works have been those by W. Soergel (1925), O. Abel (1926, 1935), O. Kuhn (1958a, 1963) and H. Haubold (1971, 1984).

France has also yielded a rich and intriguing array of dinosaur tracks, and here, as in Germany, the earliest studies were inspired by the discovery of the enigmatic footprints named *Chirotherium*. Subsequently, there were numerous finds of middle and late Triassic dinosaur tracks, many of them reviewed by L. Courel and his colleagues in 1968 (see also Demathieu and Haubold 1972).[4] Some of these were theropod tracks belonging to Hitchcock's ichnogenera *Anchisauripus*, *Eubrontes*, *Grallator* and *Otozoum* (e.g. Demathieu 1970; Plate 7, p. 118, top right), whereas other theropod tracks were referred to new and distinctive ichnogenera such as *Coelurosaurichnus*, *Saltopoides* and *Talmontopus* (e.g. Kuhn 1958b; Plate 8, p. 134, top left). The rich

assemblages of footprints described by A.F. de Lapparent and C. Montenat (1967) are of particular interest in that they may include the tracks of early ornithopods (*Anatopus*). There are few authenticated reports of dinosaur tracks in the Jurassic of France (e.g. Thaler 1962; de Lapparent and Oulmi 1964). One such report, of a trackway impressed in ripple-marked marine sediments (F. Ellenberger and Fuchs 1965), was apparently made by a bipedal dinosaur traversing a shore-line. In addition, some small three-toed footprints, possibly those of dinosaurs, have been noticed in the Upper Jurassic of Normandy (Rioult 1978: 14), though these have yet to be described in detail. Some odd-looking tracks from the Upper Jurassic of southeastern France have been attributed to hopping dinosaurs (Bernier *et al.* 1984), though this particular interpretation may not be entirely convincing (see Chapter 9). Notable among the French works on fossil tracks are an excellent review by J. Lessertisseur (1955) and a monumental study of Triassic footprints by G. Demathieu (1970), who has also conducted several important investigations into the biological significance of dinosaur tracks (e.g. Demathieu 1975, 1984).

There have been numerous reports of fossil footprints in Spain.[5] These range from early Triassic prints of 'dinosauroid' aspect (Demathieu and de Omenaca 1976; Demathieu *et al.* 1978) to late Cretaceous ornithopod tracks (Llompart 1979). Among the many examples of Jurassic footprints (e.g. de Lapparent 1966; Garcia-Ramos and Valenzuela 1977a,b) is one preserved in marine limestones. Presumably this was made by a dinosaur venturing along a shore or even into shallow sea-water (de Lapparent *et al.* 1965). In 1984,

Plate 3 A variety of dinosaur tracks. Top: An intrepid geologist stands in the tracks of a Jurassic sauropod, on steeply tilted sediments in the Chilean Andes, accessible only to skilled mountaineers. Note the small manus print, alongside the geologist's right foot, and the prominent up-welling of sediment around all the footprints. Part of a second trackway is visible at lower left. Centre: Footprint slabs on display in the gardens of the Munich Museum, about the year 1908. The footprints are from the Lower Cretaceous of Germany and were made by big ornithopod dinosaurs, possibly *Iguanodon*. Bottom: Footprints by moonlight. These ripple-marked sediments in the Lower Jurassic of Lesotho are traversed by the footprints of an ornithopod dinosaur, extending from lower right to upper left. The parallel grooves across the centre were made by a broad-bodied creature, perhaps resembling a turtle.

H. Mensink and D. Mertmann described two new ichnogenera, *Hispanosauropus* and *Gigantosauropus*, which were regarded as the tracks of large Jurassic theropods. However, the latter ichnogenus, with footprints up to 1.3 m long, is more probably the track of a sauropod. The many discoveries made in Portugal include the tracks of late Jurassic theropods (de Lapparent *et al.* 1951; de Lapparent and Zbyszewski 1957) and those of early Cretaceous sauropods, theropods and ornithopods (Antunes 1976; Madeira and Dias 1983).

Tracks of *Iguanodon* type are also known from the Lower Cretaceous of Yugoslavia (Bachofen-Echt 1926), together with some footprints attributed to theropods (Abel 1935, fig. 121) and sauropods (Leonardi 1985a). A series of isolated footprints from the Lower Jurassic of central Poland (Karaszewski 1966, 1969, 1975) was originally suspected to represent five-toed quadrupeds along with three-toed bipeds that were almost certainly dinosaurs. However, G. Gierliński and A. Potemska (1987) recently referred most of these Polish tracks to the ichnogenus *Moyenisauropus*, originally based on ornithopod trackways from the Lower Jurassic of southern Africa. Various reptile footprints, including those of coelurosaurs, are known from the Middle Triassic of Italy (von Huene 1942; Tongiorgio 1980). Tracks of similar age have also been reported from the Netherlands (Faber 1958; Demathieu and Oosterink 1983), and these may also include theropod footprints of the ichnogenus *Coelurosaurichnus* (Demathieu and Oosterink 1988).

In 1952, about 170 dinosaur footprints, including those of ornithopods and theropods, were reported from the Upper Triassic of Sweden (Bölau 1952), and subsequently theropod tracks came to light in the Lower Jurassic as well (Pleijel 1975). Discoveries of early Cretaceous dinosaur tracks on the island of Spitsbergen, well within the present-day Arctic Circle, carry important implications for dinosaurian biogeography and for the theory of polar wandering (Axelrod 1984, 1985). Some of the Spitsbergen tracks were made by carnosaurs (Heintz 1962; Edwards *et al.* 1978), but most of them represent ornithopods similar to *Iguanodon* (de Lapparent 1962; Colbert 1964). The hazardous task of obtaining plaster casts of these footprints, on vertical cliffs battered by wind and sea, has been graphically described by N. Heintz (1963; see Plate 4, p. 62, top right). Fossil footprints are known from an equally inaccessible location in the Swiss Alps, at an altitude of 2 400 m. Here, G. Demathieu and M. Weidmann (1982) investigated an area of about 350 m^2, bearing some 800 footprints of

Triassic age. These included forms similar to *Chirotherium* as well as 'dinosauroid' types that may represent coelurosaurs and ornithischians.[6]

Africa

A few dinosaur tracks are known from the Upper Cretaceous of Algeria (le Mesle and Peron 1881; Gaudry 1890: 269–71; Bellair and de Lapparent 1949). These include theropod footprints of the ichnogenus *Columbosauripus*, first defined by C.M. Sternberg (1932) for dinosaur tracks from the Cretaceous of Canada. Other Algerian footprints, of late Triassic or early Jurassic age, have been assigned to Hitchcock's ichnogenus *Grallator* (Bassoullet 1971). More frequent finds, including abundant tracks of theropods and sauropods, have been made in the Jurassic of Morocco. One of the sauropod trackways (named *Breviparopus*) could be followed for more than 90 m, and has been analysed in detail by J.-M. Dutuit and A. Ouazzou (1980). An equally long trackway made by a bipedal dinosaur, most probably an ornithopod, was reported by M. Monbaron and his colleagues in 1985. In addition to these Jurassic tracks there are unusual examples of theropod footprints from the Upper Cretaceous, some with long heel-like imprints from the metapodium (Ambroggi and de Lapparent 1954a,b). Several new and important finds of dinosaur tracks in Morocco, including the traces of swimming sauropods, have been described by S. Ishigaki (1986, 1989).[7]

Discoveries of dinosaur tracks in Niger were reviewed by P. Taquet in 1976 and 1977. The tracks, which are mostly of Jurassic age, include those of ornithopods, theropods and sauropods (Taquet 1967, 1977a,b). Most impressive of all these discoveries is a magnificent stretch of sauropod trackway, more than 60 m in length (Ginsburg *et al.* 1966; Taquet 1972; Plate 12, p. 278, bottom right). Three-toed dinosaur tracks, some made by theropods, have also been reported from the Cretaceous of Cameroon (Flynn *et al.* 1987; Congleton 1988; Jacobs *et al.* 1989).

In 1972, M.A. Raath announced the first discovery of dinosaur footprints in Rhodesia, now Zimbabwe. These were small theropod prints, of late Triassic or early Jurassic age, which were likened to those in Hitchcock's ichnogenus *Grallator*. A recently discovered trackway of 14 footprints has been attributed to a bigger bipedal dinosaur, possibly a theropod, but is of uncertain age (Broderick 1984).

Dinosaur footprints were first reported more than 50 years ago in the Lower Jurassic of Lesotho, formerly Basutoland (von Huene 1932), and lengthy research on the subsequent finds culminated in prodigious studies by P. Ellenberger (1972, 1974). In the late Triassic and early Jurassic sediments of ancient shores and lake beds Ellenberger deciphered an astonishingly rich series of trackways, testifying to the everyday comings and goings of dinosaurs and other animals (e.g. Plate 3, p. 56, bottom; Plate 6, p. 94, left). The tracks of theropods, ornithopods, prosauropods and, perhaps, of early sauropods, all occur in abundance. Among the many intriguing finds documented by Ellenberger are traces of dinosaurs that dabbled for food in the muddy sediments and the tracks of ornithopods that shifted from all fours on to their hindlegs alone. Some of the footprints, placed in the new ichnogenera *Masitisisauropus* and *Ralikhomopus*, appeared to show imprints of feather-like structures around the toes and were attributed to small dinosaurs that may have resembled birds. However, R.E. Molnar (1985) found no compelling evidence for such plume-like structures and suggested that these markings might be nothing more than invertebrate trails. Overall, Ellenberger identified more than 70 new ichnogenera, representing well over 100 new ichnospecies, though not all of these were the traces of dinosaurs.[8]

Tracks similar in age and appearance to those in Lesotho have also been identified across the border, in the South African province of Natal (van Dijk 1978; van Dijk *et al.* 1978). In addition, there are records of theropod and ornithopod footprints in the Lower Jurassic of Namibia, formerly South-West Africa (von Huene 1925; Gürich 1927; Heinz 1932).

Middle East, Asia and the Orient

Dinosaur tracks were first recorded from Israel by M. Avnimelech in 1962 (1962a,b) and were more fully described in 1966. The numerous footprints, which were discovered in a garden on the outskirts of Jerusalem, were attributed to long-limbed theropods that may have resembled ornithomimids. More recently, a Cretaceous carnosaur track was discovered in a small quarry on the occupied Jordanian West Bank, though 'the hot political situation, in June 1982, did not permit a thorough study' (Leonardi 1985a).

In 1971, A.F. de Lapparent and J. Stöcklin described two fossil footprints from the Jurassic of Aghanistan that might, perhaps, have been made by dinosaurs. P. Taquet (1977b) suspected that the track-makers were sauropods. More definite reports of dinosaur tracks, representing both theropods and ornithopods, have come from the Jurassic of Iran (de Lapparent and Davoudzadeh 1972; de Lapparent and Nowgol Sadat 1975).

So far only a single dinosaur footprint has been reported from India (Mohabey 1986). This three-toed print bears some resemblance to the tracks of ornithopods but was regarded as the forefoot impression of a sauropod on account of its association with three sauropod eggs.

There have been several finds of dinosaur tracks in the USSR, though few of these have been described in detail. Triassic theropod tracks comparable to Hitchcock's ichnogenus *Brontozoum* were reported as early as 1882 (Efremov and Vjushkov 1955), though most subsequent discoveries have been in rocks of Jurassic and Cretaceous age. The Jurassic rocks of Tadzhikistan have yielded numerous tracks, including those of theropods, ornithopods and sauropods. In addition, well-preserved tracks of ornithopods, and perhaps of theropods and sauropods, are known from the Lower Cretaceous of Georgia (Gabouniya 1951, 1952). These include tracks in the ichnogenus *Satapliasaurus*, which was originally attributed to a theropod but seems more likely to represent an ornithopod (Sarjeant 1970). Footprints from the Upper Cretaceous of Tadzhikistan were referred to a new ichnogenus, *Macropodosaurus*, by S.A. Zakharov (1964) but are of uncertain affinities. Cretaceous dinosaur tracks have also been reported from Mongolia (Namnandorski 1957).[9]

Recent reviews of dinosaur footprints in China (Zhen *et al.* 1983, 1985, 1989) list numerous discoveries, mainly in sediments of Jurassic age. Among these, the carnosaur track *Changpeipus* (Young 1960, 1979) is of particular interest in that it may show an imprint of the hand as well as the foot. An assemblage of ornithopod tracks (*Jialingpus*; Zhen *et al.* 1983) is equally interesting because it may include a rare trace of the tail. Notable among the most recent discoveries is a rich assemblage of tracks in the Lower Jurassic of Yunnan; this includes traces of theropods, both large and small, and compares closely with faunas of similar age in the Connecticut Valley, USA, and in southern Africa (Zhen *et al.* 1986). Other Jurassic discoveries include tracks of coelurosaurs (e.g. Young 1966; Plate 7, p. 118, bottom right) and ornithopods (de Chardin and Young 1929: 132; Kuhn 1958a: 24; Young 1960: 62). In addition, coelurosaur tracks

·TRACKS OF TIME·
MESOZOIC ERA, LOWER CRETACEOUS PERIOD

(*Jeholosauripus*) have been reported from the Upper Triassic or Lower Jurassic of Manchuria (Yabe *et al.* 1940a,b), and a few tridactyl footprints are known from the Cretaceous, in association with dinosaur eggs (Zhao 1979).

In recent years dinosaur footprints have been discovered in the Cretaceous rocks of Korea and Thailand. The Korean tracks represent a variety of dinosaurs, including theropods, ornithischians and sauropods (Yang 1982, 1986; Kim 1983; Lim *et al.* 1989), whereas those in Thailand have been attributed to carnosaurs (Buffetaut and Ingavat 1985; Buffetaut *et al.* 1985).

There have been a few discoveries in the Lower Cretaceous of Japan (e.g. Matsukawa and Obata 1985a,b), including the tracks of carnosaurs (Anonymous 1986b: 16) and at least one iguanodontid (Manabe *et al.* 1989).

Australia

Numerous dinosaur tracks are known from Australia, in rocks ranging in age from middle or late Triassic to early Cretaceous. The Triassic coal measures of southeastern Queensland have yielded tracks of theropods, both large and small (Staines and Woods 1964; D. Hill *et al.* 1965; Thulborn 1986). An earlier discovery of Triassic footprints, perhaps including those of small bipedal dinosaurs, was mentioned by F.S. Colliver (1956), but has not been documented in detail. Several footprint finds in the Jurassic of Queensland have been attributed to theropods,[10] while a single print illustrated by D. Hill and her

Plate 4 Dinosaur tracks in unusual settings. Top left: Arguably the most famous fossil footprint in the world. Three-year old Tommy Pendley almost manages to smile, despite his unscheduled daytime bath in a brontosaur footprint. The footprint held 18 gallons (82 l) of water and was excavated at Paluxy Creek, Texas, by R.T. Bird in 1940. Top right: Making plaster casts of *Iguanodon* footprints on the vertical cliffs of Festningen, Spitsbergen, in 1961. Centre: A footprint set to music. This track of an early Cretaceous carnosaur is mounted in the bandstand at the town of Glen Rose, in Texas. It is the type specimen of the ichnospecies *Eubrontes*(?) *glenrosensis*. Bottom left: Two footprints of Jurassic carnosaurs exposed in the roof of the Balgowan colliery, SE Queensland, Australia, in 1951. The colliery manager, Mr Godfrey, gives an indication of size. Bottom right: Early Cretaceous dinosaur tracks on sale near Paluxy Creek, Texas, in the 1940s. A good specimen fetched as much as $50.

colleagues (1966) may represent a quadrupedal dinosaur, perhaps a stegosaur. One footprint, possibly that of an ornithopod, has been reported from Lower Cretaceous rocks on the coast of Victoria (Flannery and Rich 1981), and tracks of similar age are also known from Broome, on the coast of Western Australia (Glauert 1952). These latter were originally attributed to ornithopods (McWhae *et al.* 1958) but are now regarded as the footprints of theropods (Colbert and Merrilees 1967). Undescribed tracks, possibly those of ornithopods, are also known to occur in the Lower Cretaceous of Lightning Ridge, New South Wales, famous as the source of bones and teeth preserved in precious opal.

In the late 1970s, excavations at the Lark Quarry site, in western Queensland, revealed the tracks of more than 160 Cretaceous dinosaurs, both ornithopods and theropods (Plate 14, p. 316). Many of these track-makers were small animals, no bigger than chickens, which seem to have been caught up in a stampede that was triggered by the approach of a large predatory dinosaur (Thulborn and Wade 1979, 1984; Wade 1979a,b; Knowles 1980). The Lark Quarry site has now been roofed for its protection and serves as a tourist attraction. It continues to provide a wealth of scientific data bearing on the locomotion and behaviour of dinosaurs in general (Thulborn 1984; Thulborn and Wade 1984, 1989).

NOTES

[1] Other reports of dinosaur tracks in the United States include: Roehler and Stricker 1984, Spicer and Parrish 1987 (Alaska); Riggs 1904, Farmer 1956, Bunker 1957, Brady 1960, S. Madsen 1986 (Arizona); Pittman 1984, Pittman and Gillette 1989 (Arkansas); Reynolds 1983, 1989 (California); Bird 1939a, MacClary 1939, Parrish and Lockley 1984, Lockley 1987 (Colorado); Mudge 1866, 1874, McAllister 1989 (Kansas); Woodworth 1895, Baird 1988 (New Jersey); Hunt *et al.* 1989, Lucas *et al.* 1989 (New Mexico); Langston 1974, Conrad *et al.* 1987, Pittman 1989 (Oklahoma); Marsh 1899, Anderson 1939 (South Dakota); Shuler 1917, Wrather 1922, Gould 1929, Houston 1933, Bird 1953, Herrin *et al.* 1986, Farlow *et al.* 1989 (Texas); Sanderson 1974, Stokes 1978, Miller *et al.* 1989, Parker and Rowley 1989 (Utah); Gilmore 1924, Olsen *et al.* 1978, Pannell 1986, Weems 1987 (Virginia); Branson and Mehl 1932, Shrock 1948 (Wyoming).

[2] For dinosaur tracks in South America see also: von Huene 1931a,b, Lull 1942, Casamiquela 1964, 1966, Alonso 1980, 1989, Bonaparte 1980, Alonso

and Marquillas 1986 (Argentina); Branisa 1968, Leonardi 1984a (Bolivia); Leonardi 1979a,b,c, 1980b,c, 1981, 1984a,b, 1985b (Brazil); Fasola 1966, Casamiquela and Fasola 1968 (Chile); Leonardi 1989 (all countries).

[3] For Cretaceous dinosaur tracks in Britain see also: Mantell 1851; Tylor 1862; Deck 1865; Dollo 1906; Milner and Bull 1925; Sarjeant 1974; Blows 1978; Delair 1980, 1983; Delair and Sarjeant 1985; Ensom 1983, 1984. British *Iguanodon* tracks described by the Belgian palaeontologist L. Dollo (1906) are sometimes mistakenly assumed to have originated from Belgium (e.g. Sarjeant 1987: 3).

[4] For reports of Triassic dinosaur tracks in France see also: Beurlen 1950; Heller 1952; Courel *et al.* 1968; F. Ellenberger *et al.* 1970; Demathieu 1971; Demathieu and Gand 1972; Courel and Demathieu 1976; Gand *et al.* 1976.

[5] Additional reports of dinosaur tracks in Spain include: de Lapparent and Aguirre 1956; Casanovas Cladellas and Santafé Llopis 1971, 1974; Brancas *et al.* 1979; Garcia-Ramos and Valenzuela 1979; Viera and Torres 1979; Aguirrezabala and Viera 1980, 1983; Viera and Aguirrezabala 1982; Casanovas Cladellas *et al.* 1984, 1985; Llompart *et al.* 1984; Viera *et al.* 1984; Sanz *et al.* 1985; Valenzuela *et al.* 1986; Moratalla *et al.* 1988b.

[6] For reports of Triassic dinosaur tracks in Switzerland see: Somm and Schneider 1962; Bronner and Demathieu 1977; Baud 1978; de Beaumont 1980.

[7] For additional reports of dinosaur tracks in Morocco see: Plateau *et al.* 1937; Ennouchi 1953; Lessertisseur 1955 (pl. 11, figs 4–6); Monbaron 1983; Biron and Dutuit 1981; Jenny *et al.* 1981; Jenny and Jossen 1982.

[8] Other studies on the dinosaur tracks of Lesotho include: F. Ellenberger and P. Ellenberger 1958, 1960; F. Ellenberger *et al.* 1963, 1969, 1970; F. Ellenberger and Ginsburg 1966; P. Ellenberger 1955, 1970. Fortunately, many of the ichnotaxa defined by P. Ellenberger may be synonymous with those erected by Hitchcock for tracks of similar age in the Connecticut Valley, USA (Olsen and Galton 1984). I say 'fortunately' because some of the names coined by Ellenberger are veritable tongue-twisters. Try these: *Deuterosauropodopus*, *Masitisisauropodiscus* and *Qomoqomosauropus*.

[9] For additional information on dinosaur tracks in the USSR see: Olson 1957; Gabouniya 1958; Zakharov and Khakimov 1963; Rozhdestvensky 1964; Khomizuri 1972; Novikov and Sapozhnikova 1981; Novikov and Radililovsky 1984.

[10] For Jurassic footprints in Queensland see: Ball 1933, 1934a,b, 1946; Anonymous 1951, 1952a,b; Staines 1954; Bartholomai 1966; Molnar 1982. These footprints were invariably attributed to theropods, though some might well have originated from ornithopods.

4

Dinosaur tracks in the field and laboratory

Ichnology is expanding and many papers on this subject are being written in different languages. . . methods are often different from school to school, and from country to country. . . some day we may all come to use the same methods and in this way, come to understand each other better.

G. Leonardi, *Glossary and Manual of Tetrapod Palaeoichnology* (1987)

FINDING DINOSAUR TRACKS

The search for dinosaur tracks goes on today much as it did in the nineteenth century, the basic techniques having remained essentially unaltered for 150 years. In the past, many fossil tracks were obtained from commercial stone quarries or from the roofs of coal mines, but nowadays the quarrying of sandstone for building purposes has virtually ceased in Europe and North America. The few quarries that are still in operation are usually so highly mechanized that any fossils go unnoticed and are destroyed. Opportunities for finding footprints in the roofs of coal mines have also declined with the switch from underground workings to open-cut production. Consequently, many of the more recent discoveries of dinosaur tracks have been in natural settings, such as cliffs, hillsides, shores and river beds, rather than in industrial settings such as mines or quarries. Nevertheless, a few important discoveries continue to be made in the course of commercial excavations. In the USA, for example, abundant dinosaur tracks were discovered in 1966 at Rocky Hill, Connecticut, during excavations for the foundations of a research laboratory (Anonymous 1967; Ostrom 1967), and a large series of sauropod tracks was recently uncovered in a gypsum mine near Nashville, Arkansas (Anonymous 1985; Pittman and Gillette 1989).

In searching for footprints it should be remembered that natural casts tend to be more durable than natural moulds, and that they occur on the undersides of sandstone ledges and fallen slabs. In wave-cut platforms or river beds shallow-dipping sediments may be eroded and stripped away layer by layer, revealing footprints in the form of natural moulds. The same is true in exposures of steeply dipping sediments, where large slabs of rock may slide away to expose the bedding planes (Plate 3, p. 56, top). Casts tend to be more eye-catching than moulds, not only because they are more resistant to weathering but also because they stand out against the rock surface as raised reliefs. It is also worth mentioning that some footprints are most noticeable when soaked by rain, which tends to highlight any colour differences between the footprint filling and the surrounding rock (Sarjeant 1988: 126).

Lighting conditions are important. Low-relief footprints, especially shallow or weathered moulds, are likely to be inconspicuous when the sun is directly overhead or when the skies are overcast and the lighting is diffuse. The best conditions occur at sunrise and sunset, when casts and moulds are highlighted by long shadows.

Fossil footprints may also be discovered in stone buildings and pavements. Sometimes these were built in by accident, but on other occasions they were used deliberately as architectural curiosities (e.g. Sarjeant 1974, fig. 8). The first *Chirotherium* footprint to be found in North America was noticed in a stone fireplace (Hamilton 1952), and the type specimen of the ichnospecies *Eubrontes glenrosensis* is mounted in the town bandstand at Glen Rose, in Texas (Shuler 1935; Plate 4, p. 62, centre). In 1980, G. Leonardi recorded his amazement on finding that the Brazilian town of Araquara was 'literally paved' with fossil footprints (Leonardi 1980a; see also Leonardi and Sarjeant 1986).

SITE DOCUMENTATION

When footprints are discovered it is important to document their number, sizes, shapes and distribution. This documentation is a straightforward procedure, which does not require special expertise, and it can be undertaken with a minimum of equipment. The basic equipment comprises: notebook, graph paper, pens or pencils, compass, clinometer, camera with tripod and plenty of film, a stiff brush, hammer and cold chisels, tape measure, ruler and chalk. Personal preferences will dictate the exact choice of equipment. A light (2 kg) club hammer, in conjunction with cold chisels, is often

more versatile than a standard geological hammer; and a stiff metal tape-measure is more useful than a limp one (cloth or plastic) because it can be manoeuvred with one hand. As footprints may be discovered in potentially hazardous situations, such as cliffs and abandoned quarries, it is wise to include a safety helmet in this list of basic equipment.

First of all, the rock surface should be swept clean to expose the footprints as clearly as possible. It is advisable to measure dip and strike of the rock surface and to identify the exposed rock types. If the footprints are very numerous, so that there is some risk of confusion, it may be useful to identify each one with a code number chalked alongside it. The sizes and shapes of the footprints should be carefully noted, as should measurements of pace and stride. Appropriate techniques for taking measurements are explained below. Note the compass bearing of at least one trackway or one footprint, which can then serve as a standard of reference for orienting the other tracks. It is often useful to draw up a sketch map to illustrate the distribution and orientation of the tracks at a site; this is done most easily on graph paper. If the footprints are not too numerous it is advisable to photograph each one. Otherwise it is best to photograph the whole site, area by area. The temptation to save time by photographing only a few 'selected' or 'representative' tracks will often entail the loss of important scientific data. Each photograph should include a ruler (for scale) and a deeper three-dimensional object (e.g. a matchbox) whose shadow will indicate the direction and intensity of lighting. The resulting collection of notes, measurements, sketches and photographs will provide a perfectly adequate documentation of the site.

Plate 5 Examples of footprints in Edward Hitchcock's collection. Top left: Plaster cast of footprint obtained from a living rhea; about 12.5 cm long. Note the finely impressed details, including fleshy pads beneath the toes, skin texture and claw traces. The feet of certain dinosaurs were virtually identical in structure. Top right: The tracks of two small ornithopod dinosaurs traversing an early Jurassic shore-line; the footprints, each about 10 cm long, were assigned to the ichnospecies *Anomoepus curvatus*. The rough-textured lower part of the slab was exposed to the wind and rain whereas the smooth upper part of the slab was under shallow water. Bottom: Single slab of rock split open to reveal the cast (left) and mould (right) of a single theropod footprint, *Anchisauripus tuberosus*; total width is nearly 44 cm. The two halves of the slab have been wired together in the form of a 'stone book' – a device much favoured by Edward Hitchcock.

It is rarely possible to collect dinosaur footprints, unless they occur as detached casts or on small blocks. Often, footprints occur on massive slabs or extensive bedding planes, and in such circumstances it is inadvisable to extract a single print: the removal of one print will often entail the destruction of many others, and that single example will be of much less scientific value than the original assemblage. In nearly all cases it is better to make artificial casts or moulds (see below) than to try collecting the original footprints.

EXCAVATION AND DEVELOPMENT

If a site proves to be of major scientific interest it may need to be excavated so as to reveal the full extent of the dinosaur tracks. Limited excavations can be undertaken single-handed, or by small teams, using standard rock-breaking equipment (sledgehammers, cold chisels, steel wedges, crowbars and levers). Bigger excavations, with rock saws, jackhammers and bulldozers, can be very time-consuming and are impracticable without considerable assistance in the form of manpower, equipment and financial support.

The less-arduous business of development (or preparation) entails the removal of rock adhering to footprints, be they casts or moulds. Development can be attempted in the field, though it is best undertaken in the laboratory, using standard palaeontological techniques (Kummel and Raup 1965; Rixon 1976). Obviously, development is easiest when the casts of the footprints differ from the moulds in texture and/or colour, and when there is a fairly clean parting between them. Sometimes, superfluous rock adheres so firmly to a footprint that any attempt to dislodge it might cause irreparable damage. Such partly obscured footprints may sometimes be left in the open air to weather out naturally (see Sarjeant 1975, fig. 14.2).

Museum collections sometimes contain fossil footprints that have been altered in quite misleading ways. Some examples are known to have been 'repaired' with concrete or plaster tinted to resemble the original rock, while others have been 'enhanced' for public display by artful work with hammer and chisel. One set of dinosaur tracks, now on public display in a museum, is said to have been 'improved' many years ago – by chiselling away the shallow footprints lest they should distract attention from the better-preserved ones! Finally, one should also be alert to the existence of outright forgeries, fakes and carvings, all of which are mentioned in Chapter 7.

ARTIFICIAL CASTS, MOULDS AND REPLICAS

In the field, it is often easier to make artificial casts or moulds than to try collecting the original tracks. Footprints in the form of natural moulds provide artificial casts, and natural casts are the basis for artificial moulds. Such artificial casts and moulds are valuable for several reasons. First, they are permanent three-dimensional records of footprints that occur at remote or inaccessible sites or that otherwise run some risk of being destroyed or lost. Secondly, an artificial cast may reveal details of footprint structure that are not apparent in a natural mould; artificial moulds may be equally enlightening in the study of natural casts. Thirdly, the artificial casts or moulds are easily duplicated in the laboratory and may also be used to manufacture replicas of the original footprints.

Plaster casts and moulds

One of the most widely used casting and moulding materials is plaster of Paris. The process of making a plaster cast (or mould) is straight-forward, if lengthy. First of all, any dust or debris should be swept from the footprint. The surface of the footprint and the surrounding rock should then be lightly smeared with petroleum jelly. Cut a strip of thin cardboard or stiff paper to form a wall surrounding the foot-print at a minimum distance of at least 5 cm. The wall should be at least 5 cm high, and its inner surface should also be smeared with petroleum jelly. Fix the wall in place with rolls of clay or Plasticine (modelling clay) along the base of its inner side. The plaster should be mixed with water to achieve a smooth consistency similar to that of unwhipped cream. Avoid the temptation to use sea water, as this may ultimately lead to crumbling and deterioration of the casts. The addition of a few drops of liquid detergent will ensure smooth mixing of the plaster and will minimize the formation of bubbles. A footprint 30 cm long may require as much as 5 kg of plaster. Carefully pour the mixed plaster into the well containing the footprint, trying not to create bubbles. Make sure that the plaster fills the deepest recesses of the print. In the case of a large footprint it is advisable to reinforce the cast: before all the plaster is poured into the well, insert a piece of wire netting or a piece of sacking which has been well kneaded in wet plaster. As a last resort, casts may be reinforced with a few stout twigs, providing that these have been stripped of their bark. Finally, the remaining plaster is poured in to cover the reinforcing materials. The plaster will take at least 1 h to set, but should preferably be left for as long as possible. To remove the cast, peel away the surrounding paper wall, and pull out the clay or Plasticine underlying the edges of the plaster. The overhanging edges of the plaster then allow fingertips to be inserted so as to obtain leverage. The cast should be lifted free from the print as cleanly and as smoothly as possible.

Plaster casts are economical and (with a little experience) easy to make, but they do have disadvantages: they are heavy, and rather fragile, and they need careful packaging for transport. Moreover, they cannot be obtained from footprints with undercut or overhanging walls.

Latex peels

Latex (liquid rubber) peels are much more convenient than plaster casts. They are easily made, light, durable and easy to transport, regardless of the size or shape of the original tracks. As before, the footprint should be swept clean of dust or debris. The latex is thinned with water, if necessary, and is then applied directly to the footprint with a brush or fingertips. This first layer need not be very thick – much less than 1 mm. Before this first coat dries out completely (usually in a matter of hours) it should be strengthened by applying a second coat. Additional coats can be applied in this fashion until the entire latex covering is sufficiently thick to be peeled away without tearing. Often, it is advisable to reinforce the latex, particularly if it is being used to cover a large area: lay strips of cloth on the wet latex and apply another coat over them, so as to bind the cloth in place. Deep recesses can be filled with chopped-up sponge rubber or small pieces of cloth rolled and kneaded in latex. Once the total thickness of latex is between 1 and 2 mm it can be peeled away from the rock with confidence. A newly made peel should not be folded; instead, it should be lightly dusted with talcum powder and carefully rolled up.

Such high-fidelity latex peels need not be restricted to single footprints, but may be extended to areas of several square metres, providing that they are reinforced with suitable backing materials. Peels of this size are useful as permanent records of entire trackways or large rock surfaces with numerous footprints. It is important to note that latex peels have a limited shelflife. Ideally, they should be stored in a dark cupboard because protracted exposure to light will cause the latex to crumble and deteriorate.

Latex peels have proved immensely valuable for recording finds of dinosaur tracks, as D. Baird (1951: 342) has emphasized:

> A dinosaur trackway . . . represents days of quarrying and tons of handling and shipping weight. A plaster mold of the same trackway is heavy, bulky, and fragile. A rubber mold [latex peel] can be rolled up and tossed into the trunk of the car'.

Beyond that, it is easier to photograph a latex peel in the laboratory, where lighting conditions can be adjusted, than it is to try photographing dinosaur tracks in the field.

Replicas

A cast of an artificial mould will replicate the morphology of the original natural cast; and, likewise, a mould of an artifical cast will replicate the original natural mould. Such secondary casts and moulds will, in effect, be exact duplicates of the original footprints, be they casts or moulds. Such replicas may be produced in plaster or latex, using the methods outlined above, and may also be manufactured in various combinations of plastic, resin and glass fibre, all of which share the advantages of light weight and durability. Such replicas can be astonishingly realistic, especially when tinted to match the originals, and are very satisfactory as museum exhibits. They also have another advantage: they permit leisurely study of dinosaur tracks discovered in remote or hazardous locations.

DESCRIBING DINOSAUR TRACKS

Footprint shape

A systematic description of footprint shape takes into account the number, sizes, shapes and arrangement of the **digits** (impressions of fingers or toes), along with features such as claw-marks, heel-marks, interdigital webs and skin imprints. The following introduction to the descriptive terminology refers to footprints in general: it applies equally well to prints made by the hindfoot or **pes** (that is, 'footprints' *sensu stricto*) and to those made by the forefoot or **manus** ('handprints').

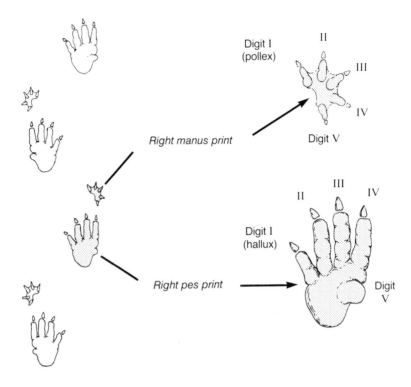

Figure 4.1 Conventional numbering of the digits in fossil footprints.

Number of digits

Many dinosaur footprints are described as **tridactyl**, meaning that they comprise the impressions of three digits. Some are **tetradactyl** (with four digits) or **pentadactyl** (with five digits). **Didactyl** (two-toed) and **monodactyl** (single-toed) footprints are found in the fossil record (e.g. Moodie 1930b; Sarjeant 1971), but they are relatively rare and unlikely to be those of dinosaurs.

In pentadactyl footprints the five digits are conventionally identified by Roman numerals from I (innermost or medial digit) to V (outermost or lateral digit; see Figure 4.1). Digit I of the hindfoot is sometimes referred to as the **hallux** (corresponding to the human big toe), while digit I of the forefoot may be termed the **pollex** (equivalent to the human thumb). Tetradactyl footprints of dinosaurs comprise digits I to IV, having lost the outermost digit (V), whereas tridactyl examples invariably comprise digits II, III and IV (Figure 4.2a–c).

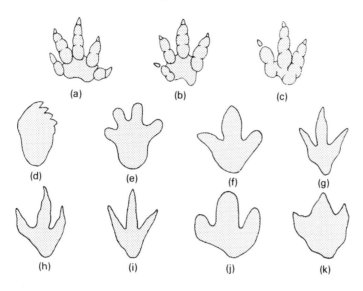

Figure 4.2 Number, size and shape of digits in dinosaur footprints.
(a) Pentadactyl print. (b) Tetradactyl print, lacking digit V. (c) Tridactyl
print, lacking digits I and V. (d–g) Variation in the length of digits, from
extremely short (d) to very long (g). (h) Curved digits. (i) Straight digits.
(j) Rounded or U-shaped digits. (k) Pointed or V-shaped digits. All the
diagrams are based on actual footprints.

There is no universal correspondence between the number of digits in
hands and feet: dinosaurs with tridactyl hindfeet may have forefeet
with as few as two digits (e.g. *Tyrannosaurus*) or as many as five (e.g.
Iguanodon).

Size of digits
The digits within a single footprint are commonly of various sizes.
Very often, digit III is the largest and probably played a major role
in supporting the animal's body weight. On account of its prominence
digit III may be regarded as the **principal digit** – a useful landmark
for making measurements of footprint dimensions. Usually digits II
and IV are somewhat smaller than digit III, and digits I and V (where
present) are smaller still. Footprints conforming to this general pattern
are described as **mesaxonic**, which means that the central digit (III)
is largest and forms the main structural axis of the foot or hand
(Figure 4.3c,d). Footprints are described as **ectaxonic** if the principal
digit is one of the outer ones (IV or V); this condition prevails in the

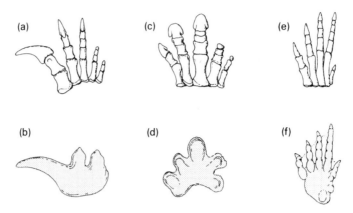

Figure 4.3 Footprint symmetry. (a,b) Entaxonic foot structure and corresponding footprint, with the largest digit on the inner side. (c,d) Mesaxonic foot structure and a comparable footprint, with digit III the largest. (e,f) Ectaxonic foot structure and corresponding footprint, with the largest digit towards the exterior.

hindfeet of lizards but is fairly uncommon in the footprints of dinosaurs (Figure 4.3e,f). Footprints are decribed as **entaxonic** if the principal digit is one of the inner ones (I or II); this condition is seen in the human foot but, again, is relatively uncommon in dinosaurs (Figure 4.3a,b). Measurements of digit size are defined below.

Shape of digits
The digits in a single footprint may be uniform in shape, though they are usually of various sizes. In a few instances one of the digits may be radically different in shape from the others (e.g. Figure 4.3a,b).

Often, the digits are straight, but sometimes one or more of them is distinctly curved (Figure 4.2h,i). As a general (though fallible) rule the convex sides of curved digits face to the exterior. The digits of dinosaur footprints are rarely sinuous or S-shaped.

In some cases the digits are U-shaped in outline, with parallel sides, and in others they are angular or V-shaped, with pointed tips (Figure 4.2j,k). Occasionally, the digits comprise a series of swellings or **nodes** (Figure 4.4). These nodes mark the presence of fleshy pads or cushions under the toes, though there has been much debate as to their exact anatomical significance (see Chapter 5). Similar **digital pads** are seen on the undersurfaces of the toes in modern birds, where they have

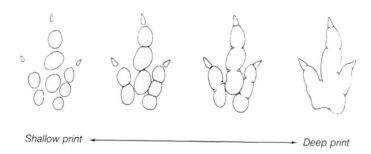

Shallow print ◀────────────────────────────▶ Deep print

Figure 4.4 Digital nodes, impressed by fleshy pads beneath the track-maker's toes. Patterns of isolated nodes, similar to that at left ('shallow print'), also appear in underprints (Figure 2.5), transmitted prints (Figure 2.6) and in some heavily weathered footprints.

sometimes been identified by the Latin name *pulvinus* ('cushion' or 'pad'; Lucas 1979, fig. 1). In deeply impressed footprints the digital nodes may be contiguous, separated only by weak constrictions. But in shallow footprints, where the dinosaur's foot did not sink so deeply into the substrate, each digit may be represented by a series of discrete nodes (Figure 4.4, left). A similar pattern of isolated nodes is sometimes apparent in underprints and transmitted prints.

At the tips of the digits there are often indications of **claws**, which may be long or short, narrow or broad, straight or curved, V-shaped or U-shaped. Sometimes, only the very tips of the claws entered the substrate, leaving puncture-marks rather than complete imprints. In the tracks of predatory dinosaurs the claw-marks tend to be sharply pointed, whereas the footprints of herbivorous or omnivorous dinosaurs reveal claws that were broader, more bluntly rounded and even hoof-like. Certain herbivorous dinosaurs, notably sauropodomorphs, had unusually large and hook-like claws on some digits, but these did not always leave an impression in fossil tracks. Sometimes, the imprint of a claw is a straightforward extension of the digit impression, but in other cases it may seem to veer off or project sideways from the end of the digit. Variations in the claws are discussed in Chapter 5.

Arrangement of digits

The digits may be spread out in a more-or-less symmetrical pattern, or one of them may be widely divergent from the others. Such a widely divergent digit is said to be **opposed** to the other digits – but

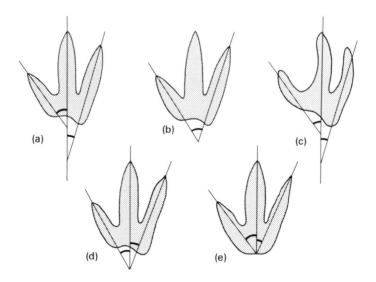

Figure 4.5 Interdigital angles, measured between the axes of two digits.
(a) Two interdigital angles (II–III and III–IV) measured between digital axes
representing lines of best fit. (b) Total divarication, measured between axes
of the innermost and outermost digits. (c) Digital axes fitted only
approximately to curved digits. (d) Interdigital angles measured on the
assumption that all digital axes radiate from a single point (compare
diagram a). (e) Interdigital angles measured on the assumption that all
digital axes radiate from a single point at the rear margin of the footprint.

this does not necessarily imply that the digit was **opposable**, with a
grasping function. In prints of dinosaur hindfeet the opposed digit is
invariably the hallux (digit I), but in handprints the opposed digit may
sometimes be the outermost one (V).

The degree to which two digits diverge may be quantified as the
interdigital angle (Figure 4.5). To measure this angle it is, of course,
necessary to specify an **axis** or midline for each of the two digits. This
is not too difficult when the digits are straight, though the placement
of the digital axes becomes an arbitrary matter when the digits are
curved. Interdigital angles are commonly cited in formal descriptions
of dinosaur tracks, though they are difficult to measure consistently
and may be so variable that they are of doubtful value (Welles 1971;
Sarjeant 1975). Small variations in the interdigital angles are unlikely
to be of great importance, so that it is usually sufficient to provide

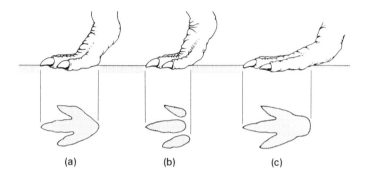

(a)   (b)   (c)

Figure 4.6 Overall shape of footprint related to foot posture. (a) Normal digitigrade posture, with the metapodium carried clear of the ground. (b) Elevated digitigrade posture, with the rear parts of the digits raised from the ground; note that the three toe-prints are separated. (c) Plantigrade or flat-footed posture, with the metapodium leaving a heel-like impression behind the toe-prints.

average or approximate values. In addition, it is useful to add a measurement of **total divarication** – that is, the interdigital angle between the innermost and outermost digits (Figure 4.5b). It is sometimes assumed that all the digital axes must radiate from a single point (Figure 4.5d,e), but some scientists do not follow that assumption and may obtain very different values for the interdigital angles (Figure 4.5a,c).

Metapodium

Most dinosaurs, like birds, were **digitigrade**, which means that they walked with the digits spread out flat on the ground. Normally, the **metapodium** ('sole' of the foot or 'palm' of the hand) did not make contact with the ground – except at its distal end, where the digits radiated from it. As a result, dinosaur footprints often comprise the imprints of the digits together with an imprint from the distal end of the metapodium (Figure 4.6a). In some cases, the rear parts of the digits were carried clear of the ground, so that there is no indication at all of the metapodium (Figure 4.6b); at the other extreme a few **plantigrade**, or 'flat-footed', dinosaurs are known to have placed the entire metapodium on the ground, resulting in a long heel-like impression behind the digits (Figure 4.6c).

In plan view, the rear margin of the footprint is often a smoothly

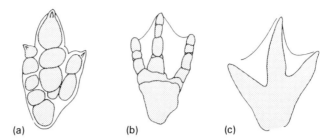

Figure 4.7 Examples of interdigital webbing in dinosaur footprints.
(a) *Otouphepus magnificus*, left footprint of a theropod dinosaur, from the
Lower Jurassic of Massachusetts, USA; about 8.3 cm long. The web-like
trace extending between the digits and round the margins of the foot was
later shown to be an artefact. (b) *Swinnertonichnus mapperleyensis*, left
footprint attributed to a coelurosaur, from the Middle Triassic of
Nottinghamshire, England; about 18.6 cm long. (c) *Talmontopus tersi*, right
footprint of a theropod dinosaur, from the Lower Jurassic of France; about
27 cm long. (Adapted from Lull 1953 (a), Sarjeant 1967 (b), de Lapparent
and Montenat 1967 (c).)

rounded convexity. Less commonly, it may be sharply angular or
concave. Prints of the hindfeet sometimes have an asymmetrical rear
margin, with a pronounced backwards bulge behind the outer toe.
This asymmetry results from the presence of a prominent fleshy pad
behind digit IV, and it is often useful in distinguishing between left
and right footprints.

Interdigital webs

The term 'interdigital web' often causes some confusion. Strictly
speaking, this anatomical term denotes any sheet of flesh connecting
the bases of two adjoining digits – even the small web between two
fingers of the human hand. In this strict sense, traces of interdigital
webs are commonplace in the footprints of dinosaurs. More often,
however, the term is taken to indicate very extensive webbing, such
as that between the toes of a duck. Such extensive webbing is rare in
dinosaur footprints (Figure 4.7; Plate 8, p. 134, top left).

Footprint dimensions

There are no universally accepted methods for measuring the dimensions of footprints, and in the past different scientists have used different methods, so that it is sometimes difficult, or impossible, to verify and compare their published findings. The methods recommended here are some of the most commonly used (Figures 4.8, 4.9). Nevertheless, it is always advisable to specify *how* measurements were obtained from dinosaur footprints. The easiest way to do this is to supplement the measurements with an explanatory sketch or diagram. Measurements should, of course, make allowance for any displacements in footprints traversed by fractures (e.g. Plate 8, p. 134, top right).

Small footprints can be measured very accurately with callipers, while larger ones are best measured with a ruler or a tape-measure. Where footprints occur in a series, or trackway, it is advisable to measure every one. This procedure may be time-consuming, but it yields valuable scientific information. Avoid the common practice of selecting and measuring one 'average' or 'representative' footprint from a whole trackway. All too often that specially selected footprint will be the largest, the best preserved and very far indeed from 'average'.

It is possible to obtain numerous measurements from a single footprint (see Leonardi 1979a, 1987), but many of these are important only for detailed taxonomic studies and are of no concern here. For present purposes it is sufficient to explain a few of the more important measurements.

Footprint length (*FL*)

The length of a footprint is best measured along, or parallel to, the axis of the principal digit. Since many dinosaur footprints are mesaxonic, their length should be measured along, or parallel to, the midline of digit III (Figure 4.8). Problems may arise where the footprint is irregular in shape or where the digits are curved: in the latter case it should be specified whether length has been measured in a straight line or along the curved axis of the digit. Measurements that include the imprint of the metapodium should be distinguished from those that do not; and, in the case of theropod footprints, measurements that include the backwardly turned hallux should be distinguished from those that exclude it. The important point is that the length, like any other dimension, should be measured unambiguously and in consistent fashion from one footprint to the next.

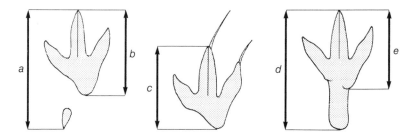

Figure 4.8 Various measurements of footprint length: *a*, maximum length of footprint including the backwardly directed hallux; *b*, length of the same footprint excluding the hallux; *c*, length of footprint excluding drag-marks at the tips of the digits; *d*, maximum length of footprint including the trace of the metapodium; *e*, length of the same footprint excluding the trace of the metapodium. All measurements are made along, or parallel to, the axis of digit III.

Footprint width (*FW*)

The width of a footprint is best measured at a right angle to footprint length. A footprint's width does not necessarily coincide with its **span**, which is the measurement from the tip of the innermost digit to the tip of the outermost one. In some of the older publications on dinosaur tracks the measurements cited for footprint width are, in fact, measurements of footprint span.

Length of digits

There are many different recommendations for measuring digit length (e.g. Demathieu 1970: 25; Leonardi 1979a: 45), some of which offer conflicting advice or are difficult to apply in the case of poorly preserved footprints. Ideally, the length of a digit should be measured along its axis or midline. In a footprint with little or no trace of the metapodium it is relatively easy to measure the length of each digit imprint (Figure 4.9a). Usually, the tips of the digits are sharply imprinted, though their bases, where they converge to join the metapodium, may be shallower and much less well-defined. A problem arises when the digits are appreciably curved: here it should be specified whether the measurement is made along the curve of the digit or in a straight line that corresponds only approximately to the digital axis.

In some cases the digits merge into an imprint made by the distal

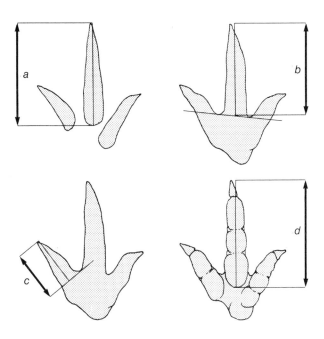

Figure 4.9 Conventions for measuring the length of digits:
a, Straightforward measurement of digit length where there is no imprint
from the metapodium; b, digit length measured to the point midway
between the hypex on each side of the digit; c, length of inner (or outer)
digit measured to the point where the digital axis intersects a perpendicular
through the single adjoining hypex; d, digit length measured to the rear
margin of the hindmost digital node. All measurements are made along, or
parallel to, the digital axis.

end of the metapodium, so that it is difficult to decide where the digits
end and the metapodium begins. If the digital nodes are obvious, the
length of each digit may be measured as far back as the limit of its
hindmost node (Figure 4.9d). Otherwise a useful convention is to
measure digit length to a point midway between the **hypex** on each
side (Figure 4.9b). (The hypex is the re-entrant or 'notch' at the junction of two digits.) However, the innermost and outermost digits are
flanked by only one hypex; in these cases, digit length may be
measured to a point where the digital axis intersects a perpendicular

through the adjacent hypex (Figure 4.9c). The distance from the tip of the digit to the level of the hypex is sometimes termed the 'free length' of the digit.

Width of digits

The digits may be so curved, tapered or irregular in shape that it is difficult to maintain a consistent technique for measuring their width. As a general guide, the width is best expressed as maximum width at a right angle to the digital axis.

Area of footprint

The area of a footprint is useful for mathematical analyses of footprint data, where the square root of footprint area provides a linear dimension of footprint size that is suitable for statistical treatment. In addition, estimates of footprint area may be used to calculate the pressures exerted on the ground by the feet of dinosaurs (Alexander 1985: 21–2). The area of a footprint can be determined by superimposing a grid of centimetre squares onto a photograph or drawing and then counting the number of squares occupied by the footprint (e.g. Dutuit and Ouazzou 1980: 100). An alternative method is to cut out a photograph or an accurate scale drawing on paper of known weight (g/cm^2) and to weigh the cut-out in a chemical balance. The area of the cut-out can be calculated from its weight, and it is then scaled up to ascertain the area of the original footprint. In using this technique it is advisable to make periodic checks on the average weight of the paper, in case it suffers hygroscopic effects.

Number and arrangement of footprints

A series of footprints made by a single animal is termed a **trackway**. The term 'trail' is sometimes applied to dinosaur trackways but is probably better confined to the continuous traces left by crawling or burrowing invertebrates. In the case of a bipedal dinosaur the term trackway is usually taken to indicate a minimum of three consecutive hindfoot prints. In the case of a quadruped the term denotes a minimum of six footprints – three consecutive prints from the forefeet along with the three associated prints of the hindfeet. Such a minimum number of consecutive footprints (sometimes termed a **set** of footprints) can provide numerous measurements, though only a few of the more important ones are explained here. Some of these

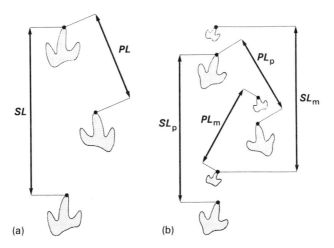

Figure 4.10 Stride length (SL) and pace length (PL). (a) Stride length and pace length in the trackway of a bipedal dinosaur. (b) Stride length and pace length in the trackway of a quadrupedal dinosaur; manus and pes must take strides of equal length (SL_m and SL_p, respectively), though their paces need not be equal in length (PL_m and PL_p, respectively). All measurements are made between corresponding points in two footprints.

measurements are defined by reference to a midline or **axis**, which is a notional line of 'best fit' separating left and right halves of the trackway.

Stride length (SL)

Studies on the locomotion of living animals define a **stride** as the distance covered by an animal during one complete cycle of limb movements. That distance can be measured directly on the trackway of a dinosaur or any other animal – as the distance between corresponding points in two successive footprints of a single foot (Figure 4.10). The measurement may be obtained from two successive prints of a single hindfoot or, in the case of a quadruped, from two successive prints of a single forefoot. Note that the measurement is made between *corresponding* points (sometimes termed 'homologous' points; Leonardi 1979a: 25). So, for example, stride length might be measured from the centre of one right hindfoot print to the centre of the next right hindfoot print. In a study of trackways made by salamanders, F.E. Peabody (1959) recommended that the base of digit

III should be selected as the reference point for making trackway measurements, but it may be difficult or impossible to apply this recommendation to poorly preserved dinosaur tracks. In practice, it is often easiest to make measurements from the tip of the principal digit, because this is usually a sharply defined and unmistakable reference point. In any event it should be noted that the choice of reference points will affect other trackway measurements, such as trackway width and pace angulation (both described below). The ratio of stride length to footprint length (**SL/FL**) is often cited in formal descriptions of fossil trackways.

Pace length (*PL*)

Pace length is the distance between corresponding points in two successive footprints (e.g. left hindfoot to right hindfoot; or right forefoot to left forefoot). In narrow trackways, where all footprints fall more or less into a single line, pace length is roughly equivalent to half stride length. In broader trackways, where left and right footprints have a definite zig-zag arrangement, pace length is somewhat more than half stride length. The forelegs and hindlegs of quadrupedal dinosaurs took strides of equal length, though they frequently took paces that were of different length. Formal descriptions of trackways often specify the ratio of pace length to footprint length (**PL/FL**).

Pace length is sometimes referred to as 'oblique pace' or 'step length', and in the older literature the terms pace, step and stride were sometimes used interchangeably, leading to much confusion. Nowadays, the terms pace and stride usually correspond to the definitions given here (Figure 4.10).

Pace angulation (*ANG*)

With measurements of two successive paces (PL_a and PL_b), and of the stride (*SL*) that they encompass, it is possible to calculate pace angulation (*ANG*) as follows:

$$\cos ANG = \frac{(PL_a)^2 + (PL_b)^2 - (SL)^2}{2 \times (PL_a) \times (PL_b)} \tag{4.1}$$

The more nearly pace angulation approaches to 180° the narrower is the trackway and the less obvious is the zig-zag arrangement of its footprints (Figure 4.11a). On a small trackway, pace angulation may

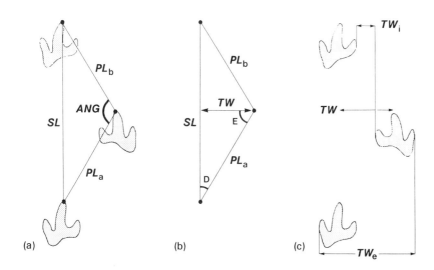

Figure 4.11 Pace angulation (*ANG*) and trackway width (*TW*) in the trackway of a biped. (a) With measurements of two successive paces (*PL*$_a$ and *PL*$_b$), and of the stride they encompass (*SL*), it is possible to calculate pace angulation (*ANG*) using equation 4.1, in text. (b) The angulation pattern, representing the triangle formed by paces and stride in diagram (a); width of the angulation pattern (*TW*) can be calculated using equations 4.2 and 4.3, in text. (c) Various measurements of trackway width, namely: minimum trackway width (*TW*$_i$), between inner margins of left and right footprints; width of the angulation pattern (*TW*), as in diagram (b); maximum trackway width (*TW*$_e$), between outer margins of left and right footprints. All measurements of trackway width are at right angles to the midline of the trackway.

be measured with a protractor, though in most instances it will be found more convenient to calculate it with the preceding equation (which is not applicable when pace angulation exceeds 180°). In the trackways of quadrupedal dinosaurs, pace angulation for the forefeet may differ considerably from that for the hindfeet.

Trackway width (*TW*)
Trackway width may be measured in several ways: as **internal** trackway width (between the inner edges of left and right footprints), as **external** trackway width (between the outer edges of the footprints), or as width of the **angulation pattern** (Figure 4.11b,c). All

these measurements are taken at a right angle to the trackway's midline, and all of them vary inversely with pace angulation. Internal and external widths are usually measured directly on the trackway, but the width of the angulation pattern can be calculated from the measurements of paces and strides (Figure 4.11b). First, calculate the angle of divergence ($\angle D$) of pace (PL_a) and stride (SL) measured from a single footprint, as:

$$\cos D = \frac{(PL_a)^2 + (SL)^2 - (PL_b)^2}{2 \times (PL_a) \times (SL)} \tag{4.2}$$

The angle of divergence ($\angle E$) of TW and PL_a may then be calculated as:

$$\angle E = 180° - (\angle D + 90°) \tag{4.3}$$

Consequently, one obtains estimates of all three angles, along with the length of one side (PL_a), in a right-angled triangle (Figure 4.11b). It is then an easy matter to calculate the length of a second side (TW) by simple trigonometry. Trackway width is not necessarily the same for forefeet and hindfeet in the case of a quadrupedal dinosaur.

Footprint rotation

Footprints may point outwards, away from the midline of the trackway, in which case they are said to show **negative rotation**, or they may point inwards, in 'pigeon-toed' style, showing **positive rotation**. The degree of rotation may be measured as the angle between the principal axis of the footprint (usually the axis of digit III) and the midline of the whole trackway. In the trackway of a quadrupedal dinosaur the pes prints and the manus prints may be rotated to different degrees, and even in different directions. The terminology adopted here follows the usage of H. Haubold (1971: 6) and W.A.S. Sarjeant (1975: 290), and it should be noted that other ichnologists (e.g. Demathieu 1970: 25; Leonardi 1979a: 37) have used the terms negative rotation and positive rotation in exactly the opposite sense. To confuse matters even more, there is a great variety of other terms, including 'pes [or manus] angulation' (e.g. P. Ellenberger 1974), 'divarication' [from the midline of the trackway], and 'toeing in' and 'toeing out' (e.g. Baird 1954). P. Ellenberger (1972: 19) has also used the medical terms 'valgus' (out-turned) and 'varus' (in-turned).

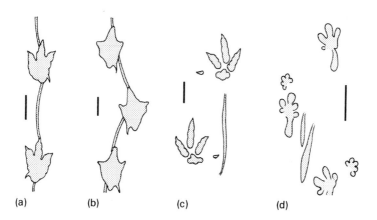

Figure 4.12 Tail marks in dinosaur trackways. (a) Continuous tail-drag in *Gigandipus*, a theropod trackway from the Lower Jurassic of the Connecticut Valley, USA. (b) An apparent tail-drag in *Hyphepus*, another theropod trackway from the Lower Jurassic of the Connecticut Valley. (c,d) Discontinuous tail-drags in *Moyenisauropus*, ornithopod trackways from the Lower Jurassic of Lesotho. Each scale bar indicates 25 cm. (Adapted from Lull 1953 (a,b), P. Ellenberger 1974 (c,d).)

Miscellaneous trackway features

Well-preserved footprints may retain indications of skin texture – not only the nodes and pads mentioned earlier, but also wrinkles, creases, scales and tubercles. Some caution should be exercised in identifying such impressions of the skin, because secondary mineral deposits sometimes infiltrate the parting between cast and mould, taking on a wrinkled, botryoidal or tuberculate habit that may bear a strong resemblance to reptilian skin texture.

Most dinosaurs seem to have walked with the tail lifted well clear of the ground, for traces of a dragging tail are relatively uncommon. Where a tail marking does occur it may be continuous or broken, straight or sinuous (Figure 4.12; Plate 6, p. 94, left). In some cases the tail was dragged or swept through the newly formed footprints, disfiguring them to some extent.

When a dinosaur rested on the ground the underside of its body sometimes left an impression between the left and right footprints. An impression located in front of the hindfeet (Figure 4.13a) probably represents the belly or the chest of the animal. In other instances

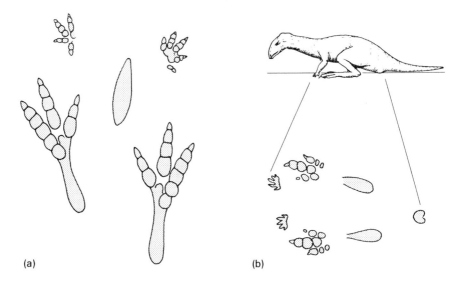

Figure 4.13 Traces of dinosaurs that squatted or lay on the ground.
(a) *Anomoepus scambus*, the trace of an ornithopod resting on all fours, from
the Lower Jurassic of the Connecticut Valley, USA; total length about
28 cm. Note the imprint made by the dinosaur's chest or belly. (b) *Sauropus
barrattii*, the trace of an ornithopod resting on all fours, also from the
Lower Jurassic of the Connecticut Valley; total length about 70 cm. Note
the heart-shaped imprint of the ischiadic callosity. Shown above is a
conjectural restoration of the track-maker in appropriate resting posture.
(All illustrations adapted from Lull 1953.)

there may be a marking between, or behind, the hindfeet (Figure
4.13b). Here, it seems that the animal assumed a sitting posture, with
its cloaca protected from abrasion by a definite swelling or callosity.
Other markings may include traces made by dinosaurs dabbling or
rooting through sediment in search of food (Figure 4.14).

ILLUSTRATING DINOSAUR TRACKS

It is notoriously difficult to produce satisfactory illustrations of
dinosaur footprints. They are often depicted in outline sketches,
silhouettes or line drawings that convey little more than the basic size
and shape. In making such a drawing it can be difficult to determine
the exact boundaries of a footprint: some prints do have sharp edges,

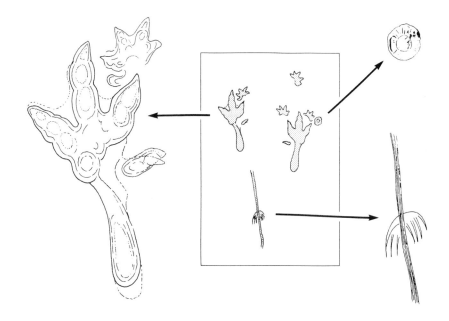

Figure 4.14 Resting and feeding traces of an ornithopod dinosaur, *Moyenisauropus natator*, from the Lower Jurassic of Lesotho; total length about 1 m. At left are the manus and pes prints, the latter with a prominent trace of the metapodium; at lower right is a fan-shaped imprint of the ischiadic callosity, bisected by the tail-furrow. At top right is a crater-like marking, apparently made by the animal's snout as it rooted in the mud. (Adapted from P. Ellenberger 1974.)

but others have rounded margins that grade imperceptibly into the surrounding rock. W.A.S. Sarjeant reported (1975: 285) that in museum collections he encountered some specimens 'in which India-ink lines, supposedly bounding a print, actually traverse the impression of a digit; drawings based on such outlines would be very misleading'. An outline drawing of a footprint should always be viewed with some reservation: it represents one person's interpretation of a complex three-dimensional object, and someone else's interpretation might differ considerably. The need for caution is demonstrated in Figure 4.15, which shows eight different interpretations of a single footprint. Despite their subjective nature, outline drawings continue to be widely used because they are easy to make and economical to publish. Unfortunately, such line drawings tend to be copied from

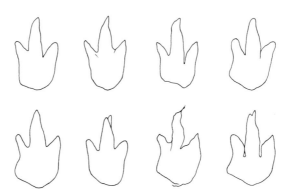

Figure 4.15 Eight different interpretations of a single footprint. Eight subjects were provided with the same photograph of a single footprint and were requested to trace the footprint's outline as carefully as possible. The photograph used in this experiment showed a theropod footprint, *Anchisauripus gwyneddensis*, about 12 cm long, from the Upper Triassic of Pennsylvania, USA (Bock 1952, pl. 44).

publication to publication until, ultimately, they may degenerate into diagrams that bear little resemblance to the original footprints. Nevertheless, outline drawings are valuable in the role of explanatory diagrams, as was stressed by Edward Hitchcock (1858: 51-2):

> In 1848 I expressed the opinion that 'for the discrimination of [footprint] species, outline sketches are better than full-shaded drawings of individual specimens, because they present more distinctly the essential characters'. . . I still remain of the same opinion.

In 1952, D. Baird introduced an ingenious method to minimize subjective interpretations of footprint shape. The footprints were outlined in ink on latex peels, and the inked outlines were then transferred to damp paper. Baird commented that this was a refinement of the direct printing method that was sometimes used by Edward Hitchcock (1858: 51):

> In the quite small tracks, however, I have sometimes touched them (when in relief,) with some coloring matter, – say a red pencil, – and then pressed upon the slabs a piece of rather thin paper, which would retain the exact form and position of the tracks.

Hitchcock also resorted to several other techniques in his efforts to obtain the most accurate representations of fossil footprints, including tracings made on paper, cloth and smoked glass. In some circumstances it is possible to use the brass-rubbing technique – by laying soft paper over the footprint and then gently rubbing the paper with graphite.

Drawings more informative than mere outlines require the judicious use of shading. Such illustrations will often demand the talents of a professional artist, or trial and error using the basic techniques recommended for scientific artwork (e.g. Staniland 1953; Zweifel 1961; Isham 1965). Even so, as D. Baird has noted (1952: 832–3), the older literature on fossil footprints is replete with shaded illustrations that should not be taken at face value:

> Many of the errors in ichnological literature can be traced to inadequate or inaccurate figuring of the types. . . As many of the early footprint types have never been photographed, subsequent students must rely on the accuracy of woodcuts, engravings, and lithographs – media in which the facts may be considerably modified by subjective interpretation.

Photographs are often the most informative of all illustrations, providing that they meet a few basic criteria. They must be of a size adequate to show all the footprint features and they must maintain adequate contrast. The source of lighting (conventionally from the upper left corner in scientific illustrations) is not so important when illustrating footprints, providing that the direction of lighting is indicated in the photograph or its caption. The easiest way to meet this requirement is to include some small three-dimensional object, such as a matchbox, in each photograph. The shadow of that familiar object will automatically reveal the source of lighting. If the direction of lighting is not specified, a photograph of a footprint can be puzzling or quite misleading. This is apparent from two photographs of a single footprint under different lighting conditions (Plate 6, p. 94, top centre and right). Stereoscopic pairs of photographs are particularly informative, but these are even more costly to publish than conventional photographs. A promising new approach is to photograph footprints through a carefully illuminated grille, so that the distorted shadowing (Moiré effect) generates a form of contour map suitable for computer analysis (Ishigaki and Fujisaki 1989).

Less satisfactory are those photographs of footprints that have been

outlined in chalk or ink, or otherwise highlighted, so as to stand out from the surrounding rock. While such practices are not too common nowadays, it may still be useful, on occasion, to emphasize the outline of a footprint mould by filling it with sand or water. In photographing wet or water-filled footprints a Polaroid lens-filter may be used to eliminate glare. If there could be any doubt about the footprint features shown in a photograph it is wise to provide a diagrammatic key.

One useful technique to illustrate large assemblages of footprints, or long trackways, is to build up a composite photomosaic, comparable to a series of very low-level aerial photographs. The footprint site, or trackway, is photographed piece by piece and the individual photographs are subsequently assembled into an overall picture (e.g. Plate 14, p. 316, bottom). In taking photographs for such a mosaic it is essential that the focal plane of the camera be absolutely parallel to the rock surface bearing the footprints; if it is not, the resulting photomosaic will appear be to be skewed or distorted. One extensive series of sauropod tracks has, in fact, been studied by means of conventional aerial photographs (Pittman and Gillette 1989).

Plate 6 Preservational and morphological features of dinosaur tracks. Left: *Moyenisauropus longicauda*, the trackway of an ornithopod dinosaur in the Lower Jurassic of Lesotho; the footprints are outlined in chalk and each is about 12 cm long. The prominent groove made by the dragging tail is an unusual feature, rarely encountered in dinosaur tracks. Top centre and Top right: Two photographs of a single coelurosaur footprint, *Skartopus australis*, from the mid-Cretaceous of Queensland, Australia; about 8 cm long. Left-hand photograph has illumination directly from right side; right-hand photograph has illumination from upper right corner. Note how a slight change in lighting has altered the appearance of the footprint. Centre: 'Squelch marks' or irregular depressions seemingly made by small dinosaurs paddling about in slushy mud, from the Upper Triassic of South Glamorgan, Wales; the area shown is about 80 cm wide. Bottom centre: Theropod footprint resembling *Anchisauripus*, filled with coarse pebbly sediment, from the Upper Triassic of South Glamorgan, Wales; about 14 cm long. Erosion of surrounding rock has left the coarse footprint filling standing on a pedestal. Bottom right: A coelurosaur footprint, *Skartopus australis*, from the mid-Cretaceous of Queensland, Australia; about 11 cm long. The prominent and sharply defined imprint of the metapodium is an unusual feature, perhaps indicating that this dinosaur used an exceptional flat-footed gait.

In photographing footprints in the field it may be necessary to reconcile the advantage of low-angle illumination with the disadvantage of low light intensity, particularly at dawn or dusk. These difficulties can often be overcome by setting up the camera and waiting for the best moment. Alternatively, the natural lighting can be supplemented – for example by vehicle headlights or an electric torch – or it can be modified by using white screens (e.g. sheets of notepaper). It is even possible to take long-exposure photographs under moonlight (Plate 3, p. 56, bottom). However, the easiest solution to all these problems is simply to take a latex peel and to photograph it at leisure in the laboratory.

INTERPRETING DINOSAUR TRACKS

Following on from the measurement and description of dinosaur tracks is the more difficult business of their interpretation. First, it is necessary to relate the footprints to existing knowledge – to identify them, to attach a formal name to them (wherever appropriate), and to accommodate them within a systematic classification of footprint types. Secondly, there is the matter of biological interpretation – to assess what significance the footprints might have for the understanding of dinosaur biology. (A third concern – the evaluation of dinosaur tracks as evidence about ancient environments – was examined in Chapter 2.) These taxonomic and biological appraisals cannot be made in isolation, but must be set in the context of what is already known about dinosaurs and their footprints. In other words, it is essential to have some familiarity with the existing literature on dinosaur tracks. These three topics – taxonomic assessment, biological assessment and the literature – are introduced here in order to provide some background for the discussions in following chapters.

Nomenclature and classification of dinosaur tracks

Nomenclature

Edward Hitchcock's pioneering work on the footprints of the Connecticut Valley, USA, began the now-standard practice of identifying fossil tracks by means of a **binomen** – that is, a combination of two formal scientific names. Each major type of track is identified by an **ichnogenus** name, corresponding to the genus names that are applied to animals and plants in the Linnean system of nomenclature. Each ichnogenus comprises one or more **ichnospecies**, equivalent to Linnean species names. A detailed code of nomenclatural practice has been developed for trace fossils, similar to the codes that exist for the guidance of zoologists and botanists (Sarjeant and Kennedy 1973; Basan 1979). A binomen was first applied to the fossil footprints of a reptile in 1835, when J.J. Kaup coined the alternative names *Chirotherium barthii* and *Chirosaurus barthii* for tracks that had been discovered in the Triassic of Germany. Ichnogenus and ichnospecies names were first applied to dinosaur footprints by Edward Hitchcock, as mentioned previously. Even so, some dinosaur tracks continue to be known by informal names. It is common to find mention, for instance, of 'ornithopod tracks' or '*Iguanodon* footprints' rather than formal ichnogenus and ichnospecies names.

Although the use of ichnogenus and ichnospecies names is widespread, a few scientists have argued that such names are inappropriate for inorganic objects such as footprints (e.g. Faul 1951; C.C. Branson 1967). An alternative system of nomenclature, employing a series of code letters and numbers, was proposed by H. Faul (1951) but failed to gain any currency. Many palaeoichnologists continue to apply formal names to fossil footprints because such names are easier to remember than numerical codes or symbols, and because they are less likely to be overlooked while searching the literature (Peabody 1955; Frey 1973: 16).

Dinosaur footprints are so variable in their appearance, even within a single trackway, that numerous 'ichnotaxa' might be established on the basis of trifling variations. Fortunately, most palaeoichnologists have adopted a conservative attitude to the naming of ichnotaxa and seem quite prepared to accept a considerable range of morphological variation within an ichnogenus or an ichnospecies (Sarjeant 1971: 347). Consequently, there has not been an excessive proliferation of names. Some of the names applied to dinosaur tracks are very appropriate and useful. For example, the ichnogenus *Coelurosaurichnus* probably does comprise the tracks of coelurosaurs; likewise, the name *Hadrosaurichnus* may well signify a hadrosaur track (Alonso 1980). Unfortunately, some other names are less appropriate or downright confusing. So, for instance, footprints in the ichnogenus *Anchisauripus* were most likely produced by theropod dinosaurs and not by the prosauropod *Anchisaurus*. Ichnogenus names have been applied not only to footprints but also to fossil eggs (e.g. Chao and Chiang 1974), coprolites and, strange as it may seem, to teeth-marks (e.g. Cruickshank 1986). No one has yet applied formal names to gastroliths or dinosaur nests.

It must be emphasized that ichnogenus and ichnospecies names apply only to trace fossils, and not to the animals that produced them. The ichnogenus name *Anchisauripus*, for example, refers to a particular type of footprint, and not to the dinosaur that made the footprint. This distinction tends to be forgotten in many works mentioning dinosaur tracks, where it is assumed or implied that the name of a footprint is also the name of the track-maker.

Classification

Although it is common practice to arrange ichnogenera and ichnospecies in a systematic classification there is, unfortunately, no consensus on which classification. Different ichnologists continue to follow quite different approaches to classification, each with their own underlying assumptions. This confusing situation is likely to persist indefinitely because the existing recommendations for footprint nomenclature make no provision for taxa above the level of ichnogenus (Sarjeant and Kennedy 1973).

Basically, there have been three approaches to the classification of dinosaurian trace fossils. (Many other approaches have been adopted in classifying trace fossils of invertebrates: see Seilacher 1953, 1964; Martinsson 1970; Vialov 1972; Frey 1973; Simpson 1975; Frey and Pemberton 1985.) The first, and perhaps the most obvious, approach is simply to deposit ichnogenera and ichnospecies in the existing Linnean classification of animals. So, for example, the theropod track *Grallator* would be placed in the order Theropoda of the subclass Archosauria. This procedure, while convenient, is open to serious objections. First, trace fossils such as footprints are not, by definition, organisms or parts of organisms; consequently, there is no justification for including them in a classification of organisms. Secondly, a trace such as a footprint might be regarded as an artefact (in the sense of being an inanimate object resulting from the activity of an organism), and the introduction of artefacts into a classification of animals would lead to a ludicrous situation: the name *Homo sapiens*, for instance, would denote not only humans, but also a veritable junk-yard, including everything from stone axes to space shuttles. Thirdly, a Linnean classification is nowadays taken to imply evolutionary relationships between taxa; and footprints do not evolve.

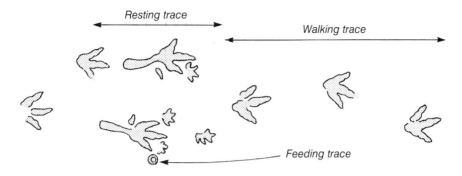

Figure 4.16 A single dinosaur trackway resulting from several different behaviours – resting on all fours, feeding, and walking bipedally. Based on *Moyenisauropus natator*, an ornithopod trackway from the Lower Jurassic of Lesotho; total length about 2.1 m. (Adapted from P. Ellenberger 1974.)

The second approach to classifying tracks and traces is to erect a Linnean-like system for these objects alone: ichnogenera might be assembled in ichnofamilies, ichnoclasses, and so on. In this way it would be possible to construct a classification of traces to mirror the classification of trace-makers. Aside from the objections mentioned already, there are some practical difficulties in implementing this proposal. Most importantly, the animals responsible for tracks and traces can rarely be identified with precision. Few traces could be allocated with confidence to their proper places within such a classification; conversely, large numbers of tracks and traces would be referred tentatively to major categories (e.g. ?Reptilia) or would simply accumulate under the uninformative, if honest, label *incertae sedis* ('of uncertain position').

The third approach is to construct an entirely independent system of classification for tracks and traces. In this case the problem is to find the most appropriate criterion to underlie the classification. Should traces be classified by function, or by morphology? Both criteria have been recommended and used on different occasions, and both present some practical difficulties. A function-based classification is convenient at higher taxonomic levels but may require subjective or arbitrary decisions to discriminate between lower taxonomic categories. It is easy to distinguish coprolites from footprints but it may be difficult or impossible to define different types of coprolites (or footprints) on a purely functional basis. Moreover, a function-based classification has difficulty in accommodating multi-functional traces. Consider, for example, the dinosaur trackway shown in Figure 4.16. In a function-based scheme one part of this trackway (a resting trace) would be classified separately from another part (a walking trace). Presumably, these two parts of a single trackway would merit recognition as different ichnogenera.

A morphology-based classification presents equally intractable problems. Here, too, the dinosaur trackway just mentioned would be dismantled into morphologically distinct ichnotaxa, which would be distributed piecemeal through the classification. This problem is exacerbated by the fact that traces are so variable in their morphology: two successive prints of a single foot may be quite different in shape, as will be explained in Chapter 5.

In practice, many works on dinosaur footprints manage to avoid these difficulties. It is widely agreed that ichnogenera and ichnospecies are to be defined on morphological criteria, and due allowance is made for the great variation that may affect the appearance of footprints. But above the level of ichnogenus there are no standard procedures. Some authorities have classified footprints in a Linnean framework (e.g. Haubold 1971) whereas others have used special ichnotaxonomic categories (e.g. Vialov 1966). Many other ichnologists avoid the problems by using informal groupings, such as 'theropod footprints' or 'ornithopod tracks'. This latter approach, which seems to be least confusing and easiest to apply, has been adopted in this book.

Biological interpretation of dinosaur tracks

Careful evaluation of a dinosaur track can provide information about the identity, size, anatomy and behaviour of the track-maker, as will be explained in following chapters. Those insights into dinosaur biology depend to a large extent on assumptions and methods that originated in studies of animals other than dinosaurs. The assessment of dinosaur locomotion, for example, draws heavily on the understanding of locomotion in living mammals (Alexander 1976; Coombs 1978a; Thulborn 1982). Similarly, the works of F.E. Peabody (1948, 1959), which were concerned primarily with tracks of salamanders and early Triassic vertebrates, introduced several techniques that were subsequently turned to advantage in the study of dinosaur tracks. Of the many and diverse lines of scientific inquiry that incidentally

Figure 4.17 The perplexing Triassic track *Chirotherium*, and attempts to visualize its maker. (a) Diagram showing part of a *Chirotherium* trackway. Note the prominent opposed digit at the outer side of the pes print; a similar digit is also present in the smaller manus print. (b) The *Chirotherium* track-maker envisaged as a 'hand-footed' amphibian. (c) The track-maker envisaged as a quadrupedal thecodontian. (d) Restoration of the thecodontian *Ticinosuchus ferox*, from the Middle Triassic of Switzerland; about 2.5 m long. (Adapted from Soergel 1925 (a–c), Krebs 1965 (d).)

advanced the understanding of dinosaur tracks few were more productive than those that were dedicated to solving the mystery of *Chirotherium*.

Chirotherium is the ichnogenus name given to unusual fossil footprints that were first discovered in the Triassic rocks of Germany in 1833. These footprints were somewhat similar in shape to a human hand, comprising four forwardly directed digits and what seemed to be a widely divergent 'thumb' or 'big toe' (Figure 4.17). But, to everyone's bewilderment, this divergent digit was clearly located on the *outer* margin of the footprint. By contrast, existing land animals have their opposed digits on the inner side of the hand or foot. In other words, there was no known animal, living or extinct, that might have produced footprints like *Chirotherium*. These perplexing tracks generated speculations and controversies for more than 130 years, during which time it was suspected that the maker of the *Chirotherium* tracks might have been an enormous ape, a bear, a marsupial, or a gigantic toad-like amphibian. H. Haubold (1984: 224–5) listed nearly 50 'classic' works on *Chirotherium*, and those represent only a sample of the enormous literature on the subject.

The mystery surrounding the origin of these hand-like tracks was fully apparent when J.J. Kaup (1835) came to invent a name for them. He proposed two ichnogenus names – *Chirotherium* (Greek, 'hand beast' or, more loosely, 'hand mammal') and *Chirosaurus* ('hand lizard') – in case the track-maker should transpire to be either mammal or reptile. In fact, the track-makers turned out to be reptilian, but since the name *Chirotherium* was listed as Kaup's first alternative it takes priority over the more appropriate name *Chirosaurus*. (The spelling *Cheirotherium*, which is used by some ichnologists, is etymologically correct but is disregarded because it does not agree with Kaup's original.)

Many of the scientists who puzzled over *Chirotherium* were reluctant to concede that the track-maker might have had an opposed outer digit; instead, they sometimes invented the most ingenious theories to explain away the problem. It was suggested, for instance, that the supposed outer digit was merely an imprint made by the ankle or the wrist (Bernhardi 1834). Another suggestion held that the supposed outer digit was nothing more than a fleshy outgrowth on the margin of hand and foot (e.g. Walther 1917). The influential anatomist Richard Owen attributed the tracks to labyrinthodont amphibians (1841), and expressed his firm conviction that the divergent digit really was the innermost one. But why, then, should this innermost digit seem to be located on the outer side of each footprint? To answer this question Owen adopted the curious notion of a creature that had lurched across the Triassic landscape, crossing and uncrossing its feet as it went. In this strange and improbable gait the track-maker was supposed to have planted its left foot on the right side of its trackway, and its right foot on the left side. In 1855, a reconstruction of this 'hand-footed labyrinthodont', modelled along the lines of an over-grown toad, duly made its appearance in the fifth edition of Charles Lyell's *Manual of Elementary Geology* (Figure 4.17b).

Chirotherium was not confined to Germany. Similar footprints came to light throughout Europe and, eventually, in North and South America, Africa and South-East Asia. Altogether, some 40 ichnospecies names (not all of them valid) were proposed for *Chirotherium* tracks of various sizes and shapes. But still the identity of the track-maker remained an open question. Eminent naturalists continued to puzzle over *Chirotherium*, suggesting that it might be the track of a crocodile, or perhaps of a dinosaur, but these speculations did little to explain the anomaly of an outwardly divergent digit. In 1925, the German scientist Wolfgang Soergel published by far the most detailed and perceptive study of *Chirotherium*. He established that the supposed 'thumb' or 'big toe' was certainly the outermost digit and emphasized that a foot of this unusual construction was known to occur in some of the thecodontian reptiles. After weighing all the evidence, Soergel concluded that *Chirotherium* track-makers were probably thecodontians that resembled crocodilians in their outward appearance and that ranged in length from about 35 cm to 8 m (Figure 4.17c).

Unfortunately, the thecodontians were then so poorly known that Soergel was unable to identify any one genus as the most likely candidate. Nevertheless, Soergel's careful judgement attracted strong support, even though the tangible remains of a *Chirotherium* track-maker remained elusive. In the following years various thecodontians were nominated as possible track-makers (e.g. von Huene 1933; Baird 1954), but none of them seemed to meet all the important criteria that had been specified by Soergel. Eventually, nearly 40 years after Soergel's investigation, the Swiss palaeontologist Bernard Krebs (1965, 1966) described a very suitable thecodontian, *Ticinosuchus*, from the Middle Triassic of Switzerland. In practically every respect *Ticinosuchus* fulfils the predictions made by Soergel in 1925 (Figure 4.17d). Even then, *Chirotherium* continued to excite speculation. As recently as 1970 G. Demathieu likened certain examples of *Chirotherium* to the feet of theropod and prosauropod dinosaurs.

The ichnogenus *Chirotherium* is unlikely to include the tracks of dinosaurs, but it deserves mention here because it attracted so many scientists to the study of fossil footprints, including those of dinosaurs. In attempting to solve the mystery of *Chirotherium* those scientists developed techniques and lines of reasoning that were subsequently turned to great advantage in the study of dinosaur tracks.

Literature

The scientific literature on dinosaur tracks is vast and scattered. Publications may be difficult or impossible to obtain and, in some cases, difficult or impossible to understand. Many of the most important sources are in languages other than English, and there are bewildering inconsistencies in terminology. For many years the study of fossil footprints was regarded as 'a minor, neglected, and somewhat disreputable branch of paleontology', with the result that many scientists, both amateurs and professionals, 'dabbled' in the study of fossil tracks and published their findings in the most unlikely places. W.A.S. Sarjeant (1975: 284) has lamented that:

> Gaining access to papers published in German high school publications . . . and 'works newspapers' . . . or by Austrian touring clubs . . . is virtually impossible; yet such places of publication are scarcely even exceptional. In Kuhn's (1963) invaluable bibliographic compilation, *Ichnia tetrapodorum*, the words 'Nicht gesehen' (*not seen*) and 'Nicht auffindbar!' (*not discoverable!*) occur frequently; one can entirely sympathize with his problems.

Despite these difficulties there are a few major starting-points for search of the literature on dinosaur tracks and traces. Works by O. Kuhn (1958a, 1963) and H. Haubold (1971, 1984) are particularly useful in this respect, as are some general reviews of vertebrate ichnology (e.g. Winkler 1886; Lessertisseur 1955; Sarjeant 1975, 1987; see also numerous papers in Gillette and Lockley 1989). Some bibliographies devoted primarily to body fossils also contain sections on dinosaur tracks (e.g. Hay 1902, 1929-30; von Nopcsa 1926, 1931), and many other sources of information are mentioned in Chapter 3.

5

Identifying the track-maker

'Tracks,' said Piglet. 'Paw-marks.' He gave a little squeak of excitement. 'Oh, Pooh! Do you think it's a-a-a Woozle?'

'It may be,' said Pooh. 'Sometimes it is, and sometimes it isn't. You never can tell with paw-marks.'

A.A. Milne, *Winnie-the-Pooh* (1926)

SOURCES OF EVIDENCE

Various pieces of evidence can provide valuable clues to the identity of a dinosaurian track-maker. The number and arrangement of footprints will indicate if the track-maker was a biped or a quadruped, while the dimensions of the footprints can be used to estimate its body size (Chapter 8). The shape of the footprints reflects the anatomy of the track-maker's feet and is a particularly important clue because each major group of dinosaurs had its own distinctive pattern of foot structure. The stratigraphic and geographic distributions of dinosaur groups, ascertained primarily from the evidence of body fossils, will indicate which sorts of dinosaurs lived where, and when, thereby narrowing down the choice of potential track-makers. And, finally, the evidence of associated fossils, be they animals, plants or other trace fossils, may indicate something about the behaviour of a dinosaurian track-maker. The animal's behaviour, in turn, can be a valuable clue to its identity.

Sometimes all these approaches will fail. No matter how assiduously one tries to identify track-makers there will always be a residue of problematical tracks. These include tracks that might have been made by animals other than dinosaurs, and those that cannot be attributed with confidence to any particular type of dinosaur. These problems are discussed in Chapter 7.

NUMBER AND ARRANGEMENT OF FOOTPRINTS

The number and arrangement of footprints should indicate if the track-maker was a biped or a quadruped. However, it should not be assumed that some types of dinosaurs were exclusively bipedal and that others were exclusively quadrupedal. Habitually bipedal dinosaurs, such as coelurosaurs, could and did walk on all fours (see Figure 6.10b), and some trackways indicate that the normally quadrupedal sauropods sometimes managed to progress by using only two feet (see Figure 6.18). All that may be said with certainty is that some sorts of dinosaurs were usually bipedal and that other sorts were usually quadrupedal. It is difficult to be more specific because there are no sharp anatomical distinctions between those dinosaurs presumed to have been bipeds and those presumed to have been quadrupeds. Some dinosaurs, such as *Tyrannosaurus*, have been described as 'obligate' bipeds because their forelegs are so much smaller than their hindlegs that it is difficult to imagine these animals walking quadrupedally. Yet even these obligate bipeds must have come down on to all fours from time to time, if only to rest, to drink or to feed (Newman 1970). At the other extreme it may be equally difficult to imagine huge sauropods rearing up on their hindlegs, though recent studies reveal that they may well have done so (Alexander 1985). Then, for some dinosaurs, such as prosauropods, the anatomical evidence is equivocal. These animals may have been basically bipedal (Cooper 1981) or basically quadrupedal (Galton and Cluver 1976), though in all probability they could switch from two legs to four, and back again, as circumstances demanded. Here the evidence of dinosaur tracks is crucially important, for it reveals the way in which dinosaurs actually moved. It is known, for example, that at least one prosauropod walked on all fours (Baird 1980; Plate 13, p. 306, top right).

The tracks of bipeds and quadrupeds look so different that it might seem impossible to confuse them. Nevertheless, the tracks of quadrupeds can be mistaken for those of bipeds, and occasionally they have been. There is a simple reason for such mistakes: quadrupeds sometimes leave tracks where the prints of the forefeet are absent or concealed, or where they are so small that they go unnoticed. In many dinosaurs the forefeet were much smaller than the hindfeet and produced correspondingly smaller and shallower footprints, or no perceptible footprints at all. If prints of the forefeet are lacking, or are overlooked, it is easy to assume that the trackway was made by a biped. A similar problem arises with the 'pseudo-bipedal' tracks that are made by quadrupeds planting their hindfeet over the prints of their forefeet. This partial or complete overlapping of hindfoot prints on to forefoot prints sometimes misleads modern naturalists. For instance, the living Himalayan bear is known to leave overlapping or 'composite' footprints that are readily mistaken for those of a man-like or ape-like biped (Napier 1972, figs 3,4). Another living quadruped, the Brazilian tapir, was long ago observed to produce pseudo-bipedal tracks along muddy river-banks (Beckles 1854: 462, footnote), and the North American beaver often places its hindfeet over the prints of its forefeet (Jaeger 1939: 228). Many other quadrupeds are capable of producing pseudo-bipedal tracks (Speakman 1954; Bang and Dahlstrom 1974; Morrison 1981; Bouchner 1982), and such potentially misleading trackways certainly do occur in the fossil record. The pseudo-bipedal trackway of a quadrupedal ornithischian was described from the Cretaceous of Brazil by F. von Huene (1931a), and recently M.G. Lockley and his colleagues (1986) discussed a similar example of a sauropod trackway in the Upper Jurassic of Colorado. There are many more examples (e.g. Plate 11, p. 208, centre), but the point should be obvious already: it sometimes needs more than a casual glance to distinguish the tracks of bipeds from those of quadrupeds.

It seems less likely that the trackway of a biped could be mistaken for that of a quadruped, since this would require the presence of 'extra' prints representing the forefeet. However, D. Baird (1964: 120) has described one such case from the Triassic of Wyoming. Here the ichnogenus *Agialopous* had been defined by E.B. Branson and M.G. Mehl (1932) as the trackway of a quadruped with tridactyl hands and feet. Baird suspected that *Agialopous* actually comprised the tracks of two bipedal dinosaurs that had travelled in the same direction, the supposed 'manus' prints being the pes prints of the smaller animal.

Occasionally, one encounters the track of an animal that seems to have been moving on three feet – with imprints of the forefoot on one side of the trackway but not on the other. Tracks of this type were sometimes formed when a quadruped crossed a sloping substrate, such as the face of a sand dune. The feet on one side carried most of the animal's body weight and were deeply impressed whereas those on the other side produced shallower footprints or none at all. Dinosaurs also used three feet in shifting between quadrupedal and bipedal gaits, because they tended to lift one forefoot from the substrate before the other, and they sometimes produced discontinuous trackways when swimming.

At some sites dinosaur tracks are so abundant that it is possible to confuse the prints of two or more animals and to assume that they constitute a single trackway. A good example of this potential pitfall is furnished by some dinosaur footprints on public display in the British Museum (Natural History), in London (see Figure 11.7b; also Charig 1979: 32; Haubold 1984, fig. 127). When they first came to light, in the Purbeck Beds (late Jurassic to early Cretaceous) of Dorset, these footprints were taken to represent the trackway of a single dinosaur that had been walking with very short steps and with its feet unusually wide apart (Anonymous 1962). It was even suggested (Swaine 1962) that the animal was taking such exceptionally short steps because it was plodding uphill! Further excavations revealed that this supposedly single trackway divided into two separate trackways of bipedal dinosaurs, each with a more typical spacing of the footprints (Charig and Newman 1962). Later still, it was found that one of these trackways was that of a dinosaur walking on all fours. The prints of the forefeet are so small and insignificant, by comparison with those of the hindfeet, that they were previously unnoticed (Norman 1980, pl. 5; see Figure 11.7c). Those small but important prints of the forefeet were not merely overlooked by a handful of palaeontologists: they escaped the attention of hundreds of thousands of museum visitors over a period of nearly 20 years.

FOOTPRINT MORPHOLOGY

Size of footprints

The size of the footprints may be used to estimate the size of the track-maker, and this, in turn, can be a valuable indication to its identity. Unfortunately, the business of predicting the track-maker's body size is so fraught with technical problems that it requires lengthy discussion at a later point (see Chapter 8). Aside from those technical problems there are two major sources of measurement error – preservation and variation. If a footprint is well preserved, with a sharply defined outline, its major dimensions can be measured with reasonable accuracy. But if a print is poorly preserved, with vague or irregular outline, it may be difficult to obtain reliable measurements. Variation in the size of footprints is the second major source of

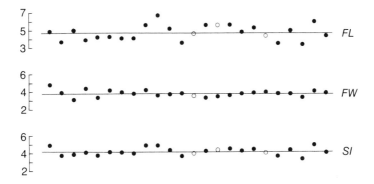

Figure 5.1 Variation in footprint length (*FL*), footprint width (*FW*) and footprint size index (*SI*) along a single dinosaur trackway. (*SI* is defined as the square root of the product of *FL* and *FW*.) The trackway of 24 footprints, ichnogenus *Wintonopus*, was made by an ornithopod; from the mid-Cretaceous of Queensland, Australia. Scale bar at left indicates percentage variation above and below the mean value (horizontal line) for each footprint variable. Open circles represent missing data points and are interpolated purely for the sake of visual continuity.

uncertainty: all too commonly the footprints in a trackway are not uniform in their size (Figure 5.1). The extent of this variation, and its causes, will be examined below, but for the moment it is sufficient to mention that such variation raises some awkward questions. If the footprints in a trackway do vary in size, which of them, if any, should be taken as the 'typical' or 'representative' example? Or is it legitimate to accept average footprint dimensions? The only way to begin answering these questions is to undertake a careful analysis of the variation that exists in dinosaur tracks. Regrettably, there have been few attempts to analyse that variation. Many works on dinosaur tracks include elementary statistics (e.g. mean footprint dimensions, with standard errors and variances) but more detailed treatments, such as correlations or analyses of variance, are uncommon (though see Demathieu and Wright 1988; Moratalla *et al.* 1988a). A comprehensive analysis of size-variation is one of the most pressing needs in the modern study of dinosaur tracks, for it could well improve the reliability of methods for predicting the size, speed and gait of a track-maker.

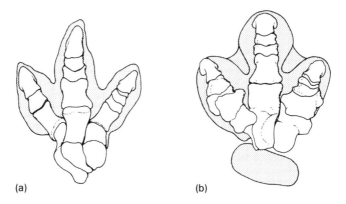

(a)   (b)

Figure 5.2 Dinosaur foot skeletons matched up against dinosaur footprints. (a) Foot skeleton of the early Cretaceous ornithopod *Iguanodon* superimposed on a footprint from rocks of similar age. (b) Foot skeleton of the hadrosaur *Hypacrosaurus* superimposed on a suspected hadrosaur footprint from the Upper Cretaceous of Alberta, Canada. (Adapted from Dollo 1906 (a), Langston 1960 (b).)

Shape of footprints

The most enlightening clue to the identity of the track-maker is the shape of its footprints. But, again, there are numerous technicalities involved in reaching an objective assessment of footprint shape: it is just as difficult to quantify and compare the shapes of footprints as it is to measure their sizes. Some standard procedures for describing footprint shape were introduced in the previous chapter, and they are important because they reduce (though never eliminate) the risk that two palaeontologists will reach different opinions about the shape of a footprint and, hence, about the identity of a track-maker.

Despite this element of subjectivity (see Figure 4.15) it remains true that each group of dinosaurs had its own characteristic pattern of foot structure and tended to leave correspondingly distinctive footprints. Dinosaurs with tridactyl feet tended to produce three-toed footprints, those with slender feet tended to produce narrow footprints, and so on. In other words, it may be assumed that each footprint is a reasonably faithful impression of the foot that made it. The obvious way to discover which sort of dinosaur was responsible for a trackway is simply to match up the footprints with dinosaur feet of the most appropriate shape and size. Such matching up of dinosaur footprints against dinosaur feet has long been standard practice (Figure 5.2). Sometimes the fossil footprint is compared directly to an actual foot skeleton; alternatively, one may work from the evidence of the footprint alone, attempting to visualize the shape of the foot that might have produced it. On occasion, it is useful to follow the reverse procedure, by predicting what sort of footprint would be produced by a particular foot skeleton. This last exercise entails something more than laying out the foot bones on a sheet of paper and tracing round them with a pencil. Instead it is necessary to arrange (or at least envisage) the foot bones in a life-like attitude, with the bases of the digits lifted from the substrate and the metapodium inclined up and backwards. Then, when they are projected on to the horizontal plane, these inclined elements will appear properly foreshortened, as they would be in an actual footprint.

Digital nodes and phalangeal formula

The **phalangeal formula** of a foot (or hand) comprises a count of the phalanges (toe or finger bones), proceeding from the innermost digit to the outermost. So, for example, the phalangeal formula of the human hand is 2:3:3:3:3. From the evidence of well-preserved skeletons it is clear that various groups of dinosaurs had their own diagnostic phalangeal formulae (see Chapter 6).

Long ago it was suspected that the swellings, or nodes, along the digits of a well-preserved footprint might bear some relationship to the phalangeal formula (e.g. E. Hitchcock 1858: 37). Theoretically, one might ascertain the phalangeal formula simply by counting the nodes; and, beyond that, the relative sizes of the nodes might reveal the relative sizes of the phalanges. Many scientists proceeded to exploit these possibilities and tried to reconstruct the foot skeletons of dinosaurs and other extinct vertebrates on the basis of their footprints. Unfortunately, the matter was not quite so straightforward, and it soon became apparent that there were some intractable problems.

In 1879, W.J. Sollas described footprints of the ichnogenus *Brontozoum* (*Anchisauripus*) from South Wales and commented that the imprints of the toes were 'contracted and swollen at intervals in correspondence with the number and position of the phalanges of the original digit' (1879: 512). When he came to check this correspondence, by examining the feet of existing birds, Sollas (1879: 513) found that there were some complications:

> On comparing the regions of the sole of the Emu's foot with its skeletal structure, one is struck with their wide divergence in details, which clearly shows the futility of too closely arguing in all cases from the skeletal structure of a foot to the impression it might make on the surface of a sedimentary deposit.

Evidently there was a need for caution, and this persists today, despite a great deal of speculation about the anatomical significance of the digital nodes.

In the early years of the twentieth century R.S. Lull suggested that the fleshy pads under the toes of existing birds were usually **mesarthral** in position (Lull 1904, 1915, 1917). That is, each pad would be located under the middle of a phalanx, and not under the joint between two phalanges (see Figure 5.3a). Such an arrangement would imply straightfoward correspondence between digital nodes and phalanges. Lull's generalization was repeated almost verbatim by O. Abel (1912) and was adopted in W. Soergel's (1925) attempt to reconstruct the *Chirotherium* track-maker.

In 1927, G. Heilmann questioned the reliability of Lull's generalization. Careful study of the feet in various birds revealed that the digital pads varied considerably in their arrangement: they were mesarthral in some birds, but **arthral** in others (i.e. enclosing the toe-joints, Figure 5.3b), and of variable location in still others. Naturally enough, Heilmann was sceptical about the prevailing view that dinosaurs had digital pads of mesarthral type. He found that there was an excellent fit between the foot skeleton of the coelurosaur *Procompsognathus triassicus* and the fossil tracks known as *Grallator tenuis*, noting that the digital nodes corresponded approximately, though not exactly, to arthral locations. Edward Hitchcock had reached roughly the same conclusion nearly 70 years earlier (1858: 37), as did F.E. Peabody at a much later date (1948: 399–402). Peabody observed that the digital pads were predominantly arthral in ground-dwelling birds such as the turkey and the rhea, and he went on to reconstruct an appropriate foot skeleton for the theropod track *Grallator cursorius*.

A few years later W. Bock (1952: 403) gave a brief but inconclusive summary of the whole problem. He, too, noted that the pattern of digital pads bore no straightforward relationship to the arrangement of phalanges, and commented that 'one takes a chance in judging a

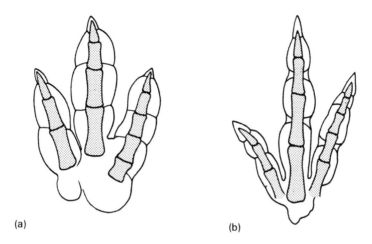

(a)   (b)

Figure 5.3 Relationship of digital nodes to foot phalanges. (a) Theropod footprint *Eubrontes giganteus*, with the foot skeleton restored on the assumption that the digital nodes are mesarthral in position; footprint about 37 cm long. Note that the joints between the phalanges coincide with constrictions between the digital nodes. (b) Theropod footprint similar to *Grallator*, with the foot skeleton restored on the assumption that the digital nodes are arthral in position; footprint about 7.5 cm long. Note that the joints between the phalanges lie in the centre of the digital nodes. (Adapted from Heilmann 1927 (a), Baird 1954 (b).)

phalangeal formula on such a rather uncertain basis'.

In an encyclopaedic study of avian skin structures A.M. Lucas and P.R. Stettenheim (1972: 71–2) also paid some attention to the distribution of digital pads:

> Our search of the literature failed to reveal names for the individual pads and spaces [on the undersides of a bird's toes]. To fill this need, we selected names that seemed appropriate; later we revised these names. Eventually we found that any set of names failed to have general applicability to species from different orders and even within orders. This is due chiefly to the fact that sometimes a pad covered parts of two phalanges and sometimes only of one.

Worse still, Lucas and Stettenheim found that the number of the pads and intervening spaces varied from individual to individual within a single species; they left no doubt that the number and arrangement of digital pads are decidedly variable in birds.

There is yet another difficulty. Successive footprints of a single animal may have different patterns of digital nodes, as was noted by W.J. Sollas (1879: 514):

> Moreover the number of phalanges indicated [in the footprint] varies with the way the foot is set on the ground; thus in one instance the Emu in our Zoological Gardens so stepped as to run the second and third phalangeal imprints of its middle toe into one.

So, if the evidence from existing birds is any guide, the pattern of nodes is likely to have varied from one ichnospecies to another. It might also have varied between trackways within an ichnospecies, and even from one footprint to the next within the trackway of a single animal. Despite these uncertainties, one generalization does remain valid: an interphalangeal joint rarely coincides with the constriction between two digital nodes. Or, to put this another way, the digital nodes are rarely mesarthral in position. However, this generalization must be regarded as a *minimal* requirement in any attempt to reconstruct the phalangeal formula of a dinosaurian track-maker. That minimal requirement does not exclude the possibility that some digital nodes were mesarthral in position. Nor does it exclude the possibility that one large node might have covered two joints.

With these qualifications in mind, the digital nodes may be interpreted with a fair degree of confidence in well-preserved footprints, particularly those of theropods and ornithopods (e.g. Plate 7, p. 118, bottom right). Many of these dinosaurs had three functional toes (II, III and IV), with a consistent phalangeal formula of 3:4:5. The pattern of nodes in a well-preserved footprint often mirrors such a formula (e.g. Figure 4.4): two nodes in digit II would correspond to the two joints between three phalanges; and three nodes in digit III would match up with the three articulations in a series of four phalanges. This correspondence between nodes and phalanges is obvious in some footprints, though in others it may be disrupted by 'extra' nodes or 'missing' nodes.

The commonest of 'extra' nodes are those impressed by pads covering the metatarso-phalangeal joints. Such a node is frequently present behind digit IV, where it forms a definite bulge at the rear of the footprint (Figure 5.4a,c,e) – a landmark that is useful for distinguishing a

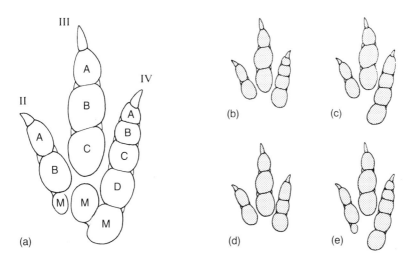

Figure 5.4 Key to the pattern of digital nodes in the footprints of bipedal dinosaurs. (a) Plan showing the maximum number of digital nodes. In each digit the nodes are identified by code-letters, beginning with the foremost. A metatarso-phalangeal node (M) is shown at rear of each digit, though this is common only in digit IV. Diagrams (b)–(e) show some variations on this plan, mainly in the nodes of digit IV. (b) Footprint without metarso-phalangeal nodes; four nodes in digit IV correspond to A–D in diagram (a). (c) Footprint with metatarso-phalangeal node behind digit IV; four nodes in digit IV correspond to A+B (amalgamated), C, D and M in diagram (a). (d) Footprint without metarso-phalangeal nodes; three nodes in digit IV correspond to A+B (amalgamated), C and D in diagram (a). (e) Footprint with metatarso-phalangeal nodes behind digits II and IV; digit IV comprises the full series of nodes, as in diagram (a).

left footprint from a right one. Less commonly there is a metatarso-phalangeal node behind digit II or digit III. Other examples of 'extra' nodes may stem from the misinterpretation of preservational features. For instance, some prints are preserved in such exquisite detail that relatively minor creases and wrinkles of the skin might be mistaken for the folds delimiting nodes. In other cases, plastic deformation of sediment in the floor of a footprint produced irregular pockets and ripples that might be mistaken for nodes and constrictions.

Often, a node seems to be missing from digit IV. This is usually because two short phalanges, immediately behind the claw, were encased in a single pad. Clearly, the pattern of nodes in digit IV demands careful interpretation: this toe had a consistent skeletal structure, comprising a string of five phalanges, but it could produce as few as three nodes or as many as five (Figure 5.4d,e).

VARIATION IN FOOTPRINT MORPHOLOGY

No two footprints are identical. Any two prints from the trackway of a single animal will differ to some extent in their size and shape, as will the cast and the mould of a single print (e.g. Sarjeant 1971: 344–5; Plate 7, p. 118, top centre and right), and in some instances there may be very striking differences between one footprint and the next (Figures 5.5 and 5.6). These variations make it difficult to ascertain the 'true' or 'normal' morphology of an animal's footprints and to deduce from that the size and shape of the track-maker's foot. To overcome those difficulties it is essential to examine all the footprints that are available, be they casts or moulds, or both. This essential requirement was justifiably stressed by D. Baird (1957: 457):

> Footprints from . . . [this Triassic horizon] reveal a sobering degree of variability both in apparent form and in manner of impression. . . Many of the footprints are so deformed by accidents of impression that they give a decidedly erroneous picture of the foot structure. Such anomalies serve to emphasize the dangers involved in any attempt to characterize and interpret footprint species without adequate quantities of well-preserved material.

Comparison of all the available footprints allows one to identify, and then disregard, some of the random variations that may affect only a few of them. Thereafter, it is necessary to decide which of the remaining footprint features are reliable indications to the track-maker's foot structure. To make that decision it is important to understand the origins and extent of morphological variation in dinosaur footprints.

A footprint gives only an imperfect idea of the track-maker's foot structure: some footprints may be incomplete because parts of the foot were not impressed into the substrate, whereas others may be obscured by adventitious or **extramorphological** features. Some of these deficiencies and extramorphological features resulted from local

Figure 5.5 Variation in the morphology of footprints along a single dinosaur trackway. Illustrations show 12 successive left footprints of *Neotrisauropus deambulator*, the trackway of a theropod dinosaur from the Lower Jurassic of Lesotho; scale bar indicates 20 cm. Note particularly the variation in footprint width and the sporadic appearance of the hallux imprint. (Adapted from P. Ellenberger 1974.)

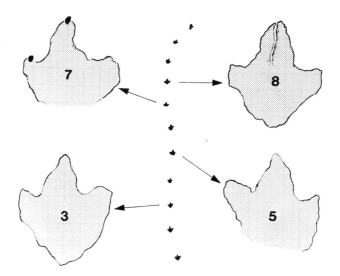

Figure 5.6 Variation in the morphology of footprints along a single dinosaur trackway. A carnosaur trackway, similar to *Tyrannosauropus*, from the mid-Cretaceous of Queensland, Australia, with four individual footprints shown at a larger scale; total length of trackway about 16.5 m. Puncture-like imprints of sharp claws are apparent in footprint 7, and digit III of footprint 8 is bisected by a crest of sediment that adhered to the underside of the toe. (Adapted from Thulborn and Wade 1984.)

variations in the texture and consistency of the substrate. Others originated from the dynamic interaction between foot and substrate – from the manner in which the foot was applied to the substrate and withdrawn from it (Thulborn and Wade 1989). These factors are responsible for seemingly random variations in the morphology of footprints. By contrast, certain other factors, such as the track-maker's taxonomic identity, age, size, sex and behaviour, are responsible for systematic or consistent variations in morphology. These systematic variations are replete with information about the identity and the behaviour of track-makers whereas the random variations are potentially misleading or confusing. The art of interpreting a fossil footprint resides essentially in distinguishing these two sorts of variations.

Sometimes, the origin and significance of morphological variations are readily apparent from the fossil footprints themselves, but more often there is a need to draw on comparative evidence from other sources. First, there is a source of information in experimental work, where living animals are persuaded to impress their tracks in selected substrate materials (Plate 5, p. 68, top left). Many sorts of animals

Plate 7 Morphological features of dinosaur tracks. Top left: *Wintonopus latomorum*, left footprint of a small ornithopod dinosaur, from the mid-Cretaceous of Queensland, Australia; width of footprint is nearly 10 cm. All three toes dragged forwards through the sediment as the dinosaur withdrew its foot; the drag-mark from the middle toe is extremely long and is slightly deflected as it runs through a coelurosaur footprint. Top centre and Top right: Natural mould (left) and cast (right) of a single theropod footprint, *Grallator variabilis*, from the Lower Jurassic of western France; about 10 cm long. Note that the cast shows details of footprint structure not apparent in the mould. Centre right: *Wintonopus latomorum*, left footprint of a small ornithopod dinosaur, from the mid-Cretaceous of Queensland, Australia; nearly 14 cm wide. The pronounced drag-mark has given the middle toe-print a Y-shaped outline. Bottom left: *Wintonopus latomorum*, another left footprint of a small ornithopod dinosaur; nearly 13 cm wide. A foreshortened print formed by the track-maker's toes plunging into the mud at a steep angle. As the toes were entering the sediment, the foot was simultaneously rotating in a clockwise direction: the undersides of the left and middle toes produced distinct slide-marks as they settled down and forwards, while the right toe pushed up a mound of sediment at the rear. Bottom right: Cast of the left footprint of a late Jurassic theropod, from Sichuan, China; scale bar is marked in cm. The superb preservation reveals narrow V-shaped claws and phalangeal nodes.

have been used in such experiments, together with a great variety of natural and artificial substrate materials (including smoked paper, clay, sand, mud, snow, plaster and dough). F.E. Peabody (1959: 3) transported salamanders to modern mudflats for 'natural recording' of their tracks, and on another occasion he went so far as to 'reconstitute' an ancient mudflat by mixing Triassic mudstone with water (1948: 303). Experimenters using artificial substrate materials often take great pains to ensure that these are as realistic as possible:

> To make a suitable bed for tracks, we cut slabs of potter's clay. . . and laid them end to end. We worked water into the surface of the clay to make a suitable mud. . . Then we conditioned the surface with a very thin coat of glycerine to simulate natural algal and bacterial growth, which acts as a natural parting medium. This prevented the substrate from sticking to the caiman's feet as he moved, the most frequent problem in producing experimental trackways. (Padian and Olsen 1984: 179)

Earlier experiments were not always so well planned:

> I went on Saturday last to a party at Mr. Murchison's house, assembled to behold tortoises in the act of walking upon dough. Prof. Buckland acted as master of the ceremonies. . . At first the beasts took it into their heads to be refractory and to stand still. Hereupon the ingenuity of the professor was called forth in order to make them move. This he endeavoured to do by applying sundry flips with his fingers upon their tails; deil a bit however would they stir; and no wonder, for on endeavouring to take them up it was found that they had stuck so fast to the piecrust as only to be removed with half a pound of dough sticking to each foot. This being the case it was found necessary to employ a rolling pin, and to knead the paste afresh. (J. Murray III in letter of January 23, 1828, cited by Murray 1919: 7–8; see also Sarjeant 1974: 269)

Experiments such as these yield precise information about the origins of footprint variations. There are no uncertainties about the track-maker's anatomy, its locomotor behaviour can be observed at first hand, and the physical properties of the substrate can be carefully monitored.

In addition, there is a wealth of equally enlightening observations on the tracks of living animals in their natural environments (e.g. Jaeger 1948; Ennion and Tinbergen 1967; Bang and Dahlstrom 1974; Murie 1975; Morrison 1981; Bouchner 1982; Triggs 1984). Then there is the evidence of variation that is already known to affect fossil tracks: the pattern of variation in one set of dinosaur tracks will obviously prove a useful guide when attempting to unravel the variation in another set of tracks. And, finally, there are significant clues to be gathered from the rocks that contain the dinosaur tracks – from the stratigraphic occurrence of the tracks, from the nature of the sediments in which they are preserved, and from associated fossils. These pieces of information may illuminate the circumstances under which dinosaur tracks were formed, thereby pointing to a range of environmental factors that might have affected footprint morphology.

At this point it will be useful to examine some common and potentially misleading variations in the morphology of dinosaur tracks.

Number of digits

Ideally, the number of digits in a footprint should correspond to the number of toes in the track-maker's foot. It might be supposed, for instance, that three-toed footprints were produced by dinosaurs with three-toed feet. In practice there is a complication: one or more toes may fail to leave a recognizable trace in the footprint. Thus, a three-toed footprint might have originated from a foot with three *or more* toes (e.g. H.C. Beasley 1896: 408). Some good examples are footprints described by G. Demathieu (1970: 176) from the Triassic of France. These three-toed prints look typically dinosaurian, but Demathieu suspected that they were actually *Chirotherium* tracks lacking imprints of the first and fifth toes (see Figure 7.3e,f).

As a general rule, the number of digits in a footprint indicates the minimum number of toes in the track-maker's foot. One or more toes may be seem to be lacking from the footprint for any of several reasons. In the first place, a digit impression may be so inconspicuous that it is simply overlooked. This is most likely to happen in the case of a small toe, often the innermost or outermost one, which supported little weight and was weakly imprinted on the substrate. For instance, D. Baird (1952: 836) remarked on two amphibian tracks where the small outermost digit had gone unnoticed in earlier studies.

Secondly, a toe may have been so short, or so high up on the foot, that it normally failed to touch the ground. Such a toe would leave

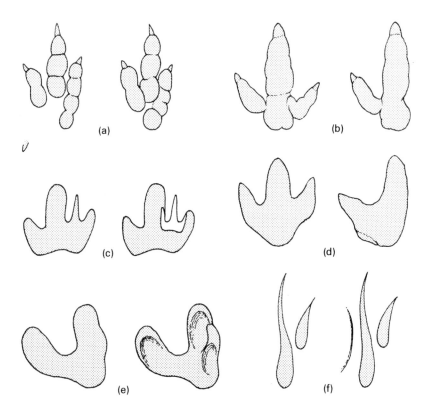

Figure 5.7 Real and apparent variations in the number of digits.
(a) Theropod footprints, *Anchisauripus*, showing four digits (left) and three digits (without hallux, right). (b) Tridactyl footprint of emu, *Dromaius*, on soft substrate (left) and didactyl variety formed on firm substrate (right). (c) Unusual four-toed footprint (left) transpires to be an amalgamation of two prints (right). (d) Normal tridactyl print of an ornithopod similar or identical to *Iguanodon* (left) and a didactyl variety where one digit has failed to leave an impression (right); compare diagram (b) above. (e) Apparently didactyl print of an ornithopod, *Wintonopus*, (left) transpires to be tridactyl, with two toe-prints amalgamated (right). (f) Apparently didactyl print of a coelurosaur, *Skartopus* (left), transpires to be tridactyl (right), with one digit represented by shallow scratch; compare Figure 5.15. (Adapted from Lull 1953 (a), photographs by author (b), Thulborn and Wade 1984 (c,e,f), Ballerstedt 1922 (d).)

an imprint only if the foot sank deeply into a soft substrate or was applied to the ground at an unusual angle. This is the case with some theropod footprints, where the hallux was located so high on the foot that 'only the imprint of the claw is ever made and that rarely' (Lull 1953: 166).

Withdrawal of the track-maker's foot was sometimes followed by slumping of the substrate, so that two or more of the digital imprints coalesced to resemble a single one. This was most likely to occur when the digits were relatively short and diverged at a low angle. Occasionally, two toes were applied to the substrate so close together that they produced a single imprint (Figure 5.7e). A similar effect resulted if the foot was planted into the substrate at an unusual oblique angle, with its undersurface facing slightly outwards or inwards rather than directly downwards. The toes on one side of the foot might fail to touch the substrate whereas those on the other side might overlap, giving the false impression of a single digit.

Much of a dinosaur's weight was borne on the principal digit, which tended to sink quite deeply into the substrate. The adjacent toes did not necessarily sink to the same depth, and if the substrate were firm enough one or more of them might instead splay out sideways, leaving only a shallow superficial marking or no impression at all. A toe was most likely to behave in this fashion if it was widely divergent from the principal digit. This effect is well seen in tracks of the living emu: on soft substrates emu footprints comprise impressions of all three toes, but on firmer ground the small inner toe fails to leave an impression and the footprints appear to be didactyl (Figure 5.7b). Similar effects have been observed in footprints attributed to the ornithopod dinosaur *Iguanodon* (Figure 5.7d).

Finally, there are tracks of dinosaurs with malformed or injured feet. A good example is furnished by a theropod trackway (*Eubrontes*) from the red sandstones of the Connecticut Valley (Figure 5.8): the left footprints comprise three digits, but the right ones lack the inner toe. Either this dinosaur's right foot was afflicted by a congenital defect or it had sustained an injury at some earlier stage in the animal's life. There are several other reports of tracks made by dinosaurs with injured or deformed feet (e.g. E. Hitchcock 1844a: 307; Jenny and Jossen 1982; Ishigaki 1986, figs 37, 38), and in a detailed study of salamanders and their tracks F.E. Peabody (1959: 23) made the following observations:

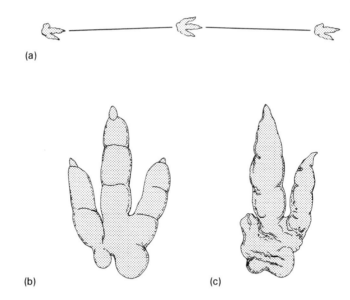

Figure 5.8 Trackway of a dinosaur with injured or malformed foot.
(a) Portion of theropod trackway, *Eubrontes*, from the Lower Jurassic of the
Connecticut Valley, USA; stride length about 1.4 m. Note that the right
footprints are consistently didactyl, lacking any impression from digit II.
(b) Normal right footprint of the ichnogenus *Eubrontes*. (c) Right footprint
from trackway shown above. (Adapted from Abel 1935 (a,c), Lull 1953 (b).)

> Abnormalities of the digits are frequently found in a study of
> a large number of individuals. Thus in [the salamander]
> *Taricha* a digit may be lacking, stunted, overdeveloped, or
> broken and healed with a right-angled bend, or the whole
> extremity may be stunted and stump-like. Also one finds
> fleshy nubbins or branches, with or without skeletal support,
> developed from either side of the digit, sometimes near its end
> so that a crude, forked tip results. The expression of such
> abnormalities in a fossil trackway is entirely possible but could
> not be expected to appear very often, and would probably not
> be identical in both feet.

Peabody's final sentence applies equally well to the tracks of
dinosaurs.

In all the cases mentioned earlier there will seem to be fewer digits
in the footprint than in the track-maker's foot. Less commonly, one

might be led to assume that a footprint has more digits than the track-maker's foot (e.g. Trusheim 1929).

First of all, one footprint may be superimposed on another, giving the appearance of a single print with an unusually large number of digits (Figure 5.7c). This may happen when two animals tread on the same spot or when a quadruped plants its hindfoot over the print of its forefoot. For example, some alleged 'sauropod' tracks in the Lower Cretaceous of Texas transpired to be amalgams produced by one bipedal dinosaur treading over the tracks of another (Farlow 1987: 3). Second, adventitious markings on the substrate may be mistaken for digit imprints. This is known to have happened in the case of an ancient amphibian track, where a pit made by a raindrop was once mistaken for the impression of a toe (see comments of Baird 1952: 836). Similarly, some 'extra digits' in examples of the theropod track *Grallator* were recognized by P.E. Olsen and P.M. Galton (1984: 99) as 'marks extraneous to the trackways'. Lastly, a tenacious substrate may have adhered to the underside of a toe, being drawn up into a longitudinal crest as the foot was lifted from the ground. As a result one digit may seem to be divided into two closely parallel ones (see Figure 5.6, print number 8). In most of these cases the appearance of 'extra' digits resulted from chance factors that were unlikely to have affected more than one or two footprints within a trackway.

Arrangement of digits

The interdigital angles are notoriously variable in dinosaur tracks, implying that the toes could be drawn together and spread apart to varying degrees. To some extent those movements of the toes would have been automatic, as they are in the feet of ground-dwelling birds. Here the toes tend to spread apart when the foot is planted on the ground to support the animal's weight; then, towards the end of the stride, when the animal begins to lift its foot from the ground, the toes draw together again. Such involuntary movements were probably most obvious in dinosaurs with long toes, and least obvious in those with elephantine feet and relatively short toes. The exact degree to which any two toes spread apart or drew together would have been dictated by chance factors including the consistency of the substrate, irregularities on the ground surface, and the angle at which the foot was applied to the substrate and withdrawn from it. It has sometimes been noticed, for instance, that dinosaurs tended to spread their toes

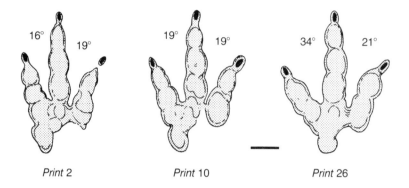

Print 2   Print 10   Print 26

Figure 5.9 Variation in the interdigital angles. The illustrations show three left footprints from a single theropod trackway, *Neotrisauropus deambulator*, from the Lower Jurassic of Lesotho; scale bar indicates 5 cm. Note the progressively broader spreading of the toes as the track-maker travelled from a firm substrate onto softer ground. (Adapted from P. Ellenberger 1974.)

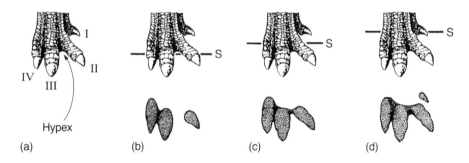

(a)   (b)   (c)   (d)

Figure 5.10 Footprint morphology related to the foot's depth of sink. (a) Hypothetical tetradactyl foot, with a single hypex indicated. (b) Foot sinks a short way into the substrate; hypex between digits II and III fails to touch the substrate surface (S), so that imprint of digit II is separate from the rest of the footprint. (c) Foot penetrates more deeply; hypex II–III sinks below the surface, so that all digital imprints are joined together. (d) Foot penetrates more deeply still, so that digit I makes an impression on the substrate surface.

more widely when they moved from a firm substrate on to softer and wetter ground (e.g. Figure 5.9; Currie and Sarjeant 1979). In the footprints of chickens, G. Gand (1976) found that the divarication of digits II and III was more variable than that of digits III and IV. By comparison, the total divarication (II–IV) was much less variable and promised to be of some value for defining and distinguishing ichnotaxa. The interdigital angles in dinosaur footprints have yet to be studied in such detail (though see Moratalla *et al.* 1988a).

The footprints within a single trackway may show different degrees of interconnection between the digital impressions. In some cases, the digits may converge to form a continuous rear margin to the footprint, evidently because the foot sank into the substrate up to, or beyond, the lower end of the metapodium (Figure 5.10c,d). In other cases, the foot did not sink so deeply, so that the digits are partly or completely separated and there is no continuous rear margin to the footprint (Figure 5.10b). If one toe-print is commonly found to be connected to its neighbour, it may be deduced that the intervening hypex was located well down the track-maker's foot. If, on the other hand, two neighbouring toe-prints are rarely connected it is likely that the hypex was so high on the foot that it rarely touched the substrate.

Size and shape of digits

In ideal circumstances the undersurface of a dinosaur's foot would be applied in parallel to the substrate, so that the resulting footprint would be identical in length. More commonly, the foot was planted into the substrate at an angle, with the toes pointing down and forwards, so that the footprint is distinctly shorter than the foot that produced it (e.g. Plate 7, p. 118, bottom left; Plate 9, p. 140, centre right). The steeper the angle at which the foot was set down, the greater was the degree of foreshortening that affected the footprint. In extreme cases, the toes plunged into the substrate almost vertically, so that the footprint comprised only a series of puncture-marks (see Figure 5.13).

Most dinosaurs seem to have been digitigrade while walking, as their footprints often comprise complete imprints of the toes, all diverging from the base of the metatarsus (see Figure 4.6a). When dinosaurs were trotting or running they frequently lifted the rear part of the foot from the ground, so that the footprints show only the distal parts of the toes (Dollo 1906; see Figure 4.6b). Humans and

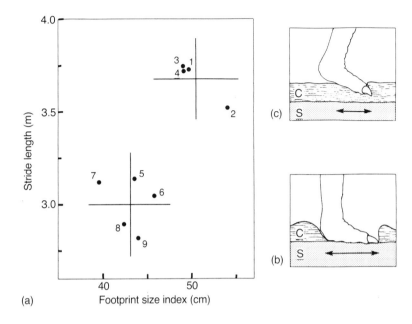

Figure 5.11 Variation in the size and shape of footprints along the trackway of a carnosaur that was apparently approaching or stalking its prey (see Figure 5.6). (a) Changes in stride length related to changes in footprint size. Footprint size index (defined as the square root of the product of footprint length and footprint width) is plotted as the mean value for the two footprints defining each stride. The first four strides are relatively long and are defined by large footprints, impressed deeply into the substrate (b). The following five strides are shorter and are defined by smaller and shallower footprints (c), almost as if the animal were walking more cautiously as it neared its prey. The substrate comprised soft mud (C) overlying firm sand (S). (Adapted from Thulborn and Wade 1989.)

ostriches have a similar tendency to rise on to the tips of their toes while running.

A good example of the way in which foot posture affected the size of the footprints is provided by the trackway of a Cretaceous carnosaur that may have been stalking its prey (Figure 5.11). The first part of this predator's trackway comprises large footprints, formed by the feet plunging deeply into the substrate. The second part of the same trackway is quite different: the animal took shorter strides, and its footprints became smaller and shallower, almost as if the animal

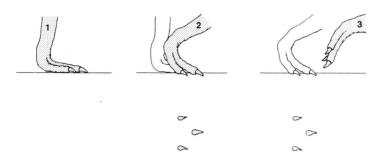

Figure 5.12 Formation of an incomplete footprint, comprising only impressions of the toe-tips. 1, At midstride, the foot maintains a large area of contact with the substrate and is too lightly loaded to make an impression. 2, At end-stride, the rear part of foot is lifted from the substrate; the entire body weight is transferred to the tips of the toes, which are forced into the substrate. 3, The foot is lifted clear to commence its next stride. Note that the toe-prints are widely splayed.

had been walking on tip-toe. Here, as in many other cases (e.g. Figure 4.16), there is clear evidence of footprint morphology being affected by the track-maker's changing behaviour.

Some dinosaurs, such as coelurosaurs, had relatively large and broad-spreading feet that may have functioned as analogues of snow-shoes when the animals traversed soft ground. Consequently, their trackways may include discontinuities or very incomplete footprints made by the tips of the toes (Figure 5.12). Similar instances of 'missing' or incomplete footprints occur in the tracks of other dinosaurs wherever these animals chanced to tread on an unduly resistant patch of sediment.

Sometimes, the foot made no impression until the very end of a dinosaur's stride, when only the tip of the longest toe remained in contact with the substrate. This toe then sank into the substrate, anchoring itself in place and allowing the shorter toes to swing forwards and touch down alongside it. The result was an extremely distinctive footprint comprising a row of punctures (Figure 5.13; Plate 8, p. 134, bottom left). Such footprints are noticeably narrower than those 'normal' examples where the toes splayed out under the weight of the track-maker (see Figure 5.12). Footprints comprising only the imprints of the toe-tips also resulted when swimming dinosaurs pushed off the bottom with their feet (as in Figure 2.8).

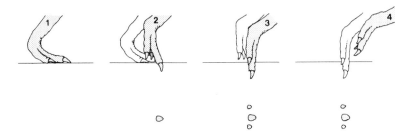

Figure 5.13 Formation of an extremely foreshorted footprint, where toes form deep punctures. 1, At mid-stride, the foot maintains a large area of contact with the substrate and is too lightly loaded to make an impression. 2, By end-stride, the dinosaur has lifted most of its foot from the substrate; the entire body weight is transferred to the tip of the longest toe, which begins to plunge almost vertically into the substrate. 3, As the longest toe sinks more deeply, the shorter toes swing alongside and also penetrate the substrate. 4, The foot is pulled clear of the substrate to commence its next stride. Note that the toe-prints are close together and side by side, unlike the footprint shown in Figure 5.12.

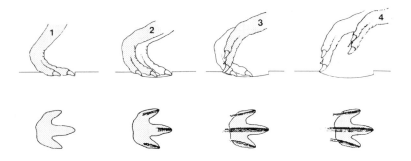

Figure 5.14 Formation of retro-scratches. 1, At mid-stride, the dinosaur's foot sinks into the substrate to form a print of normal depth. 2, At end-stride, the rear part of foot is lifted from the substrate; the tips of the toes lose their purchase and begin to slip backwards, incising grooves in the floor of the footprint. 3, In extreme cases, the tips of the toes slip back so far that they breach the rear wall of footprint. 4, The foot is lifted clear of the substrate to commence its next stride.

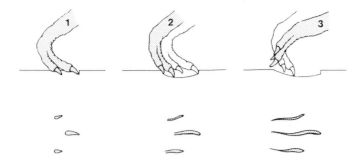

Figure 5.15 Formation of a footprint comprising furrow-like markings.
1, The foot makes no impression until end-stride, when the heavily
weighted toe-tips sink into the substrate. 2, The tips of the toes lose their
purchase and begin to slither backwards in the substrate. 3, The foot is
finally lifted clear of the substrate to commence its next stride.

Occasionally, a dinosaur's toes lost their purchase as the foot was
being withdrawn from the substrate, so that the claws slipped
backwards, incising grooves in the floor of the footprint (Figure 5.14;
Plate 8, p. 134, bottom right). In extreme cases, they slipped back so
far that they breached the rear wall of the footprint and formed
distinctive **retro-scratches**. If only the tips of the toes had entered the
substrate in the first place, these might slither backwards to produce
a footprint consisting of parallel furrows (Figure 5.15; Plate 9, p. 140,
bottom left). Retro-scratches should not be confused with **slide-
marks**, produced when the undersides of the toes slid down and
forwards into the substrate (Figure 5.16a; Plate 7, p. 118, bottom left).

On other occasions, the claws trailed through the front rim of the
footprint, leaving forwardly directed **drag-marks** or **scrape-marks**
(Figure 5.16c–e; Plate 7, p. 118, top left). The longest toe in the foot
(usually digit III in tridactyl feet) tended to produce a drag-mark most
frequently, whereas the shortest toe was least likely to do so. Drag-
marks are often quite short, though some examples are extremely
long, and they are common in the tracks of many land animals, both
living and extinct (e.g. Bang and Dahlstrom 1974; Reineck and
Howard 1978; Morrison 1981; Bouchner 1982). The development of
drag-marks was doubtless controlled by random factors such as the
consistency of the substrate and the depth to which an animal's feet
sank into it. Even so, it seems that some species of living animals are

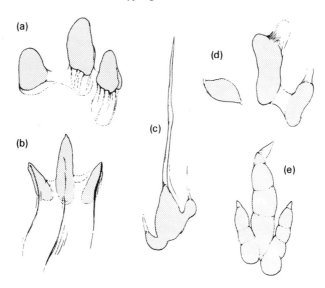

Figure 5.16 Examples of footprints with slide-marks, retro-scratches and drag-marks. (a) *Wintonopus*, an ornithopod footprint from the Lark Quarry site, mid-Cretaceous of Queensland, Australia, showing pronounced slide-marks where the digits settled down and forwards into the substrate. (b) *Skartopus*, a coelurosaur footprint, also from the Lark Quarry site, with prominent retro-scratches resulting from the backwards slippage of the toe-tips. (c) Another example of *Wintonopus*, showing an extremely long drag-mark extending from the tip of digit III; shorter drag-marks extend from digits II and IV. (d) An example of *Wintonopus* with broad drag-marks extending from digits III and IV; note the Y-shaped outline to the tip of digit III. (e) *Anchisauripus*, a theropod footprint from the Upper Triassic of Wales, with a short drag-mark extending from the tip of digit III; examples such as this are sometimes mistaken for tracks of animals with bent or deformed toes. (Adapted from photographs in Thulborn and Wade 1984 (a–d), Tucker and Burchette 1977 (e).)

more likely to produce drag-marks than are others. For instance, M.D. Leakey (1979: 454) illustrated such marks in the fossil track of a giraffe and mentioned that existing giraffes also tended to drag their toes. Consequently, the presence or absence of drag-marks might be useful for defining and distinguishing ichnotaxa, though as yet this possibility remains largely untested. However, R.A. Thulborn and M. Wade have suggested (1989) that drag-marks may be more common in ornithopod tracks than in theropod tracks, perhaps because the ornithopods had thicker and less flexible toes. In any event drag-marks should not be mistaken for the impressions of long and taper-ing digits; nor should they be mistaken for a tail-drag extending from one footprint to the next (e.g. Eberle 1933, fig. 1; F. Buckland 1875: 30; see Figure 4.12b).

A drag-mark is not necessarily a straightforward extension of the digit impression, for in many cases it veers off sideways. Most commonly, it extends forwards and outwards from the tip of the digit impression. Evidently, the foot was planted into the sediment in one direction, with the toes pointing forwards (and often inwards), but was withdrawn in a different direction, with the toes being lifted forwards and slightly outwards. In some dinosaur tracks the effect is so marked that the tip of the digit appears to be forked or Y-shaped (Figure 5.16d; Plate 8, p. 134, top right). Such effects, which are also seen in the tracks of amphibians and birds, do not indicate that the track-maker had 'split' or 'damaged' toes (Peabody 1956a). Similarly, a drag-mark veering off from the digital axis at an angle approaching 90° should not be mistaken for the impression of a 'bent' or 'deformed' toe (e.g. Tucker and Burchette 1977, fig. 3).

W.P. Coombs has distinguished several types of claws in bipedal dinosaurs (1980b: 1200):

> Bipedal dinosaurs may have one of three major ungual patterns: (i) sharply pointed, laterally compressed, or nearly circular [in cross-section], strongly hooked true claws; (ii) pointed, dorso-ventrally compressed, flat-bottomed, narrow semiclaws; and (iii) broadly rounded, dorsoventrally flattened hooves. . . Most pre-Cretaceous ornithopods have semiclaws. True claws, found only in the Theropoda, can be recognized in footprints by their tendency to form an isolated, nearly circular pit just anterior to the distalmost interphalangeal pad. Semiclaws lie flatter against the substrate, leaving an elongate, sharply pointed triangle at the end of each toe.

This threefold classification (Figure 5.17) is easily applied to many dino-saur tracks, though the footprints in a single ichnotaxon are not always consistent in the form of the claws. For instance, Coombs discovered that '*Eubrontes* prints have a diversity in the shape of the ungual [= claw] imprints, and on this feature alone such prints could not be iden-tified with certainty as theropod rather than ornithopod' (1980b: 1200).

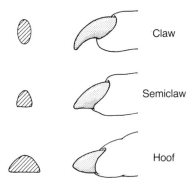

Figure 5.17 Claws, nails and hooves of dinosaurs. The small figures at left represent diagrammatic cross-sections. The term 'nail' has been applied to the flat-bottomed and relatively narrow semiclaw, as well as to structures intermediate between semiclaws and hooves.

Plate 8 Morphological and preservational features of dinosaur tracks. Top left: *Talmontopus tersi*, the natural cast of a right theropod footprint, showing clear traces of interdigital webbing, especially between the middle and left-hand toes, from the Lower Jurassic of western France; about 27 cm long. Top right: *Wintonopus latomorum*, two left footprints representing two small ornithopod dinosaurs, from the mid-Cretaceous of Queensland, Australia; the upper example is nearly 5 cm wide. The lower example is bisected by a joint that has displaced the two portions of the print; the Y-shaped tip of the middle toe-print indicates that the toe was planted into the sediment in one direction but was withdrawn in a different direction. Bottom left: Small ornithopod footprints (*Wintonopus latomorum*) and coelurosaur footprints (*Skartopus australis*), also from the mid-Cretaceous of Queensland; the matchbox indicates the scale and direction of the lighting. Several examples show secondary narrowing of the middle toe-print, caused by a suction effect on withdrawal of the foot (e.g. three prints above the matchbox, and the large print at lower right). A line of three small puncture-marks, at extreme lower-right corner, resulted when only the tips of the toes were impressed on the substrate. Bottom right: *Skartopus australis*, a right coelurosaur footprint from the mid-Cretaceous of Queensland, showing deep scratches formed by backwards slippage of the track-maker's claws; nearly 5 cm wide.

In some tracks, the claw-prints may be contiguous with the digits, while in others they may be demarcated by a constriction or transverse crease. If the claws were strongly arched, their imprints may be separated from the digit impressions by a short space of undisturbed sediment. And claws that extended downwards, rather than forwards, may be represented by puncture-like markings or steeply inclined tunnels in the tips of the digit impressions. In some dinosaur footprints the claws are deflected sideways, so that the digits seem to be oddly 'bent'. The extent of this deflection may be variable within an ichnospecies, but for reasons that remain unknown: either the claws were movable or they were fixed in slightly different positions in each track-maker. Small footprints often have sharper claws than large footprints of the same ichnospecies (e.g. Demathieu 1970: 180; Thulborn and Wade 1984: 422). This variation in sharpness might be an ontogenetic feature, or it might merely reflect a greater degree of wear and tear in bigger animals. For example, D. Baird (1957: 457) noted that 'the claws of *Grallator sulcatus* [a theropod track] are acuminate in some individuals but blunted by wear in others'. The presence of sharp claw-marks is often used to distinguish the footprints of theropods from those of ornithopods, but J.O. Farlow (1987: 11) expressed some doubt regarding this assumption:

> If the sediment across which a theropod walked were soft and/or fluid enough and/or lacking enough in cohesiveness, and the reptile's claws thin enough, talon impressions at the tips of toe impressions might collapse after withdrawal of the dinosaur's foot, even though the digit impressions themselves were preserved... At the other sedimentary extreme, if a track-registering sediment layer were thin and/or firm enough, and if the dinosaur's digital pads were thick enough, the claws might not contact the substrate.

In short, the shape, orientation, depth and even the presence of claw-traces may be highly variable, sometimes within a single trackway.

Scales or tubercles at the margin of the foot sometimes produced a series of tiny notches or grooves in the edge of the footprint (e.g. Woodhams and Hines 1989). These indentations may occur at the front of the toe-prints (if the toes slipped slightly backwards) or at the rear margin of the footprint (if the foot slid down and slightly forwards into the substrate). They also occur along the sides of the toe-prints, where the toes shifted sideways or rotated in the substrate.

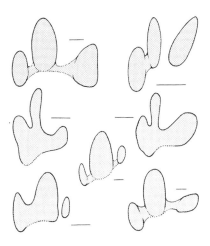

Figure 5.18 Variation in the shape of digits within a single ichnospecies. All seven ornithopod footprints (*Wintonopus*) are from a single assemblage in the mid-Cretaceous of Queensland, Australia; each scale bar indicates 2 cm. In some footprints, digit III is broadest, as it was in the track-maker's foot. In other cases, the impact of the heavily loaded digit III generated a thixotropic reaction in the substrate; on withdrawal of the foot, fluid sediment was sucked back into the print of digit III, so that it became narrower than the prints of digits II and IV (see examples at top right, centre left and centre right). (Adapted from Thulborn and Wade 1989.)

Comparable markings are found in *Chirotherium* and in the tracks of living animals (e.g. Soergel 1925, figs 13, 14; Bouchner 1982: 150, 168, 173).

Occasionally, the wall of a dinosaur footprint appears to be 'stepped' or 'terraced' (e.g. Plate 9, p. 140, top right). This feature originated when the track-maker's foot shifted its position as it settled into the substrate. At the start of a stride the forwardly extended foot would have rested on the substrate without any major supportive role, so that the footprint would, initially, have been quite shallow. At midstride the animal's centre of gravity passed forwards above the foot, which then would have sunk more deeply into the substrate. If, during this sequence of events, the foot shifted its position (by slipping backwards or sideways, or by rotating) remnants of the initial shallow footprint would persist in the form of marginal terraces. Sometimes this slippage of the foot is so marked that one or more of the toe-prints appears to be doubly imprinted (e.g. de Lapparent 1945: 270).

The stoutest toe in the foot might be expected to produce the broadest toe-print, and this is often the case. But in some instances the impact of the principal weight-bearing digit generated a thixotropic reaction in the sediment, which was sucked inwards when the toe was withdrawn. In this way the broadest toe might produce the narrowest toe-print (Figure 5.18; Plate 8, p. 134, bottom left).

Miscellaneous variations

An imprint from the metapodium is a normal feature of sauropod pes prints, and it also resulted when bipedal dinosaurs rested on the ground (see Figure 4.13) or adopted a plantigrade posture (Plate 6, p. 94, bottom right). As mentioned before, most dinosaurs were digitigrade, though some of the bigger forms may have been subdigitigrade or semiplantigrade, deriving some support from the distal end of the metapodium and from a substantial 'heel pad' of soft tissues behind it (see Figure 8.6b). It is also suggested that some of the smaller bipedal dinosaurs were in the habit of adopting a plantigrade posture (Kuban 1989b), though this possibility deserves further research. The imprint of the metapodium may be foreshortened, like the digit impressions, if the foot was applied to the substrate at an angle. In some cases the metapodium imprint is deeper than the toe-prints, while in others the reverse is true; the significance of this variation remains unknown (Thulborn and Wade 1989).

Sometimes, a dinosaur's toes entered the substrate quite crisply, so that the sediment between them was barely disturbed. In other instances the substrate seems to have had a pliable surface that buckled downwards between and alongside the toes. These buckled areas between the digits should not be mistaken for interdigital webs. Genuine traces of interdigital webs have sharply defined margins (e.g. Figure 4.7b,c), as in the footprints of birds, but appear to be uncommon in dinosaur footprints. The ichnogenus *Otouphepus*, described by J.A. Cushman (1904) from the red sandstones of the Connecticut Valley, USA, was supposed to be distinguished by very extensive traces of webbing – not only between the toes, but around the margins of the footprint as well (see Figure 4.7a). R.S. Lull (1953: 176) suspected that this web, which was defined by a dark discoloration, might have been nothing more than 'a slight wave of mud displaced by the animal's weight', but on careful examination D. Baird (1957: 507) proved it to be a thin coat of gum that was readily removed with

  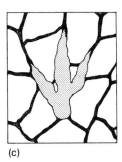

(a)   (b)   (c)

Figure 5.19 Diagrams of footprints associated with sun-cracks. (a) Footprint superimposed on previously sun-cracked substrate. (b) Sun-cracks formed after the footprint and traversing it at random. (c) Digits of footprint acted as starting-points for the subsequent development of sun-cracks.

soap and water. Once this 'web' had been removed, and other preservational peculiarities were taken into account, *Otouphepus* transpired to be an example of the common theropod footprint *Anchisauripus*.

The exposed surface of a fine-grained substrate, such as mud or clay, sometimes dried out and developed a network of shrinkage cracks (variously termed desiccation cracks, mud-cracks or sun-cracks). These may be associated with fossil footprints in any of several ways (Figure 5.19). First, the footprints may be superimposed on existing cracks. If those earlier-formed cracks were solidly filled by coarse-grained sediment, such as wind-blown sand, the superimposed footprints might be preserved in practically undisturbed form (Plate 11, p. 208, bottom left). But if those cracks remained open, the footprints might have a very different fate. Rain or floodwater might saturate the muddy substrate, causing it to re-expand; consequently, the edges of the shrinkage cracks would draw together again, thus distorting the superimposed footprints. Alternatively, the shrinkage cracks might be formed after the footprints. In some circumstances, footprints acted as starting-points for the development of shrinkage cracks (e.g. E. Hitchcock 1858: 170; Wuest 1934), because the hardening surface of the substrate was thinnest and most liable to fracture where it had been penetrated by the tip of a track-maker's toe. Incipient cracks extending from the tips of the digits should not be mistaken for drag-marks or traces of slender claws. Finally, shrinkage cracks might develop independently of a footprint and traverse it at random; in this

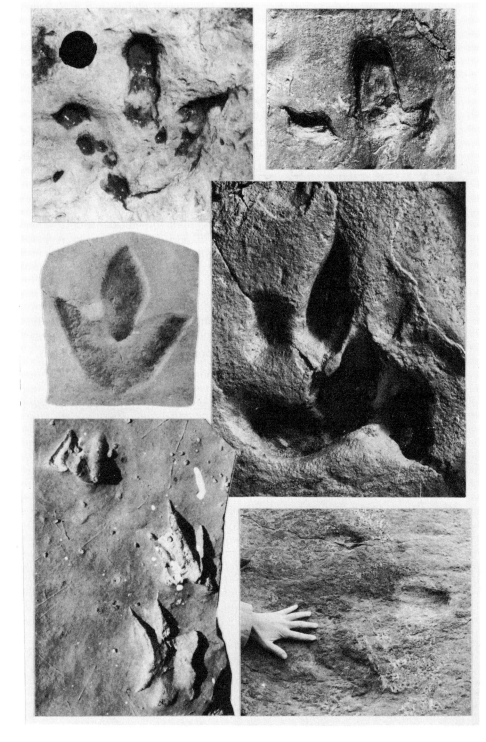

case the footprint would be divided into discrete sections that were sometimes forced wider and wider apart by the expanding cracks. Footprints may also be fragmented and dislocated by joints, or stress fractures, in a lithified substrate (e.g. Plate 8, p. 134, top right).

STRATIGRAPHIC AND GEOGRAPHIC LOCATION OF FOOTPRINTS

Each group of dinosaurs has its own definite stratigraphic range, established primarily on the evidence of body fossils, and tracks are

Plate 9 Preservational features of dinosaur tracks, and examples of ornithopod footprints. Top left: Left footprint of an ornithopod dinosaur, from the Lower Jurassic of northern Arizona; about 27 cm wide. Note that the footprint is wider than long, with a broadly rounded rear margin; the U-shaped outline of the middle toe-print is also characteristic of ornithopod tracks. Top right: *Wintonopus latomorum*, the right footprint of an ornithopod dinosaur, from the mid-Cretaceous of Queensland, Australia; 16.5 cm wide. 'Terraces' within the print reveal that the foot shifted position as it settled into the substrate. Centre left: *Gypsichnites pacensis*, left footprint of an ornithopod dinosaur, from the Lower Cretaceous of British Columbia, Canada; 29 cm long. The 'clover-leaf' shape, with the expanded middle digit, is characteristic of many big ornithopod tracks. Centre right: *Wintonopus latomorum*, two right footprints of small ornithopod dinosaurs, from the mid-Cretaceous of Queensland, Australia. The slightly bigger print, below, is about 7 cm wide and is foreshortened because the toes entered the substrate at a steep angle; on withdrawal, the middle toe smeared the sediment forwards, so that it folded over to conceal the outer part of a smaller print. Bottom left: Slab with footprint casts of small dinosaurs, from the mid-Cretaceous of Queensland, Australia; about 26 cm wide. Poorly preserved ornithopod print (*Wintonopus*) is at top left; well-preserved coelurosaur print (*Skartopus australis*) at centre right shows indications of phalangeal nodes and narrow claws. Another coelurosaur print at lower right comprises three long furrows, produced by the dinosaur's toes slithering backwards through the sediment. Numerous small markings indicate plant rootlets and invertebrate burrows. Bottom right: Footprint of an early Cretaceous ornithopod (*?Iguanodon*), exposed on the cliffs at Festningen, Spitsbergen; slightly distorted by photograph's perspective. Only the three blunt toe-tips are clearly visible, perhaps because the footprint is heavily weathered or because it might be a transmitted impression.

most convincingly attributed to particular dinosaur groups when they fall into the appropriate stratigraphic range. For example, all body fossils of hadrosaurs have been collected from rocks of late Cretaceous age. Footprints that match up with the feet of hadrosaurs might well be attributed to them if they also occur in late Cretaceous rocks (e.g. Langston 1960; Alonso 1980). Similar footprints from somewhat older rocks (e.g. Currie 1983; Lockley *et al.* 1983) cannot be attributed to hadrosaurs with the same degree of confidence.

This principle applies just as well to the geographic location of tracks. Some dinosaurs are known to have been cosmopolitan, so that it is not surprising to find their tracks distributed world-wide. This is the case, for example, with footprints of large ornithopods similar to *Iguanodon*; these have been discovered not only in Europe, but also as far afield as Australia, China, South America, North America and the island of Spitsbergen, within the present-day Arctic Circle. Other types of dinosaurs were more restricted in their distribution. Ceratopsians, for instance, are known to have existed only in the northern (Laurasian) continents and there is no certain evidence that they ever inhabited the southern (Gondwana) continents. Consequently, one would not expect to find tracks of these dinosaurs in the Cretaceous sediments of South America, Africa or Australia.

ASSOCIATED FOSSILS

Dinosaur tracks are sometimes found in association with other fossils, of which the most thought-provoking are, of course, dinosaur bones. If skeletons and footprints occur together in a single rock unit, it is only natural to suspect that the two might be related. That suspicion is easily tested by comparing the sizes and shapes of the footprints to the sizes and shapes of the dinosaur foot skeletons (e.g. Figure 5.2). A close correspondence between the two *might* indicate that the dinosaurs were responsible for the associated tracks. Notice the tentative phrasing of this conclusion: it is a statement of probability, and not a final verdict. It must be remembered that any one type of dinosaur, no matter how distinctive its foot structure, might produce a great diversity of footprint shapes, and that two quite different dinosaurs might produce very similar footprints. Unfortunately, this point tends to be forgotten when dinosaur tracks and dinosaur skeletons are discovered in the same rock unit. It is tempting to leap to the conclusion that the dinosaurs made the tracks, and even to apply the name of the dinosaur to the tracks as well. The only way to demonstrate conclusively that any one species of dinosaur was responsible for a particular type of footprint is to discover the skeleton of the animal preserved at the end of its fossil trackway. Such incontrovertible associations of tracks and body fossils are known for invertebrate animals, but have not, as yet, been reported for dinosaurs. For example, the Wealden (Lower Cretaceous) sediments of southern England contain skeletons of the large ornithopod *Iguanodon* as well as the tracks of large ornithopod dinosaurs. Some of these tracks match up very well with the feet of *Iguanodon*, but this is not conclusive proof that they are '*Iguanodon* tracks', as they are sometimes called. It is more accurate to identify them as tracks of large ornithopods that were similar or identical to *Iguanodon*. This sounds like hair-splitting, but it is actually a legitimate scientific statement of the conclusion to be drawn from the evidence. It might seem harmless to talk informally about '*Iguanodon* tracks', but all too often that casual association of dinosaur and tracks will be mistaken for a literal statement of fact.

Sometimes, footprints are associated with other dinosaur tracks, rather than dinosaur skeletons. Since some dinosaurs were anatomically adapted for social interactions, with weaponry, display structures and sound amplifiers, their tracks should be expected to occur in groups. This is true for plant-eating dinosaurs, such as sauropods (e.g. Bird 1944; Ishigaki 1986) and ornithopods (e.g. Lehmann 1978; Currie 1983), which may well have been accustomed to moving around in herds. By comparison, it is unlikely that largest predators, such as the tyrannosaurs, gathered in packs. These great carnivores are more likely to have been solitary hunters, or to have roamed in small groups (Farlow 1976), and one would normally expect to find their tracks singly or in limited numbers (though see Leonardi 1984a).

On occasion, the associated plant fossils may give an indication to the habits, and hence the identity, of dinosaurian track-makers. An excellent example is furnished by a series of tracks found recently in the roof of a coal mine in Utah (see Figure 11.3). The tracks reveal that a number of large bipedal dinosaurs threaded their way through a grove of tree-ferns, occasionally halting with their toe-tips right up against the trunks of these plants. Judging from their behaviour these dinosaurs, which were evidently stopping to investigate or browse on the tree-ferns, were more likely to have been herbivorous ornithopods rather than predatory theropods. This identification of the track-makers was substantiated by matching their footprints to the foot skeletons of ornithopods (Lockley *et al.* 1983).

6

Principal types of dinosaur tracks

An enormous three-toed track was imprinted in the soft mud before us. The creature, whatever it was, had crossed the swamp and had passed on into the forest. We all stopped to examine that monstrous spoor. . .

'Iguanodons,' said Summerlee. 'You'll find their footmarks all over the Hastings sands, in Kent, and in Sussex. The South of England was alive with them when there was plenty of good lush green-stuff to keep them going.'

Arthur Conan Doyle, *The Lost World* (1912)

CARNOSAURS

Structure of manus and pes (Figures 6.1 and 6.2)

The carnosaur manus (hand) was usually tridactyl, with three fingers conventionally identified as I, II and III. In early or persistently primitive-looking carnosaurs there was sometimes a remnant of digit IV, and in advanced forms such as *Tyrannosaurus* the outermost of the three fingers was sometimes so reduced that the manus was functionally didactyl. Overall, the hand was much smaller than the foot and seems to have had no major supportive role in locomotion. Its slender fingers terminated in strong curved claws, so large that they have sometimes been mistaken for those of the hindfoot (Russell 1970: 11). The fingers were not very widely divergent, in some instances almost parallel. In the standard tridactyl manus the phalangeal formula was 2:3:4:0:0; in the didactyl manus of tyrannosaurs the formula was 2:3:0:0:0.

The carnosaur pes (foot) skeleton resembled an enlarged version of that in modern birds. It comprised three large and forwardly spreading toes (II, III and IV), which, in many instances, were opposed by a smaller hallux (digit I); the phalangeal formula was typically 2:3:4:5:0. The three main weight-bearing toes were moderately

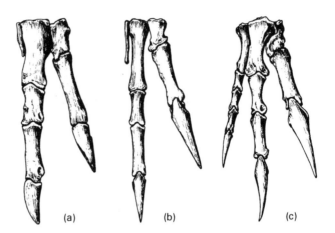

Figure 6.1 Manus skeleton in carnosaurs. (a) *Tarbosaurus*, from the Upper Cretaceous of Mongolia; about 23 cm long. (b) *Albertosaurus*, from the Upper Cretaceous of Canada; about 30 cm long. (c) *Allosaurus*, from the Upper Jurassic of the western USA; about 45 cm long. (Adapted from Maleev 1974 (a), Lambe 1917 (b), J. Madsen 1976 (c).)

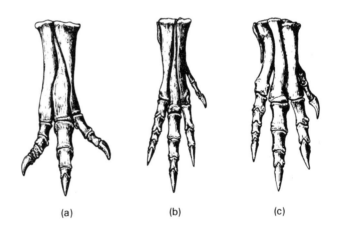

Figure 6.2 Pes skeleton in carnosaurs. (a) *Allosaurus*, from the Upper Jurassic of the western USA; about 71 cm long. (b) *Albertosaurus*, from the Upper Cretaceous of Canada; about 71 cm long. (c) *Tarbosaurus*, from the Upper Cretaceous of Mongolia; about 86 cm long. (Adapted from J. Madsen 1976 (a), Lambe 1917 (b), Maleev 1974 (c).)

divergent and arranged in an almost symmetrical pattern. In most carnosaurs these toes were quite robust, though in a few cases they were rather narrower. Digit III was the longest in the foot, while digits II and IV were slightly shorter and subequal in length. All the toes, including the hallux, terminated in large, sharply pointed and strongly curved claws. The hallux was smaller than any other digit in the foot and was sometimes located so high up on the metatarsus that it did not always touch the ground. In some carnosaurs the hallux extended medially, though in others it was directed postero-medially or straight backwards.

Tracks (Figures 6.3–6.5)

Carnosaurs were habitual bipeds, with narrow trackways in which the prints of the hindfeet often seem to be arranged in a single line. Pace angulation is commonly in the range 160°–180°, though sometimes it falls as low as 150°. Frequently, the prints of the hindfeet show slight positive rotation, and there is rarely any trace of a tail-drag. A rare example of a genuine tail-drag, with a somewhat erratic course, is apparent in Hitchcock's ichnogenus *Gigandipus* (see Figure 4.12a). The supposed tail-drag in another of Hitchcock's ichnogenera, *Hyphepus*, is more likely a series of drag-marks, each extending in consistent fashion from the tip of digit III (see Figure 4.12b). A tail-marking about 2 m long has been reported from the Jurassic coal measures of Queensland (Bryan and Jones 1946: 59), in association with large tridactyl pes prints that may well be those of theropods (e.g. Ball 1946). However, it is not known whether this tail-marking was made by a walking dinosaur or by one that was resting on the ground. The ratio of stride length to footprint length (SL/FL) is often about 5/1, though occasionally it may be as high as 8/1 or as low as 3/1.

The individual footprints are mesaxonic and tridactyl (digits II–IV) or tetradactyl (digits I–IV). In all cases, digits II, III and IV are spread out in roughly symmetrical pattern, whereas digit I, if present, extends medially or to the rear. The footprints are often a little longer than wide, though slightly broader examples are known. Those footprints with relatively slender digits (e.g. *Buckeburgichnus*, Figure 6.4j) were presumably made by gracile carnosaurs such as *Megalosaurus*; those with relatively stout digits (e.g. *Gigandipus*, Figure 6.4f) were probably produced by heavier and more robust animals resembling *Tyrannosaurus*. The total divarication of digits II–IV is often about 50°–60°,

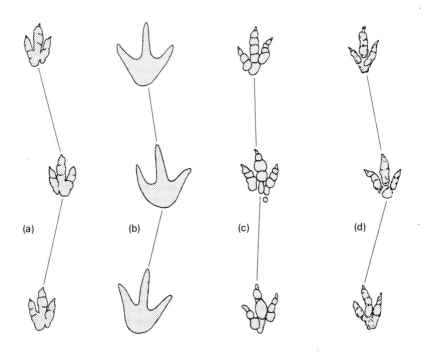

Figure 6.3 Trackways attributed to carnosaurs. (a) *Eubrontes*, from the Lower Jurassic of the Connecticut Valley, USA; total length about 2.6 m. (b) *Irenesauripus*, from the Lower Cretaceous of British Columbia, Canada; total length about 2.5 m. (c) *Eubrontes*, from the Upper Jurassic or Lower Cretaceous of Brazil; total length about 1.65 m. (d) Unnamed trackway, from the Jurassic of Morocco; total length about 2.5 m. (Adapted from Lull 1953 (a), Sternberg 1932 (b), Leonardi 1980c (c), Ishigaki 1985b (d).)

though occasionally it may be as low as 35° or as high as 75°. Interdigital angles II–III and III–IV are roughly equal.

The imprints of digits II, III and IV are usually tapered or V-shaped (Plate 10, p. 160, centre right), though their outlines are sometimes disrupted by the presence of digital nodes. In some carnosaur footprints all three digits are roughly equal in width, but in others digit III is distinctly broader than adjoining the ones. Often, there are distinct traces of large and acutely pointed claws (Plate 10, p. 160, centre left). The imprint of the hallux (where present) is narrow, sharply pointed and much smaller than the imprints of the other toes. It may appear as an oval or almond-shaped impression, a furrow, or a small puncture-mark left by the claw, though sometimes it is a

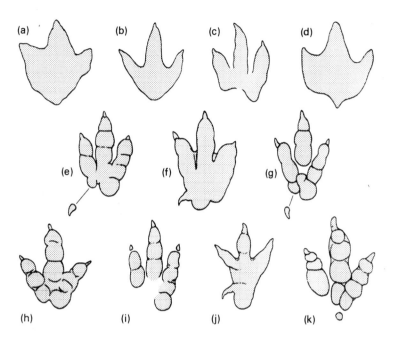

Figure 6.4 Pes prints attributed to carnosaurs. (a) Unnamed footprint resembling *Tyrannosauropus*, from the mid-Cretaceous of Australia; about 49 cm long. (b) Unnamed footprint attributed to carnosaur resembling *Allosaurus*, from the Lower Cretaceous of southern England; about 64 cm long. (c) *Eubrontes veillonensis*, from the Lower Jurassic of France; about 34 cm long. (d) Unnamed footprint attributed to carnosaur resembling *Megalosaurus*, from the Lower Cretaceous of southern England; about 25 cm long. (e) *Eubrontes giganteus*, from the Lower Jurassic of the Connecticut Valley, USA; about 37 cm long, excluding hallux. (f) *Gigandipus caudatus*, from the Lower Jurassic of the Connecticut Valley, USA; about 45 cm long. (g) *Anchisauripus minusculus*, from the Lower Jurassic of the Connecticut Valley, USA; about 31 cm long, excluding hallux. (h) Unnamed footprint attributed to carnosaur resembling *Megalosaurus*, from the Upper Jurassic or Lower Cretaceous of Brazil; about 29 cm long. (i) *Eubrontes approximatus*, from the Lower Jurassic of the Connecticut Valley, USA; about 40 cm long. (j) *Buckeburgichnus maximus*, from the Lower Cretaceous of Germany; about 70 cm long. (k) Unnamed footprint similar to *Eubrontes platypus* and attributed to carnosaur resembling *Tyrannosaurus*, from the Upper Jurassic or Lower Cretaceous of Brazil; about 29 cm long. (Adapted from Thulborn and Wade 1984 (a), Blows 1978 (b,d), de Lapparent and Montenat 1967 (c), Lull 1953 (e,f,g,i), Leonardi 1980c (h,k), Kuhn 1958a (j).)

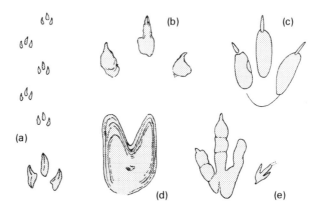

Figure 6.5 Miscellaneous traces attributed to carnosaurs. (a) Exceptionally broad trackway attributed to a carnosaur that was possibly swimming, from the Lower Cretaceous of Colorado, USA; right pes print shown enlarged (below) is about 47 cm long. (b) Deeply impressed toe-prints of a carnosaur, showing clear indications of the claws, from the Lower Cretaceous of Spitsbergen; about 23 cm wide. (c) *Dilophosauripus williamsi*, from the Lower Jurassic of Arizona, USA, showing narrow slots made by blade-like claws; about 33 cm long. (d) Trace of a sitting or squatting dinosaur, possibly a theropod, from the Lower Jurassic of Lesotho; total length about 70 cm. (e) *Changpeipus carbonicus*, from the Lower or Middle Jurassic of China; sometimes regarded as manus and pes prints of a single carnosaur, but possibly a chance association of two unrelated tracks; larger print about 39 cm long. (Adapted from Lockley 1985 (a), Edwards *et al.* 1978 (b), Welles 1971 (c), P. Ellenberger 1970 (d), Haubold 1971 (e).)

curved or sickle-shaped marking, with the convex margin facing antero-medially. The hallux may extend medially or backwards, but in no case is it directed forwards. Sometimes it runs straight backwards in the midline, affording the entire footprint an almost cross-shaped outline (Figure 6.4d).

The rear half of a carnosaur footprint tends to be triangular or wedge-shaped in outline, though the hindmost margin varies in shape: it is often smoothly rounded, though sometimes it is flatter or bluntly angular. Occasionally, one or more of the metatarso-phalangeal pads formed a definite heel-like bulge at the rear margin of the footprint, most commonly behind digit IV (e.g. Figure 6.4c,e,g–i).

Carnosaur prints are distinguished from those of coelurosaurs on the arbitrary basis of size: as a general rule the footprints of carnosaurs have a length greater than about 25 cm whereas smaller examples are conveniently attributed to coelurosaurs. The largest known carnosaur footprints are about 80 cm long. Some enormous footprints, up to 150 cm long, were attributed to a theropod by H. Mensink and D. Mertmann (1984), but might, perhaps, be the tracks of sauropods (ichnogenus *Gigantosauropus*, discussed later).

There seems to be only a single known trackway made by a carnosaur moving quadrupedally (see Figure 11.7c), and even this example is open to different interpretation. The prints of the forefeet are more widely spaced than those of the hindfeet, and are decidedly unpredictable in their placement. Pace angulation for the hindfoot prints is about 168°; for the handprints it is about 98°. Evidently, the animal must have been walking with its forelimbs spread wide apart and its head carried close to the ground (see Chapter 9). The handprints are much smaller than the prints of the hindfeet, and are roughly oval in outline. Their front margins appear to be slightly bilobed, perhaps indicating the presence of two functional digits. It is possible (or even probable) that a third digit was present in the hand, but this may have been so small that it failed to leave a definite impression (compare the hands of tyrannosaurs, Figure 6.1a,b). The prints of the hindfeet are tridactyl, with slender, sharply pointed and quite widely divergent digits. There is no trace of a tail-drag; a broad and continuous groove connects the prints of the hindfeet, but this appears to be an erosional feature rather than a genuine tail-drag.

Resting traces of carnosaurs are equally elusive, though two associated prints from the Middle Jurassic of China (*Changpeipus carbonicus*, Young 1960) have sometimes been attributed to hand and foot of a single carnosaur (e.g. Haubold 1971: 79). The two prints are side by side and might have been made by an animal that was squatting or standing on all fours (Figure 6.5e). However, it is also conceivable that these two prints were made by two different animals. Footprints from Australia (Bartholomai 1966: 149; Haubold 1971: 79) and China (Young 1979) have also been referred to the ichnogenus *Changpeipus*, but these do not resolve the uncertainties.

Two impressions made by sitting or squatting dinosaurs were discovered by P. Ellenberger (1970, 1972) in the late Triassic to early Jurassic sediments of Lesotho (Figure 6.5d). These traces show clear indications of an ischiadic tuberosity, indicating the location of the cloaca, but few other details are discernible and it is far from certain that the trace-makers were theropods.

In 1980, W.P. Coombs reported traces of swimming theropods in the early Jurassic sediments of the Connecticut Valley, USA. Some of the larger footprints, assigned to the ichnogenus *Eubrontes*, were probably made by a carnosaur resembling *Megalosaurus* (see Figure 2.8), whereas others, perhaps belonging in the ichnogenus *Anchisauripus*, were made by a smaller theropod. Each footprint comprised impressions from the tips of the three main toes (II–IV). The middle and longest toe was indicated by a claw-mark, followed by a shallow trace of the foremost digital pad. Inner and outer toes were each represented by a pit, made by the claw-tip, extending backwards into a narrow groove. Behind these grooves, which were evidently formed by backwards slippage of the toes, there was sometimes a small mound of displaced sediment. From the morphology and layout of the footprints Coombs deduced that each track-maker was swimming (or arguably wading) by kicking off the bottom with its toes. As each foot touched down, the tip of the longest toe anchored itself in the muddy sediment; then, as the track-maker pushed itself forwards, the inner and outer toes touched down and slithered back through the sediment to form distinctive grooves. One trackway with unusually long strides might have been made by a theropod gliding or floating between footfalls, and another, with regular alternation of short and long paces, led Coombs to suspect that there was a 'gallop' rhythm to the swimming strokes. Subsequently, L.C. Godoy and G. Leonardi (1985) reported similar tracks of about 40 theropods in the Lower Cretaceous of Brazil. Here the tracks tended to be aligned with ripple-marks on the substrate, perhaps indicating that the track-makers had been moving parallel to a shore-line.

Finally, some unusual trackways from the Upper Jurassic of southern France have been attributed to theropods that were hopping rather than walking or running (Bernier *et al.* 1984). These somewhat puzzling tracks, named *Saltosauropus*, are discussed in Chapter 9.

COELUROSAURS

Structure of manus and pes (Figures 6.6 and 6.7)

The coelurosaur manus usually comprised three digits, conventionally

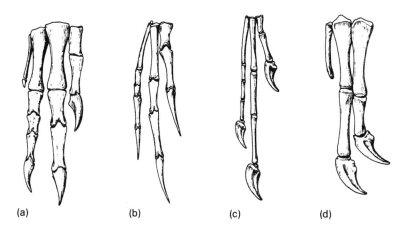

(a)   (b)   (c)   (d)

Figure 6.6 Manus skeleton in coelurosaurs. (a) *Syntarsus*, from the Lower Jurassic of Zimbabwe; about 6.6 cm long. (b) *Deinonychus*, from the Lower Cretaceous of Wyoming and Montana, USA; about 28 cm long. (c) *Ornitholestes*, from the Upper Jurassic of Wyoming, USA; about 17.5 cm long. (d) *Compsognathus*, from the Upper Jurassic of Germany and France; about 4 cm long. (Adapted from Raath 1969 (a), Ostrom 1969a (b), 1978 (d), Osborn 1917 (c).)

identified as I, II and III. The long slender fingers were only slightly divergent or subparallel, and were equipped with slim and sharply pointed claws. In some cases the inner and outer fingers were a little shorter than the middle one, so that the manus was almost symmetrical in form; in other cases the innermost finger was noticeably shorter than the other two. Often, the three fingers were about equally robust, but sometimes the outermost one was thinner and rather feeble-looking. The phalangeal formula was typically 2:3:4:0:0, though the early Jurassic coelurosaurs *Syntarsus* and *Coelophysis* were unusual in retaining the vestige of a fourth finger (formula 2:3:4:1:0). This tiny fourth finger was tucked in behind the other three and might not have been visible in the hand of the living animal (Raath 1969: 9). Despite its slender and almost trident-like structure the hand may have afforded a weak grasping action in some coelurosaurs (Galton 1971a). In all cases the hand was distinctly smaller than the hindfoot, though it was not reduced to the extreme degree seen in some carnosaurs. The hand was often about half the length of the hindfoot.

The coelurosaur pes was essentially a smaller and more delicate

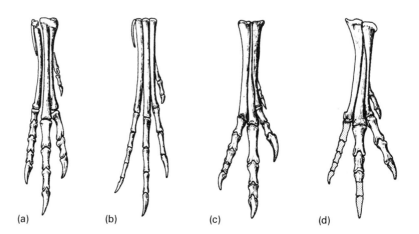

Figure 6.7 Pes skeleton in coelurosaurs. (a) *Syntarsus*, from the Lower Jurassic of Zimbabwe; about 11.5 cm long. (b) *Compsognathus*, from the Upper Jurassic of Germany and France; about 10.5 cm long. (c) *Stenonychosaurus*, from the Upper Cretaceous of Canada; about 38 cm long. (d) *Elmisaurus*, from the Upper Cretaceous of Mongolia; about 30 cm long. (Adapted from Raath 1969 (a), Ostrom 1978 (b), Kurzanov 1987 (c,d).)

version of that in the carnosaurs. It too comprised three forwardly spreading toes (II, III and IV), often opposed by a smaller hallux. The phalangeal formula was invariably 2:3:4:5:0. All three of the forwardly directed toes were relatively long and slender, with sharp claws, and were roughly equal in width. The middle toe was the longest in the foot, with the slightly shorter inner and outer toes disposed almost symmetrically to the sides. The hallux was the smallest digit in the foot; in some instances it extended no farther than the distal end of the metatarsus and could barely have touched the ground. A few early and primitive-looking coelurosaurs may have had the hallux directed forwards, alongside digit II (Raath 1969: 19), but in all other cases it was re-oriented so as to point medially or backwards.

Tracks (Figures 6.8–6.10)

Coelurosaurs were habitually, though not exclusively, bipedal. Their trackways are typically rather narrow, with pace angulation in the range 160°–170°, but occasionally as low as 150° or as high as 180°. Sometimes, the trackway is so narrow that left and right footprints fall

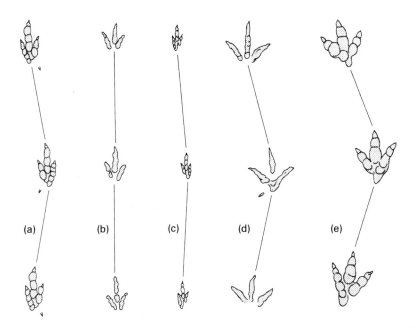

Figure 6.8 Trackways attributed to coelurosaurs. (a) *Anchisauripus sillimani*, from the Lower Jurassic of the Connecticut Valley, USA; total length about 100 cm. (b) Unnamed ichnospecies of *Grallator*, from the Upper Triassic of the Colorado-Oklahoma border, USA; total length about 46 cm. (c) *Grallator cursorius*, from the Lower Jurassic of the Connecticut Valley, USA; total length about 80 cm. (d) Unnamed trackway, from the Jurassic of Morocco; total length about 35 cm. (e) Unnamed trackway resembling both *Eubrontes* and *Grallator*, from the Upper Triassic or Lower Jurassic of Brazil; total length about 117 cm. (Adapted from Lull 1953 (a,c), Conrad *et al.* 1987 (b), Ishigaki 1985b (d), Leonardi 1980c (e).)

in a single line. The prints of the hindfeet show slight positive rotation, or no appreciable rotation at all, and there is rarely any marking left by the tail. The ratio *SL/FL* is often in the range 7/1 to 8/1, though it may fall as low as 4/1 or rise as high as 16/1 in the tracks of fast-running animals (e.g. *Skartopus*, Thulborn and Wade 1984).

Prints of the hindfeet are mesaxonic, and tridactyl or tetradactyl, with digits II, III and IV splayed out in roughly symmetrical pattern. Digit I, where present, is directed medially or to the rear. In general appearance the footprints resemble those of carnosaurs, being distinguished from them only through their smaller size; most

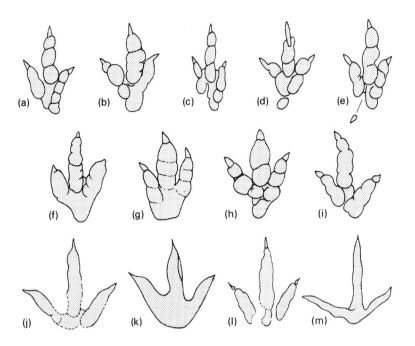

Figure 6.9 Pes prints attributed to coelurosaurs. (a) *Grallator cuneatus*, from the Lower Jurassic of the Connecticut Valley, USA; about 12 cm long. (b) Unnamed footprint attributed to a coelurosaur resembling *Syntarsus*, from the Lower Jurassic of Zimbabwe; about 11 cm long. (c) *Grallator cursorius*, from the Lower Jurassic of the Connecticut Valley, USA; about 7.5 cm long. (d) *Stenonyx lateralis*, from the Lower Jurassic of the Connecticut Valley, USA; about 3 cm long. (e) *Anchisauripus hitchcocki*, from the Lower Jurassic of the Connecticut Valley, USA; about 12 cm long, excluding hallux. (f) *Coelurosaurichnus perriauxi*, from the Middle Triassic of France; about 9.5 cm long. (g) *Coelurosaurichnus ziegelangerensis*, from the Upper Triassic of France; about 12.5 cm long. (h) Unnamed footprint resembling both *Eubrontes* and *Grallator*, from the Upper Triassic or Lower Jurassic of Brazil; about 21 cm long. (i) Unnamed footprint, from the Middle Triassic of France; about 9 cm long. (j) Unnamed footprint, from the Lower Jurassic of Sweden; about 18 cm long. (k) *Columbosauripus ungulatus*, from the Lower Cretaceous of Canada; about 12.5 cm long. (l) Unnamed footprint, from the Middle Jurassic of Buckinghamshire, England; size not specified, but probably less than 25 cm long. (m) Unnamed footprint, from the mid-Cretaceous of Israel; size not specified, but probably less than 25 cm long. (Adapted from Lull 1953 (a,c–e), Raath 1972 (b), Gand *et al.* 1976 (f,i), Kuhn 1958b (g), Leonardi 1980c (h), Pleijel 1975 (j), Sternberg 1932 (k), Delair and Sarjeant 1985 (l), Avnimelech 1966 (m).)

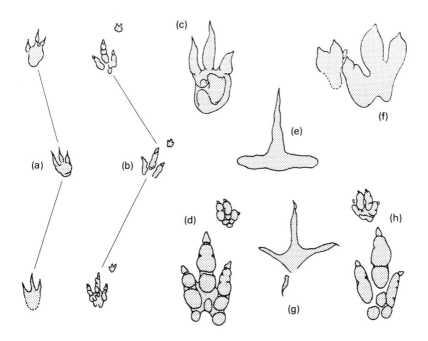

Figure 6.10 Miscellaneous traces attributed to coelurosaurs. (a) *Skartopus australis*, a bipedal trackway from the mid-Cretaceous of Queensland, Australia; total length about 80 cm. Note that the footprints are rotated positively (inwards) and show clear impressions of the metatarsus.
(b) *Atreipus milfordensis*, a quadrupedal trackway from the Upper Triassic of Pennsylvania, USA; total length about 110 cm. Note the unusual width of the trackway and negative (outwards) rotation of the pes prints.
(c) *Skartopus australis*, a pes print from the mid-Cretaceous of Queensland, Australia, showing prominent imprint of the metatarsus; about 11 cm long.
(d) *Atreipus milfordensis*, a manus-pes couple (composite) from the Upper Triassic of New Jersey, USA; pes about 10.5 cm long. (e) *Taupezia landeri*, a pes print from the Upper Jurassic of Dorset, England, showing extremely wide divarication of digits II and IV; about 15 cm long. (f) *Delatorrichnus goyenechei*, a manus-pes couple from the Middle Jurassic of Argentina; pes about 3 cm long. (g) *Ignotornis mcconelli*, a pes print from the Lower Cretaceous of Colorado, USA, showing wide divarication of digits II–IV and backwardly directed trace of the hallux; about 5.5 cm long, including hallux. (h) *Atreipus (Coelurosaurichnus) metzneri*, a manus-pes couple from the Upper Triassic of Germany, showing tetradactyl manus print; pes about 8.5 cm long. (Adapted from Thulborn and Wade 1984 (a,c), Olsen and Baird 1986 (b,d,h), Delair 1963 (e), Casamiquela 1966 (f), Mehl 1931 (g).)

coelurosaur footprints are less than 20 cm long, and the smallest examples are no more than 2 cm in length. There are reports of even smaller coelurosaur prints, down to 0.5 cm in length, but these may not be dinosaurian in origin (see Chapter 7). Coelurosaur footprints are commonly longer than wide, with FW equivalent to some 70-75 % of FL. The three main digits usually diverge at low angles, the total divarication (II-IV) being in the range 45°-50°. Occasionally, the total divarication approaches 180°, so that the footprint resembles a letter 'T', as in the ichnogenus *Taupezia* (Figure 6.10e). Interdigital angles II-III and III-IV are roughly equal.

The imprints of digits II, III and IV are narrow and slightly tapered, though their overall outlines are frequently disrupted by the presence of digital nodes (see Plate 10, p. 160, top left and top right). Digit III is longer than digits II and IV, which are usually subequal in length, and in many instances it is also a little broader as well. Normally, all three digits show traces of sharp claws. The rear half of the print is rather angular or wedge-shaped in outline, often with a prominent metatarso-phalangeal pad behind digit IV. Occasionally, there may also be traces of metatarso-phalangeal pads behind digits II and IV (Figure 6.10d,h).

The small imprint of the hallux, where present, is generally similar to that in carnosaur footprints, both in its shape and its orientation. Some coelurosaur footprints bear a close resemblance to bird tracks on account of an exceptionally long hallux print extending straight backwards from the rear margin (Figure 6.10g).

Until quite recently few traces had been attributed to coelurosaurs moving on all fours. The ichnogenus *Agialopous*, from the Upper Triassic of Wyoming, was originally interpreted as the trackway of a quadrupedal coelurosaur (E.B. Branson and Mehl 1932) but is more probably a coincidental amalgamation of two tracks made by bipeds (Baird 1964: 120). In a study of tracks from the Middle Triassic of France, G.R. Demathieu (1985: 57) mentioned that some 'scarcely quadrupedal trackways' of the ichnogenus *Coelurosaurichnus* included traces of a four-fingered hand, but these were neither illustrated nor described in detail. The presence of a fourth finger in the hand is a primitive feature for coelurosaurs and would certainly accord with the early date of these tracks. Elsewhere, M. Morales (1986: 14-15) reported an association of large and small tridactyl prints – possibly pes and manus – in the Kayenta Formation (Lower Jurassic) of Arizona. An equally questionable association of manus and pes prints, described from the Upper Triassic of Morocco under the name *Tridactylus machouensis* (Biron and Dutuit 1981), might be attributed to an ornithopod dinosaur (see Figure 6.28e).

More convincing examples, from the Upper Triassic of Germany, were described by F. Heller (1952) under the name *Coelurosaurichnus metzneri*. These included a hindfoot print of typical coelurosaur type, along with the much smaller impression of a tridactyl manus. Somewhat similar traces, from the Upper Triassic of the northeastern United States, were identified as *Anchisauripus milfordensis* and *Grallator sulcatus* by D. Baird (1957). Subsequently, these were re-examined by P.E. Olsen (1980: 43), who commented that their most distinctive feature was 'a small three-toed manus impression present in nearly every trackway'. These tracks are very similar to *Coelurosaurichnus metzneri*, and there seems little doubt that they, too, were made by small theropods. Recently, all three of the German and North American ichnospecies were referred by Olsen and Baird (1986) to a new ichnogenus, *Atreipus* (Figure 6.10b,d,h). The trackways included in this ichnogenus typically comprise tridactyl pes prints associated with small 'tulip-shaped' manus prints. The pes prints show slight negative (outwards) rotation and form a fairly broad zig-zag trackway with maximum width about two and a half times footprint width. There are prominent indications of the metatarso-phalangeal nodes, but no impression of the metatarsus itself, nor of the tail. Pace angulation is about 150°, and the ratio *SL/FL* is roughly 5/1. Each manus print, which lies just in front of a pes print, usually comprises three stubby digits, sometimes with imprints of small claws. However, one ichnospecies, *Atreipus acadianus*, has consistently tetradactyl manus prints, with an impression of a tiny fourth finger. Olsen and Baird (1986) regarded *Atreipus* as the track of a small ornithischian, though they admitted that there was considerable uncertainty surrounding the identity of the track-maker. *Atreipus* has been included here among the tracks of theropods for three reasons: (1) the coelurosaur-like morphology of its pes prints; (2) the absence of any impression from the metapodium, which occurs very commonly in the tracks of small ornithopods that travelled quadrupedally; and (3) the 'tulip-shaped' manus prints, which are quite unlike the stellate or fan-shaped prints made by small ornithopods (see Figure 6.28).

The name *Delatorrichnus goyenechei* was coined by R.M. Casamiquela (1964) for tracks of small quadrupeds in the Middle Jurassic of Argentina (Figure 6.10f; Plate 11, p. 208, centre). The prints of the

hindfoot are tridactyl, with traces of sharp claws, and would certainly be appropriate for a coelurosaur. Despite their imperfect preservation the prints of the manus, which was about two-thirds the size of the pes, show traces of at least two digits, and their heart-shaped outlines are matched in the ichnogenus *Atreipus*. In fact, these two ichnogenera, *Atreipus* and *Delatorrichnus*, seem to be remarkably similar, aside from differences in preservation and the slightly greater size of the manus prints in the latter (see Casamiquela 1966). H. Haubold (1971, 1984) listed *Delatorrichnus* among the tracks of carnosaurs, though its small size (with *FL* slightly over 3 cm) indicates that it would be better accommodated among the tracks of coelurosaurs. Small pes prints resembling those of *Delatorrichnus* occur in the Lower Jurassic of Utah and have been attributed to 'hatchling sized' theropods (Lockley 1986b: 18).

These various tracks from Europe, North America and South America are important in revealing that coelurosaurs were not exclusively bipedal and that these dinosaurs did travel quadrupedally on some occasions. And, on closer analysis, trackways such as *Delatorrichnus* and *Atreipus* may indicate that some early coelurosaurs were in the habit of moving around on all fours (see Chapter 9).

The Middle Triassic of the Netherlands has yielded questionable

Plate 10 Examples of theropod tracks. Top left: *Columbosauripus ungulatus*, a coelurosaur print from the Lower Cretaceous of British Columbia, Canada; 12.5 cm long. The V-shaped outline and narrow claw-marks are characteristic of many theropod tracks. Top right: Slab with numerous casts of coelurosaur footprints, *Grallator variabilis*, from the Lower Jurassic of western France; maximum diameter of slab is 65 cm. The footprints show a V-shaped outline, prominent phalangeal nodes and traces of narrow claws; note also the numerous casts of pits made by raindrops. Centre left: A theropod footprint attributed to *Megalosaurus*, in the Lower Cretaceous of Kvalvågen, Spitsbergen; about 23 cm wide. This is an unusually broad and short print, but has unmistakable indications of the claws. Centre right: A theropod footprint (?*Kayentapus*), showing V-shaped digits with slot-like traces of claws, from the Lower Jurassic of northern Arizona, USA; about 27 cm wide. Bottom left: R.T. Bird's excavation alongside the Paluxy River in Texas, in 1940, which revealed the trackway of a carnosaur (left) apparently pursuing a sauropod. Both trackways veer to the left in identical fashion. Bottom right: Natural cast of a coelurosaur footprint, *Skartopus australis*, from the mid-Cretaceous of Queensland, Australia; about 4.5 cm wide. The track-maker was scarcely bigger than a rooster.

coelurosaur tracks, named *Coelurosaurichnus ratumensis* by G.R. Demathieu and H.W. Oosterink (1988). These may include traces of a pentadactyl manus – which is unprecedented in theropods – but their exact identity remains an open question pending discovery of a complete pes print. The ichnospecies *Sauropus barrattii*, from the Lower Jurassic of the Connecticut Valley, has sometimes been regarded as the resting trace of a theropod (e.g. Haubold 1971: 73; 1984: 48), but Olsen and Baird (1986: 78-9) have demonstrated that this name applies to a conglomeration of unrelated tracks, including those of bipedal theropods (*Grallator*) and small ornithopods (*Anomoepus*).

From the Lower Jurassic of Rhodesia M.A. Raath (1972) recovered two small footprints, preserved side by side, which might have been made by a coelurosaur in a standing or resting pose. And among the tracks of swimming theropods described by Coombs (1980b) were some small examples that were probably made by coelurosaurs (ichnogenus *Anchisauripus*?). Tracks ascribed to hopping theropods by P. Bernier *et al.* (1984) are discussed in Chapter 9.

ORNITHOMIMIDS

Structure of manus and pes (Figure 6.11)

In ostrich dinosaurs the hand was about half as long as the foot. It comprised three long and slender fingers, all of about the same length, which were only weakly divergent or subparallel. In *Struthiomimus* the middle and outer fingers may have been enclosed in a single sheath of skin, extending to the base of the claws (Nicholls and Russell 1985: 671). All three fingers carried large claws that were distinctly more curved and hook-like than the claws on the hindfeet. These powerful hand have been likened to those of anteaters (Russell 1972: 401), and it is often suggested that they were employed in raking over the ground or digging in search of food (e.g. Osmólska *et al.* 1972: 143). The hands seems to have had little grasping ability, and the fingers probably worked as a unit rather than independently (Galton 1971a). E.L. Nicholls and A.P. Russell (1985) suggested that the hand of *Struthiomimus* might have functioned as a clamp or hook for pulling down vegetation.

The structure of the hindfoot was equally unusual, because the exaggerated length of the metatarsus endowed ornithomimids with

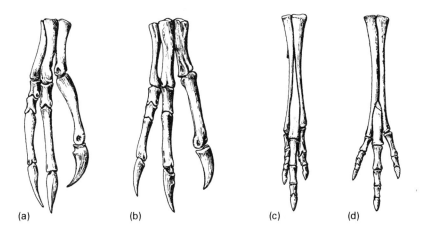

(a)   (b)   (c)   (d)

Figure 6.11 Manus and pes skeleton in ornithomimids. (a) Manus of *Struthiomimus*, from the Upper Cretaceous of Canada; about 29 cm long. (b) Manus of *Gallimimus*, from the Upper Cretaceous of Mongolia; about 31 cm long. (c) Pes of *Struthiomimus*, from the Upper Cretaceous of Canada; about 60 cm long. (d) Pes of *Gallimimus*, from the Upper Cretaceous of Mongolia; about 77 cm long. (Adapted from Nicholls and Russell 1985 (a), Osmólska *et al.* 1972 (b,d), Kurzanov 1987 (c).)

unusually long and slender feet. Digits II, III and IV were rather short and stubby, and in most cases digits I and V were entirely lacking – though *Struthiomimus* retained a tiny vestige of metatarsal V. The three functional toes diverged quite widely and carried heavy claws, reminiscent of those in ostriches and emus.

Tracks (Figure 6.12)

Ornithomimids were habitually or exclusively bipedal, and few (but distinctive) tracks have been attributed to them. The first authenticated tracks, from the Upper Cretaceous of Canada, were originally described by C.M. Sternberg (1926) under the name *Ornithomimipus angustus* (Figure 6.12a,c). These tridactyl footprints are arranged in virtually a straight line, indicating a high value for pace angulation, probably approaching 180°. The footprints show little or no rotation and in each one the middle digit is distinctly longer than the other two, which are quite widely divergent (total divarication about 78°). Digit IV is slightly more divergent than digit II, with interdigital angles about 35° (II-III) and 44° (III-IV). The middle toe-print is

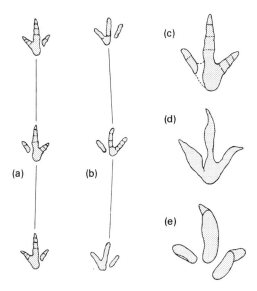

Figure 6.12 Tracks attributed to ornithomimids and similar dinosaurs.
(a) *Ornithomimipus angustus*, a trackway from the Upper Cretaceous of
Alberta, Canada; total length about 160 cm. (b) *Ornithomimipus*
(*Irenichnites*) *gracilis*, a trackway from the Lower Cretaceous of British
Columbia, Canada; total length about 155 cm. (c) *Ornithomimipus angustus*,
a pes print from the Upper Cretaceous of Alberta, Canada; about 28 cm
long. (d) Unnamed footprint from the mid-Cretaceous of Israel, not
certainly attributed to an ornithomimid; about 25 cm long. (e) *Hopiichnus
shingi*, a footprint from the Lower Jurassic of Arizona, USA, unlikely to be
the track of an ornithomimid; about 10 cm long. (Adapted from Sternberg
1926 (a,c), Sternberg 1932 (b), Avnimelech 1966 (d), Welles 1971 (e).)

slightly broader than the other two, which are about equal in width,
and all three of them show traces of a stubby claw. The imprint of
digit II is consistently separated from the conjoined imprints of digits
III and IV, implying that hypex II–III was located higher up the foot
than hypex III–IV; the ornithomimid foot skeleton does in fact show
such an arrangement of the toes (Figure 6.11d). As they are traced
backwards the imprints of digits III and IV merge into an impression
made by the distal end of the metatarsus. This impression forms a
definite bulge at the rear margin of the footprint, directly in line with
the axis of digit III. The footprints are slightly longer than broad, with

FW representing some 70 % of *FL*, and the ratio *SL/FL* is in the range 6/1 to 7/1.

At a later date Sternberg (1932) described comparable footprints from slightly earlier (Lower Cretaceous) sediments in Canada. These were given the different name *Irenichnites gracilis* but are so similar to *Ornithomimipus* that they might justifiably be included in the same ichnogenus (Haubold 1971: 75; though see Currie 1989). The footprints are smaller than those described earlier (15 cm long as opposed to 28 cm), and their chief distinction is that digit III is somewhat shorter, so that the footprint is slightly wider than long. The ratio *SL/FL* was as high as 9/1 or 10/1.

Some tridactyl footprints from the Upper Cretaceous of Algeria were regarded by P. Bellair and A.F. de Lapparent (1949) as those of ostrich dinosaurs, but these show no special resemblances to the Canadian examples described by Sternberg. Some of these Algerian footprints were referred by H. Haubold (1971: 74) to another of Sternberg's ichnogenera, *Columbosauripus*, which probably comprises coelurosaur tracks. In addition, tracks presumed to be those of long-legged theropods, perhaps resembling ornithomimids, were reported by M. Avnimelech (1962a,b, 1966) in the Upper Cretaceous of Israel.

In 1971 S.P. Welles described some footprints from the Kayenta Formation (Lower Jurassic) of Arizona under the name *Hopiichnus shingi*. Welles suspected that these were the tracks of an exceptionally long-legged dinosaur that might have resembled an ornithomimid in its body proportions. However, it is unlikely that ornithomimids were in existence at such an early date, and the tracks described by Welles are in some respects problematical. Various interpretations of *Hopiichnus* were discussed at length by R.A. Thulborn and M. Wade (1984: 453) and none of them was found to be entirely satisfactory. M.G. Lockley (1986b: 18) has since identified more tracks similar to *Hopiichnus*, but these have yet to be described. Possible ornithomimid tracks have also been reported from the Lower Cretaceous of Colorado (Lockley 1986b: 30).

There are no definite reports of traces left by ornithomimids resting on the ground or walking quadrupedally. Sternberg's original account of *Ornithomimipus* (1926: 85-6) did, however, include the following remarks:

Two faint impressions of smaller feet were shown which suggested to the observer that the animal touched the front feet on the ground while resting... Two of the tracks [= footprints] were side by side and may have been made by an animal standing with its feet very close together, but more probably were made at different times.

Unfortunately, Sternberg gave no further details or illustrations. Even so, it seems quite possible that ornithomimids did leave preservable traces of their hands, especially if these were used for digging or for raking the ground in search of food.

SAUROPODS

Structure of manus and pes (Figures 6.13 and 6.14)

The highly distinctive manus of sauropods included five columnar metacarpals grouped into a tight arc or bundle (Figure 6.13a). In some sauropods, such as *Apatosaurus*, these bones were short and thick, but in others they were tall and pillar-like, as in *Brachiosaurus*. In all cases the metacarpals were steeply inclined, with each of them supporting a short toe comprised of one or two stubby phalanges. Overall, the forefoot probably resembled that of a living hippo, with the individual toes projecting only slightly from a stout pad of tissue that encased the foot skeleton. The manus of *Tornieria*, from the Jurassic of East Africa, seems to have had a formula of 2:2:2:1:1, whereas the contemporary *Brachiosaurus* may have had 2:2:1:1:1. In *Apatosaurus* and *Diplodocus* the formula was at least 2:1:1:1:1. The innermost digit usually carried a prominent claw that curved forwards and inwards. If the other toes did possess claws these can have been no more than horny 'nubbins' along the front margin of the forefoot.

The sauropod pes is better known and was rather different in structure. It, too, comprised five digits, but the metatarsals were decidedly shorter and thicker than the metacarpals and sometimes showed a regular decrease in size towards the inner side. Moreover, the phalanges were not reduced to the extent that they were in the manus, so that the toes of the hindfoot appeared somewhat larger and more prominent. The phalangeal formula was variable: in *Apatosaurus* it was 2:3:4:2:1, in *Diplodocus* 2:3:3:2:1 and in *Camarasaurus* 2:2:2:1:1. In all cases the outer toes ended in blunt nubbin-like phalanges whereas the inner toes carried large, curved and prominently projecting claws. The number of claws on the hindfoot seems to have

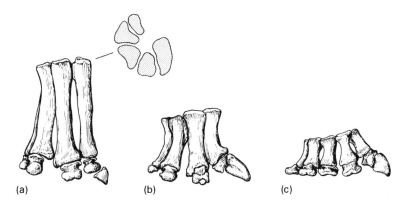

Figure 6.13 Manus skeleton in sauropods. (a) *Brachiosaurus*, from the Upper Jurassic of Tanzania; about 73 cm high. Plan view, to upper right, shows the five metacarpals grouped in a tight circular bundle. (b) *Tornieria*, from the Upper Jurassic of Tanzania; about 57 cm high. (c) *Apatosaurus* (*Brontosaurus*), from the Upper Jurassic of the western USA; about 23 cm high. (Adapted from various sources listed by Farlow 1987.)

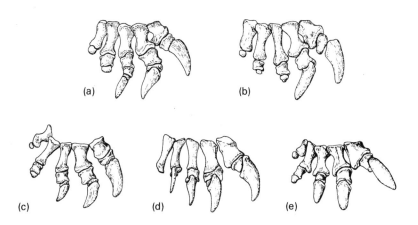

Figure 6.14 Pes skeleton in sauropods. (a) *Apatosaurus* (*Brontosaurus*), from the Upper Jurassic of the western USA; about 80 cm wide. (b) *Barosaurus*, from the Upper Jurassic of Tanzania; about 50 cm wide. (c) *Diplodocus*, from the Upper Jurassic of the western USA; about 65 cm wide. (d) *Pleurocoelus*, from the Lower Cretaceous of Texas, USA; about 47 cm wide. (e) *Tornieria*, from the Upper Jurassic of Tanzania; about 60 cm wide. (Adapted from various sources listed by Farlow 1987.)

varied. *Rhoetosaurus*, a primitive-looking sauropod from the Jurassic of Australia, had claws on digits I to IV, but in more advanced sauropods the claws were restricted to digits I–III (e.g. *Apatosaurus, Tornieria*) or digits I and II (e.g. *Barosaurus*). Regardless of their number these claws decreased in size from innermost to outermost. The metatarsals were not so steeply inclined as the metacarpals, so that the sole of the hindfoot spread out over a larger area than the sole of the forefoot. Roughly speaking, the hindfoot resembled that of an elephant, with the 'heel' region underlain by a large wedge-like cushion of supporting tissue.

Tracks (Figures 6.15–6.18)

The enormous trackways of sauropods have been reported from Jurassic and Cretaceous sediments in many parts of the world[1], and very early sauropods may be represented among the numerous tracks described by P. Ellenberger (1970, 1972, 1974) from the Upper Triassic of southern Africa. Some of these latter are, aside from their relatively small size, 'almost identical with the very characteristic footprints attributed to the true sauropods of the Jurassic' (Charig *et al.* 1965: 202). Footprints of the ichnogenus *Agrestipus*, described by R.E. Weems (1987) from the Upper Triassic of Virginia, USA, might also represent an early sauropod. Unfortunately, few sauropod trackways have been described in great detail, so that the following account is somewhat generalized (see also Farlow *et al.* 1989; Pittman and Gillette 1989).

Typical sauropod trackways are eye-catching on account of their immense size (Plate 3, p. 56, top; Plate 4, p. 62, top left). The individual footprints are large basin-like depressions, often surrounded by a prominent rim of displaced sediment. Despite their size some sauropod trackways are surprisingly narrow, as are those of elephants (e.g. H. Allen 1888, fig. 1; Sikes 1971, pl. 8): for instance, J.M. Dutuit and A. Ouazzou (1980) described one example from Morocco in which prints of the hindfeet overlapped the midline of the trackway. In other trackways the prints of the hindfeet are not quite so close to the midline, though they are very far indeed from widely spaced. Often, it seems, the maximum width of trackway left by the hindfeet is equivalent to some two and a half to three times the width of a hindfoot print.

In poorly preserved trackways the individual prints of the hindfeet appear as oval or subcircular basins without much detail, but better-preserved examples reveal that the hindfoot print is roughly oval in

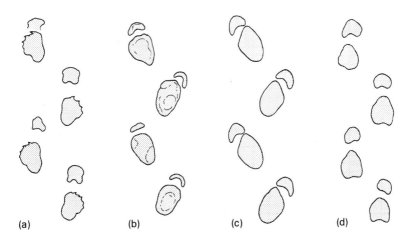

(a) (b) (c) (d)

Figure 6.15 Trackways attributed to sauropods. (a) *Brontopodus birdi*, from the Lower Cretaceous of Texas, USA; total length about 4.9 m. (b) Unnamed trackway from the Jurassic of Morocco; total length about 7 m. (c) *Breviparopus taghbaloutensis*, from the Upper Jurassic or Lower Cretaceous of Morocco; total length about 6.2 m. (d) Unnamed trackway from the Lower Cretaceous of Arkansas, USA; total length about 5 m. (Adapted from Farlow 1987 (a), Ishigaki 1985a (b), Dutuit and Ouazzou 1980 (c), Pittman 1984 (d).)

outline (narrower behind), with a series of notches representing the claws along the front margin (Plate 12, p. 278, centre left). The hind-foot prints are slightly longer than wide and show definite negative rotation, pointing outwards from the midline of the trackway at an angle between 20° and 30°. The outwardly curved imprints of the clawed toes are so very short that it is practically impossible to measure them accurately or to define the interdigital angles. The remainder of the print is usually a featureless oval depression formed by the large and spreading footpad, and in some cases the inner margin of the foot is much more deeply impressed than the outer margin (Bird 1939b: 259, top; Farlow 1987, fig. 16C). Footprint length is often about 90–100 cm, though it is sometimes as little as 50 cm (e.g. *Elephantopoides*, Kaever and de Lapparent 1974) or as much as 150 cm (e.g. *Gigantosauropus*, Mensink and Mertmann 1984). For the hindfeet, the ratio SL/FL is about 7/1 or 8/1, and pace angulation is often in the range 120°–140°.

Prints of the manus are completely different in appearance (Plate 12,

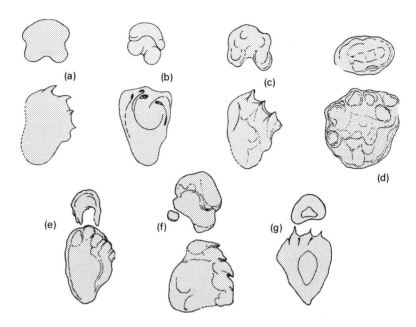

Figure 6.16 Manus-pes couples attributed to sauropods. (a) *Brontopodus birdi*, from the Lower Cretaceous of Texas, USA; pes about 87 cm long. (b) Unnamed track (*Brontopodus?*), from the Lower Cretaceous of Texas, USA; size not specified, but pes probably 80–90 cm long. (c) *Brontopodus birdi*, from the Lower Cretaceous of Texas, USA; pes about 90 cm long. (d) Unnamed track, from the Jurassic of Morocco; pes about 75 cm long. (e) Unnamed track, from the Jurassic of Morocco; pes about 80 cm long. (f) Unnamed track (*Brontopodus?*), from the Lower Cretaceous of Texas, USA; pes about 70 cm long. (g) *Breviparopus taghbaloutensis*, from the Upper Jurassic or Lower Cretaceous of Morocco; pes about 110 cm long. (Adapted from Farlow 1987 (a,c,f), Beaumont and Demathieu 1980 (b), Ishigaki 1985a (d), Ishigaki and Haubold 1986 (e), Dutuit and Ouazzou 1980 (g).)

p. 278, top left). They are semicircular or horseshoe-shaped impressions (convex forwards) that are about half the size of the pes prints, and usually they show no clear indications of separate digits, nor even of the massive claw on the pollex. W. Langston mentioned that he had 'never seen any clear indication of a claw in front foot tracks' of sauropods (1974: 97), though L. Ginsburg and his colleagues (1966) suggested that there might be evidence of the pollex claw in some sauropod tracks from Niger. The absence of claw-marks or of separate

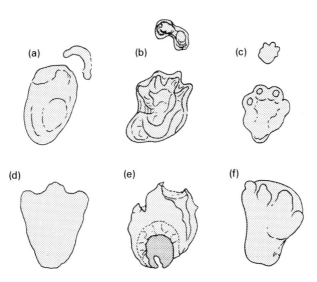

Figure 6.17 Miscellaneous tracks attributed to sauropods, showing variation in the morphology of manus prints. (a) Poorly defined manus-pes couple in an unnamed track, from the Lower Jurassic of Morocco; pes about 140 cm long. (b) Unnamed manus-pes couple, from the Jurassic of Morocco; pes about 97 cm long. (c) *Deuterosauropodopus major* (type A), from the Upper Triassic of Lesotho; pes about 40 cm long. (d) Supposed right manus print, from the Upper Cretaceous of India; about 22.5 cm long. (e) *Gigantosauropus asturiensis*, from the Upper Jurassic of Spain; about 132 cm long. Originally attributed to a gigantic theropod dinosaur, this track seems more likely to comprise a sauropod manus print overtrodden by the pes. (f) Supposed manus print of a small sauropod, from the Upper Jurassic of Colorado, USA, showing definite indications of the digits; about 18 cm long. (Adapted from Ishigaki 1985a (a), Ishigaki and Haubold 1986 (b), P. Ellenberger 1972 (c), Mohabey 1986 (d), Mensink and Mertmann 1984 (e), Lockley *et al.* 1986 (f).)

digit imprints has generated some speculation about the exact posture of the sauropod forefoot during locomotion (see Chapter 9). Despite these uncertainties the manus impressions are invariably wider than long and show distinct negative rotation, though this is not so pronounced as in the hindfeet. In some trackways the manus prints are about as deep as the pes prints, but in others they are noticeably shallower (Dutuit and Ouazzou 1980). The stride of the forefeet is 8

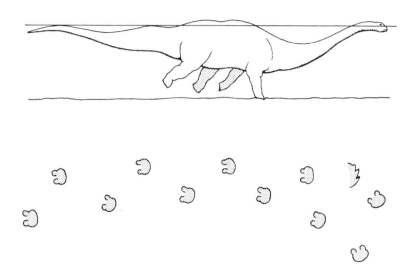

Figure 6.18 Trackway attributed to a floating sauropod, from the Lower Cretaceous of Texas, USA, with a conjectural restoration of the track-maker (above). Note that the track-maker shoved off at one point with its left hindfoot, thus changing direction to the right. (Adapted from Bird 1985.)

to 15 times the length of a forefoot impression, and pace angulation is slightly less than that for the hindfeet (100°–125°).

A few sauropod tracks do not conform to this general pattern. An isolated pes print, reported by J.B. Hatcher (1903) from the Upper Jurassic of Colorado, USA, appears to show four or five narrow, widely divergent and rather sharply pointed digits. This odd-looking print might have been disfigured by backwards slippage of the track-maker's toes, or it might have originated from some dinosaur other than a sauropod; D. Baird has suggested (in Farlow *et al.* 1989) that it might be an ornithopod manus print. Another strange-looking sauropod print, illustrated by O. Kuhn (1963, pl. 4, fig. 6), owes its exaggerated claw-marks to deep shadows combined with oblique perspective (see Bird 1954: 708, lower left). A manus print reported from the Upper Jurassic of Colorado has indications of four stubby digits, with a pronounced heel-like extension at the rear, and thus bears superficial resemblance to a miniature hindfoot print (Figure 6.17f). And a somewhat similar print, but with clear indications of only three digits, was recently described by D.M. Mohabey (1986) from the Upper Cretaceous of India. This looks something like an ornithopod pes print (Figure 6.17d), but was attributed to the manus of a sauropod dinosaur on account of its association with three sauropod eggs. A superb example of a typical horseshoe-shaped manus print from the Upper Jurassic of Spain was illustrated by J.C. Garcia-Ramos and M. Valenzuela (1977a, fig. 8), though it was not specifically attributed to a sauropod. The diminutive trackway of a juvenile sauropod, with pes prints about 20 cm long, was recently identified in the Lower Cretaceous of Korea (Lim *et al.* 1989).

The relative placement of manus and pes prints varies a great deal, even within a single trackway. Usually, the pes was planted behind the manus, though in some instances there was partial or complete overlap; in other cases the hindfoot was placed well behind the forefoot, so that there is a considerable space of undisturbed sediment between the two prints. The manus print lies to the exterior of the pes print in some trackways (e.g. *Breviparopus*, Dutuit and Ouazzou 1980), but directly ahead of the pes print in others. Despite its great size the tail rarely left any trace in sauropod trackways; either it was carried clear of the ground, or sauropods lived in shallow waters where the tail floated free of the substrate. Rare traces of a tail-drag were reported by R.T. Bird (1941: 77; 1944: 67).

There are no known traces made by sauropods resting on the ground or walking bipedally, though one famous series of footprints described by R.T. Bird (1944: 66) appears to have been made by a sauropod swimming in shallow water. Here the track-maker was steadily and repeatedly pushing off the bottom with its forefeet (Figure 6.18), though at one point it shoved off with its left hindfoot, thus changing its direction to the right. Similar traces of swimming sauropods were recently described by S. Ishigaki (1986, 1989) from the Jurassic of Morocco.

PROSAUROPODS

Structure of manus and pes (Figures 6.19–6.21)

The pentadactyl manus of prosauropods had an extremely distinctive entaxonic structure, with the inner fingers more strongly developed than the outer ones. The pollex was by far the thickest finger in the hand, with a stout block-like metacarpal, nearly as wide as long,

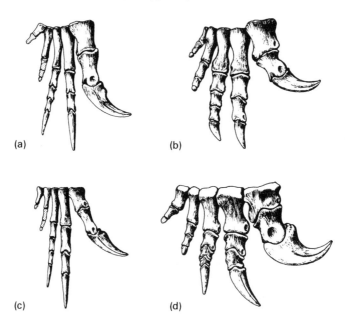

Figure 6.19 Manus skeleton in prosauropods. (a) *Anchisaurus*, from the Lower Jurassic of the eastern USA, about 10 cm long. (b) *Plateosaurus*, from the Upper Triassic of Germany; about 12 cm long. (c) *Efraasia*, from the Upper Triassic of Germany; about 14 cm long. (d) *Massospondylus*, from the Lower Jurassic of southern Africa; about 15 cm long. Drawings (a) and (c) represent 'narrow-footed' prosauropods; (b) and (d) represent 'broad-footed' prosauropods. (Adapted from Galton 1976 (a), Galton and Cluver 1976 (b), Galton 1973 (c), Broom 1911 (d).)

which supported a massive phalanx and an enormous curved claw. The second and third fingers were more 'normal-looking', with claws of moderate size. The second finger was narrower than the pollex, but was the longest in the hand, while the third finger was basically a smaller version of the second. By comparison, the outer two fingers were rather feeble; they terminated in rounded nubbins of bone and probably lacked claws. The phalangeal formula was usually 2:3:4:3:2, as in *Gryponyx*, *Lufengosaurus* and *Massospondylus*, though the outermost finger was sometimes variable in structure. In *Plateosaurus*, for instance, the phalangeal formula ranged from 2:3:4:3:0 to 2:3:4:3:3 (Jaekel 1914), and some individuals may also have had an extra phalanx in digit IV (e.g. *Pachysaurus ajax*, figured by von Huene (1932)

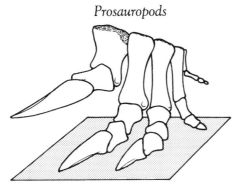

Figure 6.20 Left manus of the prosauropod *Plateosaurus* in its normal walking posture. The dinosaur's weight is supported principally on digits II and III, while the clawed pollex is retracted medially and carried well clear of the ground. (Adapted from Galton 1971b.)

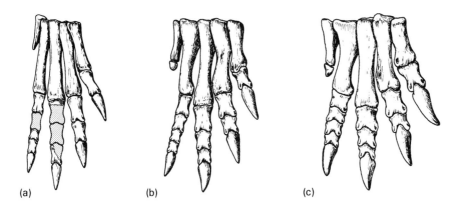

(a)   (b)   (c)

Figure 6.21 Pes skeleton in prosauropods. (a) *Anchisaurus*, from the Lower Jurassic of the eastern USA; about 21 cm long. (b) *Massospondylus*, from the Lower Jurassic of southern Africa; about 32 cm long. (c) *Lufengosaurus*, from the Lower Jurassic of China; about 35 cm long. Drawing (a) exemplifies a 'narrow-footed' prosauropod; (b) and (c) are 'broad-footed' prosauropods. (Adapted from Galton and Cluver 1976 (a), Broom 1911 (b), Young 1947 (c).)

with a formula of 2:3:4:4:2). By manipulating the fossil hand bones (Galton 1971b) it has been possible to investigate the mobility of the prosauropod pollex. When this finger was straightened its huge claw would have pointed inwards, towards the midline of the trackway; and when it was flexed its claw would have swung across the palm of the hand, perhaps affording a reasonably strong grasp. This mobile claw might have served as a weapon (Galton 1971b) or as a device for grooming (Cooper 1981) or feeding – perhaps to pull down vegetation. Whatever its function, the pollex claw was probably retracted out of harm's way when a prosauropod walked on all fours. Unlike the claws of cats, which are retracted upwards and backwards, the pollex claw would have been retracted upwards and inwards, towards the midline of the trackway (Figure 6.20). The manus was between one-third and one-half the size of the pes, though its proportions varied considerably: some prosauropods (e.g. *Anchisaurus*) had a rather slender manus, which was longer than wide, whereas others (e.g. *Massospondylus*) had a much broader forefoot.

The prosauropod hindfoot was pentadactyl, with the phalangeal formula 2:3:4:5:1 or, less commonly, 2:3:4:5:0. All five toes extended forwards in a slightly divergent pattern, with toes I, II and III being successively longer. Digit I was about two-thirds the length of digit III, which was often rivalled in size by digit IV. These four strong-clawed toes probably supported most of the animal's body weight, implying that the foot was functionally tetradactyl. By contrast, the outermost toe was short, feeble-looking and clawless; usually, it comprised a short metatarsal with one vestigial phalanx, and in normal circumstances it might not have touched the substrate. Although prosauropods can be separated into those forms with relatively broad hindfeet and those with relatively narrow hindfeet, this distinction is not so well marked as in the forefeet. Overall, the prosauropod pes was very distinctive in structure, being functionally tetradactyl, with digit I directed forwards alongside digit II, and with digits III and IV roughly equal in size. It was quite unlike the symmetrical feet of theropods and ornithopods, where the middle toe (III) was noticeably longer than the adjoining ones. In its general appearance the prosauropod pes bore a strong resemblance to the primitive foot-structures of thecodontians and early crocodilians.

Tracks (Figures 6.22 and 6.23)

Most of the prosauropod tracks described in the scientific literature were not made by prosauropods at all. This confusing state of affairs arose because some old misconceptions were unquestioningly perpetuated in standard works on dinosaur tracks.

First, it became common practice (following Lull 1904) to regard tracks such as *Anchisauripus*, from the Connecticut Valley sandstones,

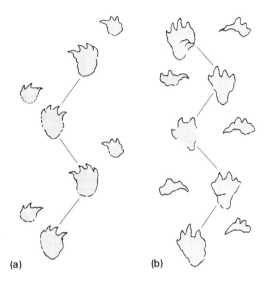

Figure 6.22 Trackways attributed to prosauropods. (a) *Tetrasauropus unguiferus*, from the Upper Triassic of Lesotho (reconstituted from illustrations of separate manus-pes couples); total length about 3.22 m. (b) *Navahopus falcipollex*, from the Lower Jurassic of Arizona, USA; total length about 0.8 m. (Adapted from P. Ellenberger 1972 (a), Baird 1980 (b).)

as those of prosauropods. Yet *Anchisauripus* is obviously the track of a theropod: it has a mesaxonic pattern of foot structure, with three forwardly directed digits (II, III, IV) and an opposed hallux. Footprints of this type correspond to the foot skeletons of theropods, but they cannot be matched to the feet of prosauropods – where there are *four* forwardly directed digits, the hallux is not opposed, and digit IV is nearly as large as digit III. D. Baird (1980: 228) has commented that:

> . . . in order to adapt the foot of *Anchisaurus* or *Ammosaurus* [both prosauropods] to the *Anchisauripus* footprint, Lull (1915, pp. 141, 154) had to dislocate and rotate the hallux (metatarsal and all) and do procrustean violence to the lengths of the other three digits. His correlation is thus doubly untenable, and footprints of the *Anchisauripus* type have nothing to do with Prosauropoda.

Even so, the tridactyl footprints of theropods and ornithopods

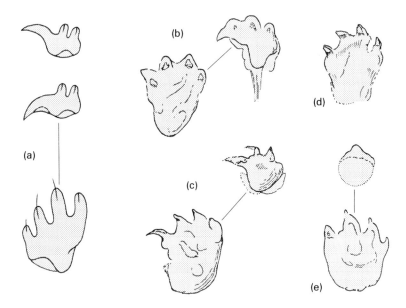

Figure 6.23 Manus and pes prints attributed to prosauropods.
(a) *Navahopus falcipollex*, a manus-pes couple from the Lower Jurassic of
Arizona, USA; pes about 13 cm long. Additional manus print (above)
shows the pollex claw in a more extended position. (b) *Pseudotetrasauropus
jaquesi*, a manus-pes couple from the Upper Triassic of Lesotho; pes about
49 cm long. (c) *Tetrasauropus unguiferus*, a manus-pes couple from the Upper
Triassic of Lesotho; pes about 44 cm long. (d) *Pseudotetrasauropus acutunguis*,
a (left?) pes print from the Upper Triassic of Lesotho; about 32 cm long.
This might represent the track of a prosauropod walking bipedally.
(e) *Paratetrasauropus seakensis*, a manus-pes couple from the Upper Triassic
of Lesotho; pes about 28 cm long. (Adapted from Baird 1980 (a),
P. Ellenberger 1972 (b–e).)

continued to be attributed to prosauropods (e.g. Haubold 1971, fig. 52;
Cooper 1981, fig. 89).

A second complication arose when prosauropods were identified as
the source of *Otozoum*, another ichnogenus from the Connecticut
Valley (Lull 1953). However, some examples of *Otozoum* include traces
of a pentadactyl manus that is totally unlike the entaxonic manus of
prosauropods. In fact *Otozoum*, like *Chirotherium*, might actually

comprise tracks of thecodontians or early crocodilians (Baird 1980: 228; Haubold 1984: 47), and perhaps of ornithopod dinosaurs as well (see later).

Third, it must be recalled that prosauropods had hindfeet so similar to those of certain thecodontians that these two sorts of animals would probably have produced identical tracks while walking bipedally. Consequently, it is possible that some of the tracks ascribed to bipedal thecodontians might actually be those of prosauropods. This ichnological dilemma was neatly summarized by P.E. Olsen and P.M. Galton (1984: 95), who concluded that some trackways of thecodontians, 'crocodiliomorphs' (crocodile-like reptiles) and prosauropods could be differentiated only through the form of the manus prints.

In short, much of the existing literature on dinosaur footprints is more of a hindrance than a help when one attempts to identify the tracks of prosauropods. This is because the tracks of theropods and other reptiles have been attributed to prosauropods whereas authentic prosauropod tracks may have been misidentified as those of thecodontians.

Trackways are confidently ascribed to prosauropods only when they include clear traces of the entaxonic manus that is unique to these dinosaurs. Such trackways include *Navahopus*, from the Lower Jurassic of North America (Baird 1980; Plate 13, p. 306, top right), and some examples of *Tetrasauropus* and *Pseudotetrasauropus*, from the Upper Triassic of southern Africa (P. Ellenberger 1970, 1972).[2] *Navahopus* is based on the track of a prosauropod that traversed the sloping surface of an ancient sand dune; it may not be representative of all prosauropod tracks, as the animal's gait might have been modified to suit the unusual substrate and because the footprints were somewhat obscured by slumping sand. The other trackways, from southern Africa, are less certainly attributed to prosauropods. For these reasons the following description is somewhat generalized and tentative.

When a prosauropod walked on all fours it produced a rather broad trackway, with maximum width about two and a half times the width of a single pes print. The pes prints are roughly oval in outline and are turned outwards from the midline of the trackway at an angle between 16° and 22°. However, this out-turning of the pes prints is often masked by the strong inwards curvature of the claw-traces. Footprint width is equivalent to some 75–85 % of footprint length, which is known to range from about 12 cm in *Navahopus* to 50 cm in

Tetrasauropus. Pace angulation for the hindfeet is between 108° and 130°, and the ratio SL/FL ranges between 2.6/1 (*Navahopus*) and 4.6/1 (*Tetrasauropus*). Digits I–IV of the pes diverge at low angles and terminate in impressions of stout claws (represented by scratch-like markings in *Navahopus*). Digits I, II and III are successively bigger, while digit IV is fractionally shorter than III. The traces of the claws are forwardly arched, with their tips directed medially.

Imprints of the forefeet are smaller and somewhat shallower than those of the hindfeet, and they form a trackway that is about three and a half or four times the width of a single manus print. They are basically oval or elliptical in outline and wider than long – the width being exaggerated by the large, medially directed claw on the pollex. There are definite traces of only three digits (I, II and III), each equipped with a claw. All three digits are relatively short; numbers II and III have their claw traces directed forwards, or curved somewhat medially, whereas that of digit I is directed straight inwards to the midline of the trackway. This large pollex claw was fully impressed only when the manus sank rather deeply into the substrate. The manus prints show slightly greater rotation than those of the pes, being turned outwards from the midline of the trackway at an angle between 19° (*Navahopus*) and 34° (*Tetrasauropus*). Invariably, the manus imprint is located in front of, and decidedly lateral to, the imprint of the hindfoot. Pace angulation for the manus impressions ranges from 60° to 69° in *Navahopus* but is not recorded for *Tetrasauropus*. There are no traces of a tail-drag.

There are no convincing reports of tracks made by prosauropods that were moving bipedally. *Pseudotetrasauropus acutunguis*, the trackway of a biped discovered by P. Ellenberger (1972) in the Lower Jurassic of Lesotho, comprises pes prints that resemble those made by prosauropods walking on all fours (Figure 6.23d). However, these footprints appear to differ from those of other prosauropods in having claw-traces that extend outwards rather than inwards, towards the midline of the trackway. It is doubtful that the changeover from a quadrupedal gait to a bipedal gait could have effected such dramatic reorientation of the claws, and there remains considerable doubt as to the identity of the track-maker. There are no definite reports of traces left by prosauropods that were swimming or resting on the ground.

SMALL ORNITHOPODS

Structure of manus and pes (Figures 6.24 and 6.25)

In smaller ornithopods the hand was pentadactyl, rather unspecialized-looking and much smaller than the foot. As conventionally shown (Figure 6.24a), the manus of *Hypsilophodon*, from the Lower Cretaceous of England, has fingers I–IV spreading forwards and digit V extending sideways. The phalangeal formula was probably 2:3:4:3:1, though there was possibly an additional phalanx in each of the outer two digits. Digits II and III were the longest and stoutest in the manus, with III being only a little longer than II. Digit I was about half the length of digit III, and digit IV was slightly shorter still. The fifth digit is not completely known but seems to have been short and feeble-looking. Digits I–IV terminated in small claws, but the tip of digit V remains unknown. In other small ornithopods, such as *Fabrosaurus* and *Dryosaurus*, the hand is poorly known but is presumed to have resembled that of *Hypsilophodon*. The manus of *Heterodontosaurus*, from the Lower Jurassic of southern Africa, was slightly different in structure. Its phalangeal formula was 2:3:4:3:2 and the outer two digits were reduced in size and diverged quite widely from the inner three (Figure 6.24b). These inner three digits were fairly thick, with robust and bluntly tipped claws. In *Hypsilophodon*, the hand was about one-third the size of the foot, though in other animals it was somewhat smaller (about a quarter the size in *Fabrosaurus*) or larger (about half the size in *Heterodontosaurus*).

Even today there remains some uncertainty about the exact arrangement of the five fingers. Early illustrations of the manus in the small Jurassic ornithopod *Laosaurus consors* (later known as *Othnielia rex*) showed digit V parallel to the other four (e.g. Marsh 1896) but, in fact, there is little evidence for such an arrangement. The only well-preserved example of the hand skeleton is that of *Hypsilophodon*, which P.M. Galton (1974) reconstructed with a widely divergent digit V (Figure 6.24a). Restorations of the manus in *Fabrosaurus* (Thulborn 1972, fig. 7R) and *Scutellosaurus* (Colbert 1981, fig. 20B) were modelled after the pattern in *Hypsilophodon*. Recently, P.E. Olsen and D. Baird (1986) reconstructed the hands of *Hypsilophodon* and *Lesothosaurus* (*Fabrosaurus*) with a less widely divergent digit V. Consequently, the five fingers form a radiating or fan-like pattern that is closely matched in the manus prints attributed to small ornithopods (Figures 6.24c and 6.28).

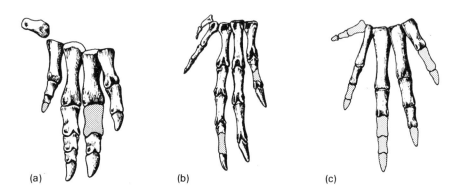

(a)   (b)   (c)

Figure 6.24 Manus skeleton in small ornithopods. (a) *Hypsilophodon*, from the Lower Cretaceous of southern England; about 5 cm long. (b) *Heterodontosaurus*, from the Lower Jurassic of southern Africa; about 7 cm long. (c) *Fabrosaurus* (*Lesothosaurus*), from the Lower Jurassic of southern Africa; about 3 cm long. (Adapted from Galton 1974 (a), Santa Luca 1980 (b), Thulborn 1972 (c), Olsen and Baird 1986 (c).)

The hindfoot was fairly uniform in structure throughout the smaller ornithopods. *Hypsilophodon* had a long, slender and functionally tetradactyl foot, with all four toes directed forwards and with the phalangeal formula 2:3:4:5:0 (Figure 6.25c). Digit I was rather short, reaching down to about the distal end of the metatarsus, and digit V was represented only by a vestigial splint of its metatarsal, tucked in at the rear of the foot. In their arrangement the three strongest toes (II, III and IV) resembled their counterparts in theropod dinosaurs. Digit III was longest and stoutest, while the adjoining ones were slightly smaller and roughly equal in size. However, the toes of ornithopods differed from those of theropods in one important detail: their claws were relatively thick and blunt, and might be termed semiclaws or nails (see Figure 5.17), whereas those of theropods were slim and sharply pointed. The foot of *Heterodontosaurus* (Figure 6.25a) was somewhat narrower than that of *Hypsilophodon*, with digit II distinctly shorter than digit IV, and digit I more widely divergent.

The main structural variation was in the size of digit I. In some forms, such as *Hypsilophodon*, this digit was long enough to reach the ground when the animal was in a normal standing pose, but in others, such as *Fabrosaurus* and *Othnielia* (*Laosaurus*), it was shorter

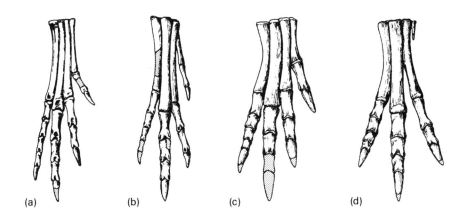

(a) (b) (c) (d)

Figure 6.25 Pes skeleton in small ornithopods. (a) *Heterodontosaurus*, from the Lower Jurassic of southern Africa; about 13.5 cm long. (b) *Fabrosaurus* (*Lesothosaurus*), from the Lower Jurassic of southern Africa; about 12 cm long. (c) *Hypsilophodon*, from the Lower Cretaceous of southern England; about 17 cm long. (d) *Dryosaurus* (*Dysalotosaurus*), a composite based on specimens from the Upper Jurassic of western North America and Tanzania; between 13 and 32 cm long. (Adapted from Santa Luca 1980 (a), Thulborn 1972 (b), Galton 1974 (c), Galton 1977 (d).)

and might have touched down only when the foot sank into the substrate. Reduction of digit I was carried to completion in *Dryosaurus*, where the foot was functionally tridactyl (Figure 6.25d).

Tracks (Figures 6.26–6.28)

In their bipedal gait small ornithopods produced fairly narrow trackways, with pace angulation in the range 150°–170°. Often, the footprints show slight but distinct positive (inwards) rotation, and in rare instances there is a tail-drag. The ratio *SL/FL* is commonly between 4/1 and 8/1, though it may reach 20/1 in the tracks of fast-running animals (*Wintonopus*, Thulborn and Wade 1984).

Each footprint is mesaxonic and tridactyl (digits II–IV) or tetra-dactyl (I–IV), and in all cases digits II, III and IV spread out in a more or less symmetrical pattern. Digit I, where present, is usually represented by a forwardly directed imprint alongside the base of digit II. Sometimes, the imprint of digit I points medially, though it never extends backwards, in contrast to the situation in some theropods.

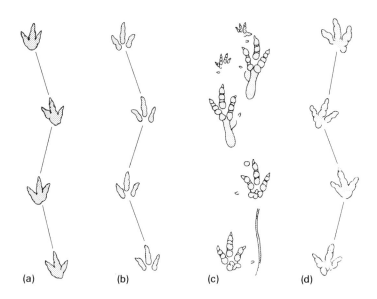

Figure 6.26 Trackways attributed to small ornithopods. (a) *Anomoepus curvatus*, from the Lower Jurassic of Massachusetts, USA; total length about 1.1 m. (b) *Anomoepus*, from the Lower Jurassic of Lesotho (extrapolated from sequence of three footprints); total length about 1.1 m. (c) *Anomoepus*, from the Lower Jurassic of the Connecticut Valley, USA; total length about 67 cm. Note the tail-drag and the changeover to quadrupedal resting posture, with traces of the ischiadic callosity and manus. (d) *Moyenisauropus natator*, from the Lower Jurassic of Lesotho; total length about 1.6 m. (Adapted from photograph (a), F. Ellenberger and P. Ellenberger 1958 (b), Lull 1953 (c), P. Ellenberger 1974 (d).)

The footprints may range in length from 2 cm (some examples of *Wintonopus*) up to about 25 cm.

The ratio *FL/FW* is variable. Often, these two dimensions are roughly equal, with footprint width representing something between 90 % and 115 % of footprint length. Footprints with slender digits were presumably derived from gracile ornithopods, such as *Fabrosaurus*, whereas those with broader digits were probably made by heavier and more robust dinosaurs resembling *Hypsilophodon*. The total divarication of digits II–IV is commonly about 60°, though it is sometimes as low as 40° or as high as 80°. Interdigital angles II–III and III–IV are often subequal.

The imprints of toes II, III and IV are parallel-sided or slightly

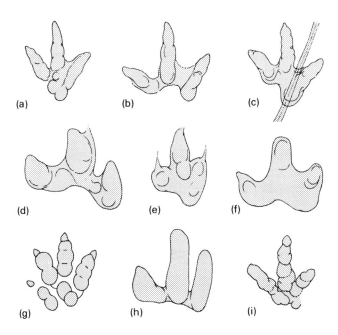

(a) (b) (c)

(d) (e) (f)

(g) (h) (i)

Figure 6.27 Pes prints attributed to small ornithopods. (a) *Moyenisauropus dodai*, from the Lower Jurassic of Lesotho; about 14.5 cm long.
(b) *Moyenisauropus dodai*, pes print of a running individual, from the Lower Jurassic of Lesotho; about 11.5 cm long. Note the widely divergent digit II.
(c) *Moyenisauropus natatalis*, a pes print intersected by tail-drag, from the Lower Jurassic of Lesotho; about 16.5 cm long. (d) *Wintonopus latomorum*, from the mid-Cretaceous of Queensland, Australia; about 9.6 cm long.
(e) *Wintonopus latomorum*, from the mid-Cretaceous of Queensland, Australia; about 4 cm long. (f) *Moyenisauropus*, from the Lower Jurassic of Poland; about 7.5 cm long. (g) *Anomoepus isodactylus*, from the Lower Jurassic of the Connecticut Valley, USA; about 11.5 cm long. (h) *Wintonopus latomorum*, from the mid-Cretaceous of Queensland, Australia; about 9.5 cm long. (i) Unnamed (right?) footprint resembling *Anomoepus*, from the Middle Triassic of France; about 7 cm long. (Adapted from P. Ellenberger 1974 (a–c), Thulborn and Wade 1984 (d,e,h), Karaszewski 1969 (f), Lull 1953 (g), Gand *et al.* 1976 (i).)

tapered and sometimes show definite nodes. All three are roughly equal in width, though in some cases III is slightly broader than IV. Digit IV is fractionally shorter than III, and digit II may be slightly shorter still. Well-preserved pes prints include traces of bluntly rounded claws, often with a U-shaped or broadly V-shaped outline.

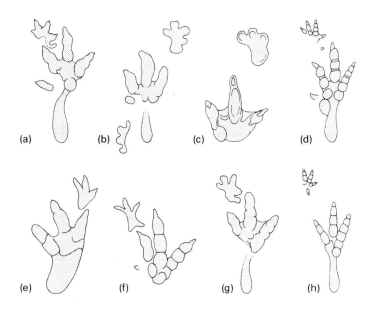

Figure 6.28 Manus-pes couples attributed to small ornithopods. (a)
Moyenisauropus natator, from the Lower Jurassic of Lesotho; pes about
20 cm long, excluding metatarsus. (b) *Moyenisauropus natator*, from the
Lower Jurassic of Lesotho; pes about 22 cm long, excluding metatarsus. An
additional manus print is shown at lower left. (c) *Moyenisauropus levicauda*,
from the Lower Jurassic of Lesotho; pes about 11 cm long. (d) *Anomoepus
intermedius*, from the Lower Jurassic of the Connecticut Valley, USA; pes
about 10.5 cm long, excluding metatarsus. (e) *Tridactylus machouensis*, from
the Upper Triassic of Morocco; pes about 18 cm long, excluding
metatarsus. Not certainly attributed to an ornithopod, and possibly a
chance association of two unrelated tracks. (f) *Apatichnus circumagens*, from
the Lower Jurassic of the Connecticut Valley, USA; pes about 7.5 cm long.
(g) *Moyenisauropus natatalis*, from the Lower Jurassic of Lesotho; pes about
16 cm long, excluding metatarsus. (h) *Anomoepus scambus*, from the Lower
Jurassic of the Connecticut Valley, USA; pes about 9.5 cm long, excluding
metatarsus. Couples (a), (d) and (h) are from the tracks of animals resting
on all fours; other couples are from the tracks of animals walking
quadrupedally. (Adapted from P. Ellenberger 1974 (a–c,g), Lull 1953 (d,f,h),
Biron and Dutuit 1981 (e).)

Frequently, the pes prints are asymmetrical on account of a prominent metatarso-phalangeal node behind digit IV (e.g. Figure 6.27a–e).

Tracks of ornithopods that moved quadrupedally are well known from the Lower Jurassic rocks of the Connecticut Valley and southern Africa (Figure 6.28; Plate 11, p. 208, bottom right), and have also been reported from rocks of similar age in central Poland (Gierliński and Potemska 1987). The Polish examples comprise only isolated pes and manus prints, both of which may show indications of quite extensive interdigital webbing. (The cosmopolitan ichnogenus *Atreipus*, regarded by Olsen and Baird (1986) as the track of a quadrupedal ornithopod, has been discussed here among the tracks of coelurosaurs.) In many respects the pes prints resemble those in the tracks of ornithopods that walked bipedally, though it is more common to find traces of the metapodium and the hallux when the animals moved quadrupedally. Pace angulation for the hindfeet seems to have decreased when ornithopods shifted on to all fours, and is frequently in the range 120°–150°. In addition, the switch from bipedal gait to quadrupedal gait usually entailed a changeover from positive (inwards) rotation to negative (outwards) rotation for the hindfeet. The ratio SL/FL (excluding metapodium) is often about 5/1, and there is sometimes an obvious tail-drag (Plate 6, p. 94, left).

The star-shaped or fan-shaped manus prints are very distinctive. They usually lie ahead of the much larger pes prints, and sometimes slightly medial or lateral to them, and they form a trackway with pace angulation about 100°. In some trackways the manus prints show no rotation, so that digit III points directly ahead, but more commonly they are rotated outwards so that, in extreme cases, digit III points directly sideways. Some manus prints show five fingers arranged in a stellate pattern, with total divarication between 120° and 150°; others are rather more fan-shaped, comprising foreshortened impressions of all five fingers. Each finger impression is parallel-sided or slightly tapered, with a bluntly rounded tip or the trace of a tiny claw. There is little or no indication of the metacarpus, and the imprint of the outermost finger is sometimes faint or lacking. Figure 6.28 shows some characteristic variations in the appearance of the manus prints.

Resting traces are generally similar to the trackways left by small ornithopods walking on all fours, except that the pes prints are side by side (see Figures 4.13, 4.14). In their sitting posture small ornithopods frequently left a trace of the ischiadic tuberosity, which appears as a small circular, heart-shaped or brush-like marking midway between the rear ends of the pes prints (e.g. Figure 4.14). The traces of ornithopods lying in a prone position show a larger basin-shaped imprint of the animal's belly (see Figure 4.13a). A recent study by P.E. Olsen and D. Baird (1986) concluded that Edward Hitchcock's ichnospecies *Sauropus barrattii*, which is often regarded as a typical example of an ornithopod resting trace (e.g. Lull 1953, fig. 161), is probably an artificial conglomeration of several unrelated tracks.

There are no definite reports of traces made by small ornithopods that were swimming. However, it is likely that such traces do exist, because they are known for bigger ornithopods and for theropods.

IGUANODONTS

Structure of manus and pes (Figures 6.29–6.31)

In all these larger ornithopods the manus was pentadactyl and considerably smaller than the foot. Beyond that it is difficult to make generalizations because the details of manus structure varied so much from species to species. Nevertheless, the middle three digits (II, III and IV) were invariably the strongest in the hand, and the phalangeal formula of digits I–III was always 2:3:3. Digit IV had two or three phalanges whereas digit V had as few as two or as many as four. Thus the entire phalangeal formula might range from 2:3:3:2:2 (*Camptosaurus*, *Tenontosaurus*) to 2:3:3:3:4 (*Iguanodon bernissartensis*). Digit I was smaller than digit II and usually diverged from it at a low angle. *Iguanodon* was exceptional in having this digit developed into a solid bony spike that projected medially from the hand. There might also have been a similar thumb-spike in the African *Ouranosaurus* (Taquet 1976) and the Australian *Muttaburrasaurus* (Bartholomai and Molnar 1981), though the fossil evidence is inconclusive. The outermost finger was slender, feeble-looking and widely divergent (except, apparently, in *Camptosaurus*). Usually, the inner three fingers terminated in blunt claws whereas the outer two were completed by rounded nubbins of bone. *Iguanodon*, again, is an exception because it had curiously expanded and hoof-like tips to digits II and III (Figures 6.29b, 6.30). The overall size of the manus varied a good deal. In *Camptosaurus* it was scarcely one quarter the size of the foot, whereas in *Iguanodon* it was well over half the size.

The mesaxonic pes was basically a more robust version of that in smaller ornithopods, though, once again, there are some minor

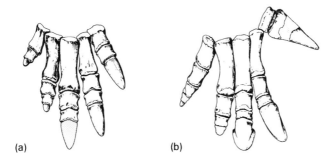

Figure 6.29 Manus skeleton in iguanodonts. (a) *Camptosaurus medius*, a small species or possibly a juvenile, from the Upper Jurassic of Utah, USA; about 11 cm long. (b) *Iguanodon bernissartensis*, from the Lower Cretaceous of Belgium; about 36 cm long. (Adapted from Gilmore 1925 (a), Norman 1980 (b).)

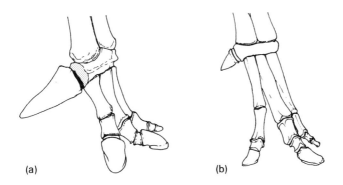

Figure 6.30 Manus of iguanodonts in quadrupedal walking posture. (a) *Iguanodon bernissartensis*, from the Lower Cretaceous of Belgium. (b) *Iguanodon mantelli*, from the Lower Cretaceous of southern England. In each case the dinosaur's weight is supported principally on digits II and III; digit V is concealed. In (a) the first disc-like phalanx of digit I is indicated by solid shading; metacarpal I (paler shading) is fused with the wrist bones to form a single slab-like element. (Adapted from Norman 1980.)

variations. In some cases the foot was tetradactyl, with the phalangeal formula 2:3:4:5:0, but in others it was tridactyl, with the formula 0:3:4:5:0. There was considerable variation in the size of digit I: *Tenontosaurus* had a slender foot with strongly-developed digit I, but in

(a)   (b)   (c)

Figure 6.31 Pes skeleton in iguanodonts. (a) *Tenontosaurus*, from the Lower Cretaceous of the western USA; about 33 cm long. (b) *Camptosaurus dispar*, from the Upper Jurassic of Utah, USA; about 38 cm long. (c) *Iguanodon bernissartensis*, from the Lower Cretaceous of Belgium; about 55 cm long. (Adapted from Ostrom 1970 (a), Galton and Powell 1980 (b), Norman 1980 and Casier 1960 (c).)

Camptosaurus this digit was reduced in size, and in *Iguanodon* it was deleted altogether, leaving a functionally tridactyl foot. The functional digits, whether three or four in number, always terminated in broad and somewhat flattened claws with rounded tips.

Tracks (Figures 6.32–6.34)

In their bipedal gait the iguanodonts produced moderately narrow trackways with prints of the hindfeet arranged in a definite zig-zag series (Plate 11, p. 208, bottom left). Pace angulation is commonly in the range 130°–150°, the footprints are rotated positively (inwards), and there is rarely any trace of a tail-drag. The ratio *SL/FL* often falls between 3/1 and 6/1. The mesaxonic pes prints are nearly always tridactyl (digits II, III and IV), and commonly between 40 and 50 cm long, though unusually large examples may reach nearly 70 cm (e.g. tracks attributed to *Iguanodon*; Haubold 1971: 86). Even bigger ornithopod footprints are known (e.g. Brown 1938; Leonardi 1980b; Lockley *et al.* 1983), but these might have originated from hadrosaurs rather than iguanodonts.

Footprint width is only a little less than footprint length, and in some instances it may be slightly greater. The three toe-prints are typically parallel-sided or slightly tapered, with U-shaped or broadly V-shaped traces of claws. Occasionally, the toes were so thick and

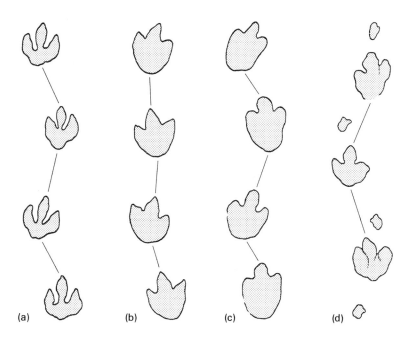

Figure 6.32 Trackways attributed to iguanodonts. (a) Unnamed trackway of an ornithopod similar or identical to *Iguanodon*, from the Lower Cretaceous of Sussex, England; total length about 4.44 m. (b) Unnamed iguanodon trackway, from the Lower Cretaceous of Colorado, USA; total length about 3.65 m. (c) Unnamed trackway resembling *Gypsichnites*, from the Upper Jurassic of Oklahoma, USA; total length about 2.98 m. (d) *Caririchnium leonardii*, a quadrupedal trackway from the Lower Cretaceous of Colorado, USA; total length about 2.8 m. (Adapted from Delair and Sarjeant 1985 (a), Lockley 1987 (b,d), Lockley 1986b (c).)

fleshy that the resulting footprint has the outline of a clover-leaf or trefoil (e.g. Figure 6.33d,e,k; Plate 9, p. 140, centre left). There may be traces of phalangeal nodes, though these tend to be more obvious in smaller footprints. The imprint of digit III is generally longest and widest, while that of digit IV is only slightly shorter and may be nearly as broad. Digit II is often smaller still. The total divarication of digits II–IV is commonly about 60°, though it may be as little as 30° or as much as 70°. Interdigital angles II–III and III–IV are roughly equal; these angles are often about 30°, though they may be as small as 20° or as great as 40°.

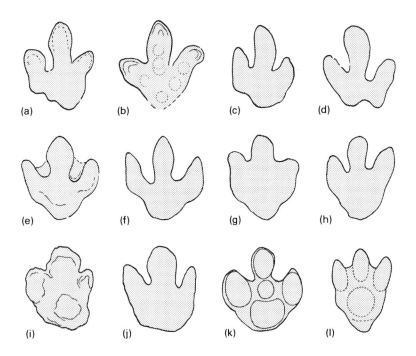

Figure 6.33 Pes prints attributed to iguanodonts. (a) *Pseudotrisauropus maserui*, from the Lower Jurassic of Lesotho; about 29 cm long.
(b) *Gyrotrisauropus planus*, from the Lower Jurassic of Lesotho; about 48 cm long. (c), (d) Unnamed footprints resembling *Gypsichnites*, in a single trackway from the Upper Jurassic of Oklahoma, USA; both about 45 cm long. (e) Unnamed footprint, from the Lower Cretaceous of Colorado, USA; about 34 cm long. (f) Unnamed footprint attributed to an ornithopod resembling *Iguanodon*, from the Lower Cretaceous of Sussex, England; about 55 cm long. (g) Unnamed footprint resembling *Gypsichnites* and attributed to an ornithopod resembling *Camptosaurus*, from the Upper Jurassic of Colorado, USA; about 45 cm long. (h) Unnamed footprint attributed to ornithopod resembling *Iguanodon*, from the Lower Cretaceous of Spain; about 68 cm long. (i) Unnamed footprint attributed to ornithopod resembling *Iguanodon*, from the Lower Cretaceous of Portugal; about 45 cm long. (j) *Ornithopodichnites magna*, from the Upper Cretaceous of Spain; about 43 cm long. (k) Unnamed footprint, from the Upper Jurassic or Lower Cretaceous of Mexico; about 30 cm long. (l) *Sousaichnium pricei*, from the Lower Cretaceous of Brazil; about 56 cm long. (Adapted from P. Ellenberger 1972 (a,b), Lockley 1986b (c,d,g), Lockley 1987 (e,f), Casanovas Cladellas and Santafé Llopis 1971 (h), Antunes 1976 (i), Llompart *et al.* 1984 (j), Ferrusquia-Villafranca *et al.* 1978 (k), Leonardi 1980b (l).)

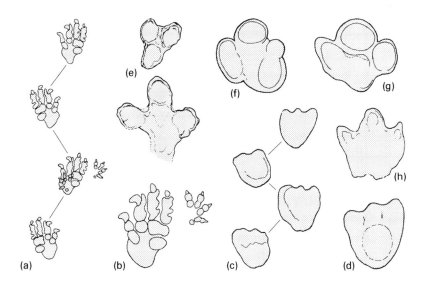

Figure 6.34 Miscellaneous tracks attributed to ornithopods. (a) *Otozoum moodii*, a trackway from the Lower Jurassic of the Connecticut Valley, USA; total length about 2.75 m. Note the pair of manus prints (left one almost completely overtrodden by pes). (b) *Otozoum moodii*, a manus-pes couple from the Lower Jurassic of the Connecticut Valley, USA; pes about 49 cm long. (c) Unnamed trackway attributed to large ornithopod, from the Jurassic of Brazil; total length about 2.3 m. Note the exceptionally great width of the trackway. (d) Unnamed pes print attributed to large ornithopod, from the Jurassic of Brazil; about 55 cm long. (e) *Caririchnium leonardii*, a manus-pes couple from the Lower Cretaceous of Colorado, USA; pes about 34 cm long. (f) *Paratrisauropus bronneri*, a pes print from the Upper Triassic of Switzerland; about 12 cm long. (g) *Paratrisauropus latus*, a pes print from the Upper Triassic of Switzerland; about 13 cm long. (h) *Paratrisauropus mendrezi*, a pes print from the Lower Jurassic of Lesotho; about 18 cm long. (Adapted from Lull 1953 (a,b), Leonardi 1980b (c,d), Farlow 1987 (e), Demathieu and Weidmann 1982 (f,g), P. Ellenberger 1972 (h).)

Although there seems little doubt that iguanodonts could, and did, walk on all fours, there are few definite reports of trackways testifying to such behaviour. Perhaps to be included here are some of the Connecticut Valley tracks that were assigned to the ichnogenus *Otozoum* by Edward Hitchcock (Figure 6.34a). *Otozoum* is often regarded as the track of a prosauropod, though in reality it does not correspond at all closely to the distinctive hand and foot structures of these dinosaurs; instead, some examples of *Otozoum* seem to find a much closer match in the tracks of ornithopods. The tetradactyl prints of the hindfeet are up to 50 cm long, with a relatively large imprint from digit I. This digit is usually lacking from ornithopod footprints of Jurassic and Cretaceous age, but might not be unexpected in a track-maker of such early date (late Triassic or early Jurassic). The pes prints of *Otozoum* have broad parallel-sided digits with indications of prominent fleshy nodes. Digit III is the longest in the foot, as in ornithopods, and all four digits show traces of thick and rather blunt claws. The prints of the hindfeet are rotated negatively (outwards) at about 10° from the midline of the trackway, and they include prominent impressions of the metatarso-phalangeal pads and, perhaps, of the metatarsus. The ratio *SL/FL* is about 3/1, and pace angulation is about 120°. The small manus prints comprise five narrow digits splayed out in fan-shaped pattern, and they show pronounced outwards (negative) rotation, with digit III sometimes pointing directly sideways. Occasionally, there is a trace of a tail-drag. In most of these trackway features, except its overall size, *Otozoum* resembles the tracks of small ornithopods that moved quadrupedally. Tracks similar to *Otozoum* may also be present in footprint assemblages described by P. Ellenberger (1972, 1974) from the Upper Triassic and Lower Jurassic of southern Africa.

Footprints reported by A. Bartholomai (1966: 150) from supposed Lower Cretaceous (actually Lower Jurassic) sediments at Mt Morgan, in Queensland, Australia, included 'not only the large three-toed prints of the hind foot, but also much smaller five-toed prints presumably made by the forefeet of the dinosaur'. This combination of tridactyl pes prints and pentadactyl manus prints is quite characteristic of ornithopods in their quadrupedal gait. However, there remains some doubt because these tracks have never been illustrated, and because footprints from the same site have been described as 'just what would be expected of Lower Jurassic theropod tracks' (Molnar 1982: 617).

Trackways of the ichnogenus *Caririchnium*, from the Lower Cretaceous of Brazil (Leonardi 1984a) and Colorado, USA, were attributed by M.G. Lockley (1986b: 27) to ornithopods that travelled quadrupedally. *Caririchnium* displays large tridactyl pes prints of standard ornithopod type, which show slight positive (inwards) rotation and form a fairly narrow zig-zag trackway with maximum width between 1.6 and 1.8 times footprint width. Pace angulation ranges from 145° to 160°, and the ratio *SL/FL* is between 4/1 and 5/1. There is no trace of a tail-drag. The small manus prints are located just in front of the pes prints and have an irregular elliptical or bean-shaped outline, with no consistently clear indications of separate digits (Plate 11, p. 208, top right). In the examples illustrated by Lockley (1986b) digit I is represented by a small spur that points medially. The manus prints have pace angulation between 145° and 170°, and the ratio of stride length to manus length is between 12/1 and 14/1.

A few poorly preserved footprints from the mid-Cretaceous of Queensland have been attributed to medium-sized ornithopods that were swimming by pushing off the bottom with their hindfeet (Thulborn and Wade 1989). Such behaviour is not surprising, in view of the fact that theropods and hadrosaurs seem to have done likewise (Coombs 1980b; Currie 1983).

There appears to be only a single report of a big ornithopod's resting traces. In the Lower Cretaceous of Portugal M.T. Antunes (1976) discovered two hindfoot prints of iguanodont type, side by side, with a trough-like tail trace some distance behind them. Antunes attributed these traces to a large ornithopod, possibly *Iguanodon*, that was sitting down kangaroo-style, with its tail functioning as a prop. S.Y. Yang (1982) has reported the 'seat places' of bipedal dinosaurs in the Upper Cretaceous of Korea, but it is not clear whether these originated from ornithopods or theropods.

HADROSAURS

Structure of manus and pes (Figure 6.35)

The tetradactyl hadrosaur manus lacked any trace of digit I and had the phalangeal formula 0:3:3:3:4. Its outermost finger diverged quite widely from the other three, which were stouter, subequal in length and apparently bound together in a tight symmetrical bundle. There was a sizeable hoof at the tip of digit III, and a smaller one on digit II, whereas digit IV terminated in a bluntly rounded nubbin. Evidently, the animal's weight was carried on digits II and III, and to a lesser extent on digit IV, with the slender and divergent digit V having only a minor supportive role. The orientation of the hoof-like extremities makes it clear that the fingers were steeply inclined, so that a quadrupedal hadrosaur would, literally, have walked on its fingertips (as did *Iguanodon*; see Figure 6.30).

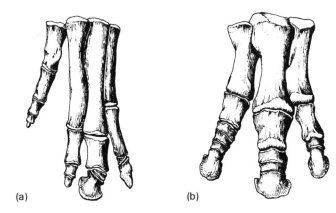

(a)   (b)

Figure 6.35 Skeletons of manus (a) and pes (b) in the hadrosaur *Edmontosaurus* (*Anatosaurus*), from the Upper Cretaceous of western Canada; manus about 45 cm high, pes about 72 cm high. (Adapted from Lull and Wright 1942.)

The hindfoot resembled that of iguanodontids in practically every respect. It was tridactyl and mesaxonic, with three stout and broad-spreading toes ending in blunt hoof-like claws. The phalangeal formula, as in *Iguanodon*, was 0:3:4:5:0. The hindfoot of hadrosaurs is often thought to have differed from that of iguanodonts in having extensive interdigital webs. Unfortunately, this possibility remains unsubstantiated, despite the evidence of skin imprints preserved in so-called 'mummified' hadrosaurs (e.g. Brown 1916; Lull and Wright 1942; Horner 1984a): these skin imprints have been discovered on just about every part of the hadrosaur body – except the hindfoot.

Tracks (Figures 6.36 and 6.37)

Hadrosaurs had hands and feet so similar to those of certain iguanodonts that it is difficult to identify their tracks with certainty. In practice, tracks have been attributed to hadrosaurs, rather than iguanodonts, for four reasons: first, because they are of appropriate age (late Cretaceous or slightly earlier); secondly, because they occur in geographic regions known to have been frequented by hadrosaurs; thirdly, because they are sometimes found in association with hadrosaur body fossils (e.g. Langston 1960); and, finally, because

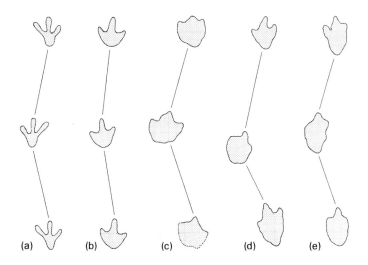

Figure 6.36 Trackways attributed to hadrosaurs. (a) Unnamed trackway, from the Upper Cretaceous of Brazil; total length about 2.8 m. (b) *Gypsichnites pacensis*, from the Lower Cretaceous of British Columbia, Canada (extrapolated from a sequence of only two footprints); total length about 2.1 m. (c) *Amblydactylus kortmeyeri*, from the Lower Cretaceous of British Columbia, Canada (trackway of a small individual, with first footprint partly restored); total length about 80 cm. (d) *Hadrosaurichnus australis*, from the Upper Cretaceous of Argentina; total length about 2.6 m. (e) *Hadrosaurichnus australis*, from the Upper Cretaceous of Argentina; total length about 2.4 m. Trackways of early Cretaceous age (b,c) are not certainly attributed to hadrosaurs. (Adapted from von Huene 1931a (a), Sternberg 1932 (b), Currie and Sarjeant 1979 (c), Alonso 1989 (d,e).)

they might in some instances show traces of interdigital webbing.

The idea that hadrosaurs had distinctively webbed hands and feet arose from the discovery of a 'mummified' specimen that showed an impression of the skin covering the hand. Subsequently, this specimen was interpreted in several different ways, leading to divergent opinions about the habits and habitats of hadrosaurs and about the morphology of hadrosaur tracks. According to one interpretation the hand was encased in a 'mitten' of skin, thus resembling a paddle. This notion accorded with the widespread assumption that hadrosaurs were aquatic or amphibious in their habits. It also prompted a suggestion that the hindfoot was webbed as well, though, as mentioned

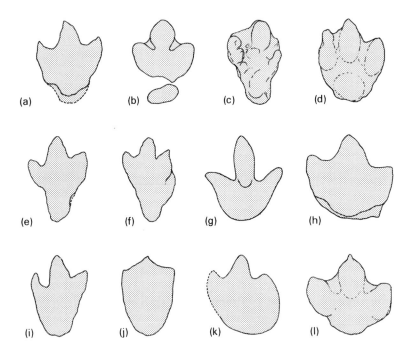

Figure 6.37 Pes prints attributed to hadrosaurs. (a) *Amblydactylus gethingi*, from the Lower Cretaceous of British Columbia, Canada; about 64 cm long. (b) Unnamed footprint, from the Upper Cretaceous of Alberta, Canada; about 62.5 cm long, excluding 'heel pad'. (c) Unnamed footprint, from the Upper Cretaceous of Utah, USA; about 140 cm long, including metatarsus. (d) *Amblydactylus gethingi*, from the Lower Cretaceous of British Columbia, Canada; about 62 cm long. (e) *Dinosauropodes bransfordii*, from the Upper Cretaceous of Utah, USA; about 73 cm long, including metatarsus. (f) *Dinosauropodes magrawii*, from the Upper Cretaceous of Utah, USA; about 136 cm long, including metatarsus. (g) *Gypsichnites pacensis*, from the Lower Cretaceous of British Columbia, Canada; about 29 cm long. (h) *Amblydactylus kortmeyeri*, from the Lower Cretaceous of British Columbia, Canada; about 11 cm long. (i) *Hadrosaurichnus australis*, from the Upper Cretaceous of Argentina; about 37 cm long. (j) *Taponichnus donottoi*, from the Upper Cretaceous of Argentina; about 57 cm long. (k) *Telosichnus saltensis*, from the Upper Cretaceous of Argentina; about 52 cm long. (l) *Amblydactylus kortmeyeri*, from the Lower Cretaceous of British Columbia, Canada; about 42 cm long. Footprints of early Cretaceous age (a,d,g,h,l) are questionably attributed to hadrosaurs. (Adapted from Sternberg 1932 (a,g), Langston 1960 (b), Bird 1985 (c), Currie 1983 (d), Lockley and Jennings 1987 (e,f), Currie and Sarjeant 1979 (h,l), Alonso 1980 (i), Alonso and Marquillas 1986 (j,k).)

before, this has never been substantiated by the discovery of webbing on a 'mummified' specimen. If these assumptions were valid one would expect to find obvious traces of interdigital webbing in hadrosaur tracks. However, a second interpretation maintains that the fingers were separate and that the supposed 'mitten' is merely a fold of skin that slipped down, like a loose sleeve, from the wrist and palm of the animal's carcass. A third interpretation suggests that the folds of skin overlying the fingers are the collapsed remnants of prominent pads on the palm of the hand. At present, it seems impossible to decide whether or not the hadrosaurs did have webbed hands and feet. Even so, two points may be stated with certainty. First, the strongest fingers in the hadrosaur hand terminated in flat hoof-like claws. This might indicate that the animals were accustomed to walking on all fours on dry land, though it does not rule out the possibility that hadrosaurs could swim. Secondly, some – though not all – of the pes prints attributed to hadrosaurs do seem to show definite traces of interdigital webs (Figure 6.37b,l).

In their bipedal gait hadrosaurs produced fairly narrow trackways with prints of the hindfeet arranged in a slightly zig-zag pattern. Pace angulation was often about 160°, though sometimes as low as 140°. The pes prints are turned inwards at about 18° to the midline of the trackway, and there is rarely any trace of a tail-drag.[3] The ratio SL/FL is usually about 6/1, but it may be as little as 4/1 or as much as 7/1. Each tridactyl and mesaxonic footprint is about as wide as long, though in some instances the length is exaggerated by an imprint from the metatarsus (Figure 6.37a–f). The three toe-prints are short, broad and strongly rounded in outline, sometimes affording the footprint a clover-leaf shape. Digits II and IV are slightly shorter than III, and are roughly similar in size, shape and angle of divergence. The total divarication of digits II–IV is commonly about 65°, though it may be as little as 55° or as much as 80°. Often, the hind margin of the footprint is smoothly rounded in outline, but in some examples there is a definite rearwards bulge formed by the metatarso-phalangeal node behind digit IV. One very fine hadrosaur footprint described by W. Langston (1960) showed an oval imprint of a 'heel pad' behind the impression of the conjoined digits (Figure 6.37b), and in a few footprints there appear to be traces of webbing between the digits (e.g. Figure 6.37b,c,d,l). Hadrosaur pes prints commonly have a length between 40 and 50 cm, with the largest examples reaching more than 85 cm (e.g. Brown 1938; Russell and Béland 1976) or even 100 cm

(Lockley *et al.* 1983). Smaller footprints, down to a length of 10 cm, have sometimes been attributed to juveniles (e.g. Currie and Sarjeant 1979; Currie 1983).

In 1983, P.J. Currie reported numerous tracks of quadrupedal hadrosaurs (ichnogenus *Amblydactylus*) in the Lower Cretaceous sediments of British Columbia, Canada. Although these were not described in detail, Currie did remark that the handprints were roughly crescent-shaped and lacked any division into separate digits (1983: 64). In addition he mentioned the trackways of hadrosaurs that were swimming (1983: 67):

> The trackway ... appears to indicate that hadrosaurs were efficient swimmers. Here an animal was walking on the muddy bottom of a quiet body of water. As the water became deeper, its strides decreased, and it appears to have been pushing off the bottom with its toes because the mark for the heel pad is very shallow and poorly defined. At one point, the midline of the trackway shifts more than a metre to the right, and several steps later, it shifts to the left again. It would be difficult to explain these shifts unless the three or four tonne weight of the body was buoyed up by water.

Currie also suggested that some discontinuous trackways were made by swimming hadrosaurs that were only occasionally touching down with their feet. On the other hand, J.O. Farlow (1987: 14) considered that the tracks described by Currie were rather too ancient to be those of hadrosaurs and suspected that they most probably originated from iguanodonts.

Tracks of the ichnogenus *Caririchnium*, from the Lower Cretaceous of Colorado, were originally attributed by M.G. Lockley (1985: 136) to a hadrosaur moving on all fours, though it seems equally probable that the track-maker was an iguanodont (Lockley 1986b: 27). There are no definite reports of resting traces left by hadrosaurs.

STEGOSAURS

Structure of manus and pes (Figure 6.38)

The stegosaur manus is poorly known, but seems to have been only slightly smaller than the pes and functionally tetradactyl, with four stubby digits arranged in a semicircular or radiating pattern. Despite

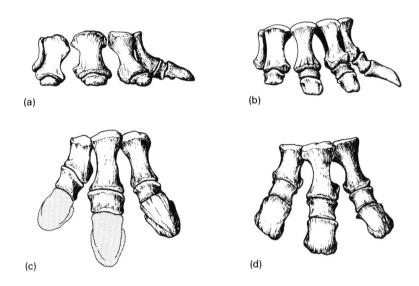

(a)

(b)

(c)

(d)

Figure 6.38 Manus and pes skeleton in stegosaurs. (a) Manus of *Stegosaurus*, from the Upper Jurassic of western North America; about 37 cm wide. (b) Manus of *Kentrosaurus*, from the Upper Jurassic of Tanzania; about 21 cm wide. (c) Pes of *Stegosaurus*, from the Upper Jurassic of western North America; about 33 cm wide (not from same individual as manus, (a)). (d) Pes of *Kentrosaurus*, from the Upper Jurassic of Tanzania; about 19 cm high. (Adapted from Gilmore 1914 (a,c), Galton 1982 (b,d).)

the presence of only four functional digits there were five metacarpals, though the outermost one seems to have borne no phalanges and had no direct supportive role. The middle three metacarpals were roughly equal in size whereas the inner and outer ones were fractionally shorter. In *Stegosaurus*, from the Upper Jurassic of North America, the phalangeal formula was probably 2:1:1:1:0. Digit I terminated in a flat and hoof-like claw while digits II to IV ended in small rounded nubbins. *Kentrosaurus* (also known as *Kentrurosaurus*), from the Upper Jurassic of East Africa, apparently had a phalangeal formula of 2:2:2:1:0. Here, there were flat hooves on digits I–IV, decreasing in size from innermost to outermost. In all cases the intermediate phalanges, between the metacarpals and the terminal hooves or nubbins, were compressed into short, disc-like bones.

The stegosaur hindfoot is a little better known. It was tridactyl and mesaxonic, with the formula 0:2:2:2:0, and in some instances there persisted a non-functional vestige of the fifth metatarsal. The digits were not quite so stubby as those of the manus, and each of them terminated in a relatively large hoof. In its general appearance the stegosaur pes resembled a shortened version of that seen in certain iguanodonts. The middle toe (III) was the longest in the foot, and the adjoining two, which were roughly similar in size and shape, diverged from it at a relatively low angle.

Tracks (Figure 6.39)

Stegosaurs are usually envisaged as habitual or obligate quadrupeds, and their combination of relatively large tetradactyl manus and stubby tridactyl pes would probably render their tracks unmistakable. Even so, there are no unambiguous reports of tracks made by these dinosaurs.

In 1950 F. von Huene speculated that trackways of the ichnogenus *Rigalites*, from the Middle Triassic of Argentina, might represent a 'pre-stegosaur', though originally he had suspected them to be the tracks of an ankylosaur (von Huene 1931b). Such an extremely early occurrence of stegosaurs, or their identifiable ancestors, does seems unlikely, and *Rigalites* has since been regarded as the track of an ornithopod (Lessertisseur 1955: 113) or a crocodilian (Haubold 1971: 63).

One trackway from the Lower Cretaceous of Brazil, described under the name *Caririchnium* by G. Leonardi (1984a), was originally attributed to a stegosaur but is more likely to be the trackway of an iguanodontid ornithopod moving on all fours (as discussed before). The manus prints are considerably smaller than the pes prints, as would be appropriate for an ornithopod rather than a stegosaur.

In a report on dinosaur tracks discovered in the Upper Jurassic of Chile, J.T. Gregory included the following remarks (in Dingman and Galli 1965: 28):

> I am greatly puzzled by the rather elongate footprints which appear to have perhaps three toes at the anterior end as shown in your photographs. . . It is conceivable from the shape of the foot that they might be [from] some quadrupedal ornithischian dinosaur, perhaps a *Stegosaur*, although the size seems very large for that group. The shape of the foot and the arrangement of the trackway is quite unlike what would be expected in one of the ornithopods.

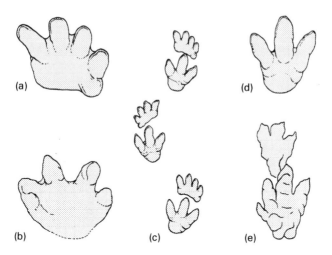

Figure 6.39 Tracks attributed to stegosaurs. (a) Conjectural manus print, based on the manus skeleton of *Stegosaurus*. (b) Unnamed print, possibly representing the manus of a stegosaur, from the Lower or Middle Jurassic of Queensland, Australia; about 19 cm wide. (c) Conjectural trackway pattern for a stegosaur, based on the manus and pes skeleton in *Stegosaurus*. (d) Conjectural pes print, based on the pes skeleton of *Stegosaurus*. (e) *Rigalites ischigualastensis*, a manus-pes couple from the Middle Triassic of Argentina; pes about 35 cm long. This is sometimes regarded as the track of a 'pre-stegosaur', but it is more probably derived from a crocodilian or an ornithopod dinosaur. (Adapted from D. Hill *et al.* 1966 (b), von Huene 1931b (e).)

Unfortunately, no further details were given, and the published photographs (Dingman and Galli 1965, figs 7 and 9) are not detailed enough to allow identification of the track-maker. Tridactyl pes prints would be certainly be appropriate for a stegosaur, though more definite proof would come from the size and shape of the manus prints.

One isolated print from the Lower Jurassic coal measures of south-eastern Queensland (D. Hill *et al.* 1966, pl. 15, fig. 5) originated from a quadrupedal dinosaur of some sort and is quite possibly the manus print of a stegosaur. The print is relatively short and broad-spreading, with four stubby digits arranged in a radiating pattern, and it conforms quite closely to the structure of the manus in stegosaurs (Figure 6.39b).

However, it is difficult to substantiate this identification because no skeletal remains of stegosaurs have ever been reported from Australia.

For the sake of completeness illustrations of *conjectural* stegosaur tracks, based only on the evidence of the hand and foot skeleton, are included in Figure 6.39.

ANKYLOSAURS

Structure of manus and pes (Figure 6.40)

Ankylosaurs had a pentadactyl manus that was only slightly smaller than the pes. Its five stubby digits were arranged in a semicircular or radiating pattern and terminated in flat, broad-spreading and rather spade-like claws. In *Talarurus*, from the Upper Cretaceous of Mongolia, the phalangeal formula was 2:3:3:3:2; in *Sauropelta*, from the Lower Cretaceous of Wyoming and Montana, USA, it was about 2:3:4:3:2, with perhaps an additional phalanx in either or both of the outer two digits. There was no great disparity in size among the inner four digits, though numbers II and III were the biggest. By comparison, digit V appeared relatively weak.

The pes comprised four functional digits (I–IV) and, in at least some cases, a vestige of metatarsal V. The digits were not so abbreviated as those of the manus, and their spade-like claws were slightly less obtuse in outline. As in the manus, the digits were disposed in a radiating pattern. Digits II, III and IV were about equally thick, though digit III was a little longer than the other two. Digits II and IV were about equal in length, and digit I was only about two-thirds the length of digit III. The phalangeal formula was 2:3:4:5:0, or perhaps 2:3:4:4:0 in some cases.

Tracks (Figures 6.41 and 6.42)

Only a few tracks have been ascribed to ankylosaurs, which were undoubtedly habitual quadrupeds. A trackway in the Upper Cretaceous of Brazil, originally attributed to a 'dinosaurio plantigrado' by L.J. de Moraes (1924), was later interpreted by F. von Huene (1931a) as the work of an ankylosaur. However, this somewhat problematical track might perhaps have been produced by a ceratopsian. The ichnogenus *Rigalites*, reported by von Huene (1931b) from the Middle Triassic of Argentina, was originally considered as a

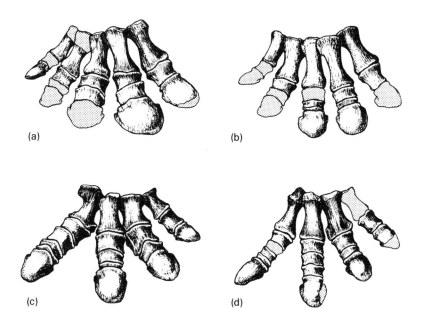

(a)

(b)

(c)

(d)

Figure 6.40 Manus and pes skeleton in ankylosaurs. (a) Manus of
Talarurus, from the Upper Cretaceous of Mongolia; about 24 cm wide.
(b) Manus of *Sauropelta*, from the Lower Cretaceous of the western USA;
about 34 cm wide. (c) Pes of *Talarurus*, from the Upper Cretaceous of
Mongolia; about 35 cm wide. (d) Pes of *Nodosaurus*, from the Upper
Cretaceous of the western USA; about 40 cm wide. (Adapted from Maleev
1956 (a,c), Ostrom 1970 and Coombs 1978a (b), Lull 1921 (d).)

possible ankylosaur track, but has since been attributed to a variety
of reptiles, including 'pre-stegosaurs' (von Huene 1950), ornithopods
(Lessertisseur 1955) and crocodilians (Haubold 1971).

   More certainly of ankylosaurian origin is the ichnogenus *Tetrapodosaurus*, founded by C.M. Sternberg (1932) on a trackway in the Lower
Cretaceous of British Columbia, Canada (Plate 11, p. 208, top left).
Sternberg suggested that *Tetrapodosaurus* was made by a forerunner of
the late Cretaceous ceratopsians, but K. Carpenter (1984) has
demonstrated that it matches very closely to the hands and feet of the
early Cretaceous anklyosaur *Sauropelta*. The manus prints are
pentadactyl and only a little narrower than the tetradactyl pes prints.
The best-preserved examples are broader than long, and show clear
indications of five short, blunt and widely divergent fingers. The pes

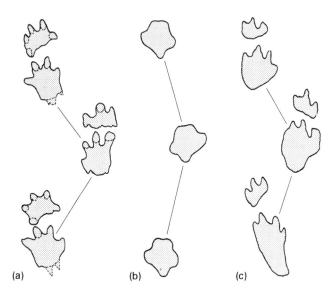

(a)   (b)   (c)

Figure 6.41 Trackways attributed to ankylosaurs. (a) *Tetrapodosaurus borealis*, from the Lower Cretaceous of British Columbia, Canada; total length about 2.05 m. (b) Unnamed trackway, from the Upper Cretaceous of Brazil; total length about 2.03 m. Possibly a pseudo-bipedal track, with manus prints overtrodden and obliterated by pes prints. (c) *Metatetrapous valdensis*, from the Lower Cretaceous of Germany; total length about 2.13 m. In all three cases there is considerable uncertainty about the track-maker's identity. (Adapted from Sternberg 1932 (a), von Huene 1931a (b), Haubold 1971 (c).)

prints, by contrast, are longer than wide, with traces of four slightly longer, though equally blunt, toes. Both manus and pes show pronounced negative (outwards) rotation. The entire trackway is relatively broad, being about two and a half times the width of a pes print. Pace angulation ranges between 100° (manus prints) and 115° (pes prints), and there is no trace of a tail-drag. The stride is relatively short, about four times the length of a pes print and about six times as long as a manus print.

Less certainly of ankylosaurian origin is *Metatetrapous*, the trackway of a quadruped from the Lower Cretaceous of Germany (Figure 6.41c). In its size and general appearance *Metatetrapous* is not greatly different from *Tetrapodosaurus*, though the pes prints are somewhat more elongate, the trackway is slightly narrower, with pace angulation

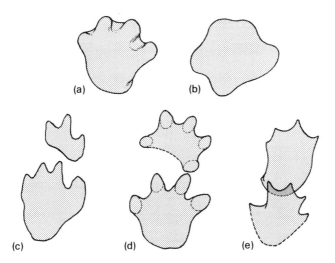

Figure 6.42 Miscellaneous tracks attributed to ankylosaurs. (a) Unnamed pes print, from the Upper Jurassic of Dorset, England; about 77 cm long. (b) Unnamed pes, print from the Upper Cretaceous of Brazil; about 33 cm long. (c) *Metatetrapous valdensis*, a manus-pes couple from the Lower Jurassic of Germany; pes about 44 cm long. (d) *Tetrapodosaurus borealis*, a manus-pes couple from the Lower Cretaceous of British Columbia, Canada; pes about 34 cm long. (e) *Meheliella* (*Walteria*) *jeffersonensis*, a manus-pes couple from the Lower Cretaceous of Colorado, USA; pes about 14 cm long. All illustrations except (a) are composites, based on several prints, and in no case is the track-maker identified with certainty as an ankylosaur; *Meheliella* (e) may not be dinosaurian in origin. (Adapted from Anonymous 1987 (a), von Huene 1931a (b), Haubold 1971 (c), Sternberg 1932 (d), Mehl 1931 (e).)

about 130°, and the number and shape of digit impressions, in both manus and pes, are less clearly established. In the tetradactyl pes of *Metatetrapous* the digital impressions are less divergent and slightly sharper than those of *Tetrapodosaurus*. The manus has indications of three or four digits at most, and these, too, are more sharply pointed. However, Sternberg did note (1932: 74) that imperfect manus prints of *Tetrapodosaurus* sometimes showed only three digits instead of five. A four-toed footprint discovered in the Jurassic coal measures of Queensland (Figure 6.39b) also bears some resemblance to *Tetrapodosaurus*. This Australian footprint was tentatively attributed

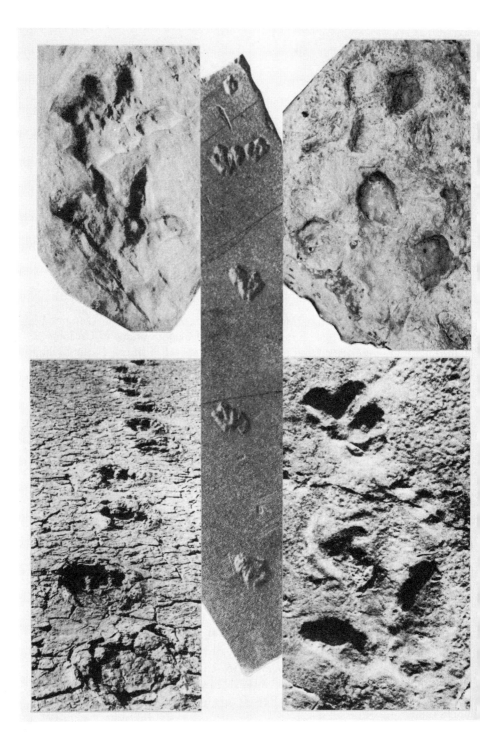

to a stegosaur but might just as well have originated from an early ankylosaur.

Some quadrupedal dinosaur tracks recently discovered in the Upper Jurassic of Dorset, southern England, might conceivably represent an ankylosaur such as *Sauropelta*. However, the great size of tracks, with pes prints up to 77 cm long, seems more appropriate for a sauropod track-maker, as does the apparently horseshoe-like outline of the manus prints (Anonymous 1987, fig. 2b).

CERATOPSIANS

Structure of manus and pes (Figure 6.43)

The small and primitive-looking protoceratopsians had hands and feet rather similar to those of certain ornithopods, such as *Thescelosaurus*. The manus, which was about half the size of the pes, was pentadactyl, with the phalangeal formula 2:3:4:3:1 (*Leptoceratops*). The five fingers were quite slender and widely divergent, and they terminated in blunt claws. Digits II and III were the strongest in the manus, whereas digit I was slightly smaller and digits IV and V were noticeably reduced in

Plate 11 Examples of dinosaur trackways. Top left: *Tetrapodosaurus borealis*, left manus and pes prints of an ankylosaur, from the Lower Cretaceous of British Columbia, Canada; manus (above) is 29 cm wide. Top right: *Caririchnium leonardii*, manus and pes prints of an iguanodont, from the Lower Cretaceous of Colorado, USA; pes print (below) is 34 cm long. Note that the digits of the manus appear to have been bound into a tight bundle. Bottom left: *Sousaichnium pricei*, the trackway of a bipedal iguanodont, traversing a sun-cracked surface in the Lower Cretaceous of the Rio do Peixe Basin, Brazil; each pes print is about 56 cm long. The seemingly regular alternation of long and short paces is an inconsistent feature of dinosaur trackways in general. Bottom right: *Moyenisauropus natator*, right manus and pes prints of an ornithopod dinosaur, from the Lower Jurassic of Lesotho; pes print is about 20 cm long. Fan-like arrangement of digits in the manus is typical for trackways of early ornithopods walking on all fours. A prominent drag-mark accounts for the seemingly 'bent' middle digit of the pes. Centre: *Delatorrichnus goyenechei*, natural cast of the trackway of a small coelurosaur travelling on all fours, from the Middle Jurassic of Argentina; each pes print is about 3 cm long. The tiny heart-shaped manus prints are in most cases overlapped by pes prints; an isolated example is preserved at the extreme top of the slab.

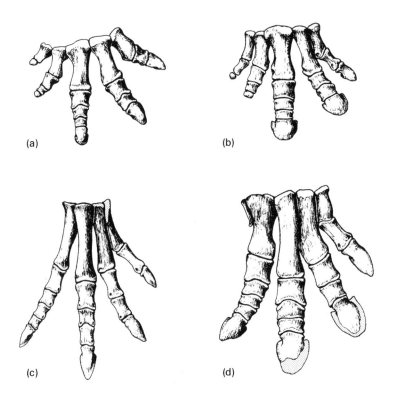

(a)   (b)

(c)   (d)

Figure 6.43 Manus and pes skeleton in ceratopsians. (a) Manus of *Leptoceratops*, from the Upper Cretaceous of Alberta, Canada, and Wyoming, USA; about 11 cm high. (b) Manus of *Styracosaurus* from the Upper Cretaceous of Alberta, Canada; about 28 cm high. (c) Pes of *Leptoceratops*; about 22 cm high. (d) Pes of *Styracosaurus*, about 44 cm high. (Adapted from Brown 1914 (a), Lull 1933 (b,d), Brown and Schlaikjer 1942 (c).)

size. The pes was even more reminiscent of that in some ornithopods. It comprised four long and quite broadly spreading toes (I–IV) with, in some cases, the fifth represented by a vestige of its metatarsal. The phalangeal formula was 2:3:4:5:0. In *Leptoceratops* the four functional toes were quite long and slender, with fairly sharp claws; in *Protoceratops* the toes were stubbier, with broader and blunter claws resembling small hooves. Digits II and IV were roughly similar in their size, shape and angle of divergence from the slightly longer digit III. Digit I was the shortest in the foot, about two-thirds the length of digit III.

The hands and feet of the bigger ceratopsians, such as *Triceratops* and *Styracosaurus*, were correspondingly shorter and heavier. In addition, they seem to have been flat and broad-spreading, rather than pillar-like. In *Triceratops*, for example, H.F. Osborn (1933: 8) noted that:

> . . . the fore feet have rather the short, flat character of the tortoise feet than the round, compact, cylindrical form of the mammalian quadruped; it is quite impossible to throw the metapodials into any such sharply convex form as those of the elephant, nor would the distal ends of the radius and ulna admit of it.

Of the hind feet, Osborn remarked that they 'seem to have been comparatively broad, short and spreading as in tortoises, not compactly rounded as in proboscideans' (1933: 12).

The manus of the larger ceratopsians was about two-thirds the size of the pes. It comprised five robust digits, arranged in a moderately divergent pattern, and invariably had a phalangeal formula of 2:3:4:3:2. In its general appearance the manus resembled that in protoceratopsians, though the digits were stubbier and terminated in broad flat claws that might justifiably be termed hooves. In addition, the phalanges (excluding the terminal hooves) were somewhat shorter and thicker than those in protoceratopsians, though they were not compressed into the disc-like bones that are found in stegosaurs and ankylosaurs. The pes was rather similar in appearance to the manus, except in being slightly larger and in having only four functional digits (I–IV). Its phalangeal formula was 2:3:4:5:0, and in some instances the fifth digit was represented by a vestige of its metatarsal. Digits II, III and IV were the strongest in the foot. Digit I, though somewhat shorter, was not very widely divergent and terminated, like the others, in a broadly rounded hoof.

Tracks (Figure 6.44)

There are few and questionable reports of tracks that might have been made by ceratopsians. *Tetrapodosaurus*, from the Cretaceous of Canada (Figure 6.41a), is sometimes regarded as the trackway of an early ceratopsian but, as mentioned previously, it is more convincingly matched to the hands and feet of ankylosaurs. (H. Haubold (1971: 103) suggested that *Meheliella* (*Walteria*), a problematical track from

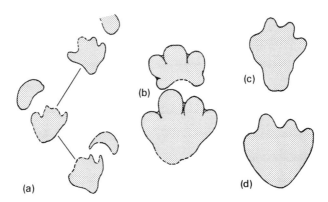

Figure 6.44 Miscellaneous tracks questionably attributed to ceratopsians. (a) Unnamed trackway, from the Upper Cretaceous of Colorado, USA; total length about 2.13 m. (b) Unnamed manus-pes couple, from the Upper Cretaceous of Colorado, USA; pes about 36 cm long. (c), (d) Unnamed pes prints, from the Upper Cretaceous of Utah, USA; both about 39 cm long. (Adapted from Lockley 1986b (a,b,d), Lockley and Jennings 1987 (c).)

the Cretaceous of Colorado (Mehl 1931), might in some respects be comparable to *Tetrapodosaurus*; and the trackway sketched under the ceratopsian *Chasmosaurus* by R.T. Bakker (1986b: 212–3) also appears to be *Tetrapodosaurus*.) More puzzling is an unnamed trackway from the Upper Cretaceous of Brazil (Figure 6.41b). According to F. von Huene (1931a) this is a pseudo-bipedal track in which the quadrupedal track-maker consistently planted its hindfeet over the impressions left by the forefeet. The pes prints are broadly pentagonal in outline, with indications of four extremely short, blunt and widely divergent digits. Each footprint is slightly longer than wide (33 cm as opposed to 30 cm), and the average pace length is 93 cm, implying that the ratio SL/FL was about 5.5/1. The footprints are laid out in a definite zigzag arrangement, with pace angulation about 155°, and the ratio of maximum trackway width to footprint width is about 2/1. Unfortunately, there is no reliable information about the manus prints, which appear to have been entirely obliterated by the prints of the hindfeet. The track-maker was tentatively identified as an ankylosaur by F. von Huene (1931a), though H. Haubold suggested (1971: 89) that it might perhaps have been a ceratopsian. There are, indeed, a couple of points to support Haubold's idea. First, the footprints bear some resemblance to a ceratopsian track that was recently illustrated by M.G. Lockley (1986b; see below); and, second, the overlapping of pes print onto manus print might imply that the forefoot was lifted from the ground just before the hindfoot was set down. In other words, the ipsilateral forefoot and hindfoot might have been clear of the ground simultaneously – and this seems more likely to have occurred in ceratopsians than in the relatively short-legged ankylosaurs. Nevertheless, one important fact counts against these ideas: there are no conclusive reports of ceratopsian body fossils from South America. The same objection applies to *Ligabueichnium bolivianum*, a suspected ceratopsian trackway from the Upper Cretaceous of Bolivia (Leonardi 1984a, 1989).

Recently, M.G. Lockley (1986b) published some illustrations of possible ceratopsian tracks in the mid-Cretaceous of Colorado. Interestingly, the pes prints appear to be very similar in size and shape to those in the Brazilian trackway just described. More importantly, the Colorado tracks include clear indications of the manus (Figure 6.44a). Each manus print has the shape of a forwardly arched crescent and lies slightly in front of, and lateral to, the pes print. One of Lockley's illustrations (1986b, fig. 13) shows the manus prints to be rather featureless, but another (pl. 2) betrays indications of four, or possibly five, stubby digits. Similar footprints, named *Dinosauropodes osborni*, occur in the Upper Cretaceous of Utah (Parker and Rowley 1989).

NOTES

[1] For examples of sauropod trackways see: Dingman and Galli 1965 (Chile); Malz 1971, Kaever and de Lapparent 1974, Hendricks 1981 (Germany); Lim *et al.* 1989 (Korea); Dutuit and Ouazzou 1980, Monbaron 1983, Ishigaki 1985a, 1986, 1989 (Morocco); Ginsburg *et al.* 1966, Taquet 1972, 1976 (Niger); Antunes 1976 (Portugal); Mensink and Mertmann 1984 (Spain); Bird 1939b, 1941, 1944, 1954, 1985, Langston 1974, Prince 1983, Lockley *et al.* 1986, Farlow 1987, Farlow *et al.* 1989, Pittman 1989, Pittman and Gillette 1989 (USA); Gabouniya 1951, 1952, Zakharov 1964 (USSR).

[2] Comments about *Tetrasauropus* are based on P. Ellenberger's accounts (1970, 1972) of the ichnospecies *T. jaquesi*. Two other of Ellenberger's ichnotaxa may also be the tracks of quadrupedal prosauropods: *T. unguiferus* (1970) and *Pseudotetrasauropus jaquesi* (1972). Another two ichnospecies – *T. gigas* (1970) and *T. seakensis* (1970) – are of uncertain

status and were regarded by Olsen and Galton (1984) as examples of *Brachychirotherium*.

[3] Sternberg mentioned that the pes prints of *Gypsichnites pacensis* 'point slightly outward from the line of march' (1932: 70). My restoration of the trackway (Figure 6.36b), which is an extrapolation from only two footprints, shows such an arrangement, though I suspect it to be a misinterpretation based on identification of the left footprint as a right one and vice versa. The *Gypsichnites* footprint in Figure 6.37g is deliberately reversed, as it would appear in a trackway with footprints showing positive (inwards) rotation of normal type for bipedal dinosaurs.

7

Problematical and anomalous tracks

Upon a long layer of the slaty stone were marks of ripplings of some now waveless sea; mid which were tri-toed foot-prints of some huge heron, or wading fowl.

Pointing to one of which, the foremost disputant thus spoke:– 'I maintain that these are three toes.'

'And I, that it is one foot,' said the other.

'And now decide between us,' joined the twain.

Herman Melville, *Mardi* (1849)

THECODONTIAN OR DINOSAUR?

Dinosaurs are distinguished from their primitive-looking relatives, the thecodontians (*sensu* Benton and Clark 1988: 300), by a series of relatively minor anatomical differences, many of which are found in the hindlimb bones and the hip skeleton and seem, in retrospect, to have been correlated with an overhaul or 'improvement' in the posture of the animals. Briefly, the first thecodontians had a primitive sprawling posture, somewhat like that of crocodiles and lizards, where the belly rested on the ground and the knees stuck out sideways. By contrast, the dinosaurs developed an erect posture, like that of mammals and birds, with the feet well under the body, the knees tucked in alongside the flanks, and the belly lifted clear of the ground (Figure 7.1). Such postural improvement looks mechanically efficient, and it might also have allowed dinosaurs to maintain uninterrupted ventilation of their lungs while they were walking and running (Carrier 1987).

The development of this erect posture, which was thoroughly documented by A.J. Charig (1972) and J.M. Parrish (1986), entailed reorientation of the hindlimb and considerable remodelling of the major joints at hip, knee and ankle. These structural changes occurred rather gradually among the thecodontians, many of which managed to achieve only an intermediate or 'partly improved' condition.

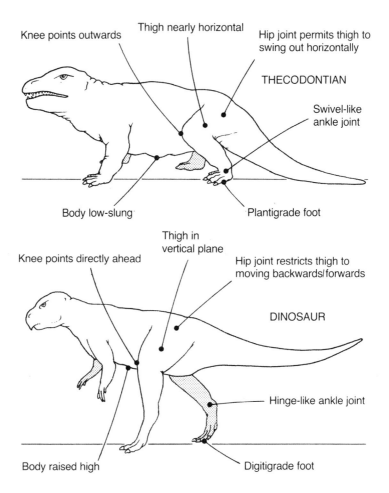

Figure 7.1 The sprawling posture of a thecodontian (above) contrasted with the erect or 'improved' posture of a dinosaur (below). The thecodontian is *Erythrosuchus*, from the Lower Triassic of South Africa; about 4.5 m long. The dinosaur is a primitive ceratopsian, *Psittacosaurus*, from the Lower Cretaceous of Mongolia; about 2 m long.

Consequently, one would expect to find an equally gradual transition between the tracks of thecodontians and those of dinosaurs.

The tracks of primitive thecodontians are unlikely to be confused with those of Triassic theropods and ornithopods. Primitive thecodontians with a sprawling posture had plantigrade hindfeet with five toes and an ectaxonic pattern of foot structure, whereas theropods and

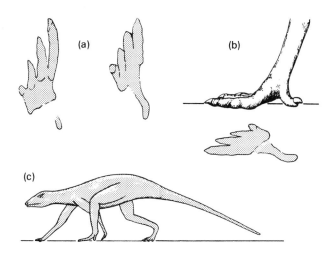

Figure 7.2 *Rotodactylus* and its interpretation. (a) Two right pes prints of the ichnogenus *Rotodactylus*, from the Lower Triassic of the southwestern USA; between 4 and 5 cm long. (b) Restoration of the track-maker's left foot, in lateral view, showing backwardly turned digit V acting as a prop. Subsequent reduction of digit V would leave a digitigrade foot of dinosaurian type. (c) Conjectural restoration of *Rotodactylus* track-maker. (Adapted from Peabody 1948 (a,b), Haubold 1984 (c).)

ornithopods had digitigrade hindfeet with three or four toes and a mesaxonic pattern of foot structure. On the other hand, some advanced thecodontians approached so closely to dinosaurs in their locomotor anatomy that their tracks might be indistinguishable from those of early dinosaurs (Baird 1954: 184). Of particular interest here is the ichnogenus *Rotodactylus*, comprising the tracks of reptiles with digitigrade hindfeet, where the backwardly turned digit V acted as a 'prop' to support the elevated rear part of the foot (Figure 7.2). *Rotodactylus* track-makers were probably small and agile quadrupeds that may have been capable of bipedal progression, and they were envisaged by F.E. Peabody (1948: 340) as:

> ... reptiles of dinosaur-like form with relatively long slender fore limbs as compared with purely bipedal dinosaurs. The foot structure is unique and perhaps is prototypic for the specialized running foot so successful among dinosaurs and birds.

H. Haubold (1966) visualized the *Rotodactylus* track-maker as an elegant little quadruped looking something like the reptilian equivalent of a greyhound (Figure 7.2c) – a creature that would match very closely to the skeleton of the late Triassic archosaur *Lagosuchus*, which was either an early dinosaur or a thecodontian so advanced that it was on the brink of achieving true dinosaur status (Bonaparte 1975; Thulborn 1980).

In practical terms, tracks in Triassic sediments may be attributed to theropod or ornithopod dinosaurs, rather than thecodontians, if they meet three criteria:

1. they are mesaxonic;
2. digit V is strongly reduced or entirely absent;
3. the track-maker was consistently digitigrade when walking.

According to these criteria *Rotodactylus* is not the track of a dinosaur: the track-maker was consistently digitigrade, but it still had an ectaxonic foot with a prominent digit V. By contrast, the tracks of early theropods (e.g. *Coelurosaurichnus*) and ornithopods (e.g. *Anomoepus*) do meet all three criteria.

In spite of these distinctions there are still cases where thecodontian tracks might easily be confused with those of Triassic dinosaurs. Good examples are known from the Middle Triassic of France, where L. Courel and G. Demathieu (1976) reported small theropod tracks, *Coelurosaurichnus*, along with those of a second and somewhat problematical ichnogenus named *Sphingopus*. Well-preserved examples of *Sphingopus* comprise five digits and, for that reason, are unlikely to be mistaken for the hindfoot prints of dinosaurs, but imperfect examples sometimes lack the innermost and outermost digits, thus resembling tridactyl footprints similar to *Coelurosaurichnus* (Figure 7.3a,b). In effect, some *Sphingopus* tracks might be classified as dinosaurian whereas others would not, depending entirely on the quality of their preservation.

The three criteria listed above are not so easily applied in the case of prosauropod dinosaurs, where the hindfoot retained traces of the primitive ectaxonic structure (with digits III and IV nearly equal in length) and a moderately large digit V (see Figure 6.21). Despite this difficulty the tracks of prosauropods that travelled quadrupedally may be distinguished with confidence from those of thecodontians by the entaxonic structure of the manus. Moreover, no thecodontians seem to have survived beyond the close of the Triassic period, so that

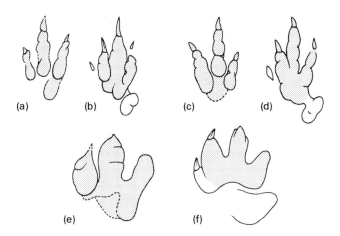

Figure 7.3 Tridactyl footprints of Triassic dinosaurs compared with contemporary footprints of thecodontians. (a) Theropod footprint, *Coelurosaurichnus*, compared with (b) footprint of the ichnogenus *Sphingopus*. (c) An unnamed 'dinosauroid' print resembling *Anchisauripus* and *Grallator*, compared with (d) another example of *Sphingopus*. (e) An unnamed 'dinosauroid' print, compared with (f) a 'crocodiloid' thecodontian print, probably *Brachychirotherium*. In all cases an imperfect thecodontian print, lacking digits I and V (unshaded), might readily be misidentified as a tridactyl dinosaur print. (Adapted from Courel and Demathieu (a), Demathieu 1970 (b–f).)

ectaxonic pes prints encountered in Lower Jurassic sediments are very likely to be those of prosauropods.

ORNITHOPOD OR THEROPOD?

There is usually no problem in distinguishing ornithopod pes prints from theropod pes prints, providing that they are well preserved and show clear indications of the claws and the digital outlines. However, as J.O. Farlow remarked (1987: 12), such well-preserved examples represent 'the clear endpoints of a rather blurred continuum' wherein even the experts may disagree over the identification of a particular footprint. In the past, different investigators have relied on their own, and sometimes arbitrary, criteria for distinguishing theropod prints from ornithopod prints. Some of those criteria may be valid, but

Table 7.1 Comparison of footprint proportions in ornithopod and theropod tracks*

Dinosaurs	Mean ratio FW/FL	Ratio of maximum width to length in digit III
Coelurosaurs	0.73 ± 0.19 (N = 40)	0.32 ± 0.03 (N = 12)
Carnosaurs	0.77 ± 0.14 (N = 44)	0.36 ± 0.05 (N = 11)
Small ornithopods	0.91 ± 0.18 (N = 43)	0.44 ± 0.01 (N = 13)
Large ornithopods[†]	0.90 ± 0.15 (N = 44)	0.61 ± 0.19 (N = 10)

* Based on a sample comprising (1) published mean values for ichnotaxa and (2) mean values for unnamed natural associations of tracks; all published identifications and measurements were accepted uncritically.
[†] Iguanodonts and hadrosaurs.

others are of questionable value, and none of them is infallible. Nevertheless, it is worthwhile to review those criteria, since some of them may prove useful in dealing with poorly preserved footprints.

Footprint proportions
Theropod footprints tend to be a little longer than wide, whereas ornithopod footprints tend to be nearly as wide, or even wider, than long (Table 7.1; see also Moratalla *et al.* 1988a).

Shape of digits
Theropod footprints comprise tapering digits, often with a V-shaped outline, while ornithopod footprints typically have parallel-sided digits with a more U-shaped outline (compare Plates 9, p. 140 and 10, p. 160). These differences in shape tend to be most pronounced in digit III, which was usually planted very firmly into the substrate. Some footprints of bigger ornithopods comprise oval digits, broadened towards their tips, and thus resemble a clover-leaf in outline (see Figure 6.33; Plate 9, p. 140, centre left).

Digital extremities
Impressions of narrow claws are commonly regarded as diagnostic features of theropod tracks, though they are not necessarily consistent in appearance (Coombs 1980b: 1200) and are sometimes lacking altogether (Farlow 1987: 11). By contrast, the toe-prints of

ornithopods are bluntly rounded – a feature that is most obvious in digit III, and particularly in the bigger footprints attributed to hadrosaurs and iguanodonts (e.g. Plate 9, p. 140, top left). The tracks of smaller ornithopods may show indications of moderately narrow claws (semiclaws or 'nails'), but these are neither so slender nor so sharply pointed as the claws of small theropods.

Length of digit III

It has been suggested that tridactyl footprints with the middle digit longer than the other two are 'characteristic of carnivorous dinosaurs' (Sanderson 1974: 234). This generalization is probably unsound, bearing in mind the many factors that may affect footprint shape, and the problems that may arise in obtaining consistent measurements (see Figure 4.9). In addition, it is worth noting that the relative length of digit III varies according to the overall size of theropod footprints in the ichnogenus *Grallator* (*sensu* Olsen and Galton 1984: 97): the middle toe is relatively long in small footprints but relatively short in bigger examples. Similar variation is apparent in ornithopod tracks.

Width of digits

The toe-prints of theropods tend to be distinctly narrower than those of ornithopods (Table 7.1; Moratalla *et al.* 1988a). Presumably this difference indicates that the theropods had more flexible toes, allowing the foot to be used as an efficient grasping device. In attempting to apply this criterion it should be borne in mind that the width of the toe-prints is sometimes affected by preservational factors (see Figure 5.18).

Curvature of digits

The impressions of digits III and, to a lesser extent, IV are sometimes appreciably curved in the footprints of theropods. Such curvature (usually convex to the exterior) is most clearly expressed in small footprints such as those in the ichnogenera *Anchisauripus* and *Grallator*. The thicker digits of ornithopod footprints show little or no curvature.

Orientation of hallux

An impression of the hallux is sometimes present in the footprints of theropods and small ornithopods, but is relatively uncommon in the tracks of bigger ornithopods. The theropod hallux extended postero-

Table 7.2 Comparison of interdigital angles in ornithopod and theropod tracks (sources of data and conventions as in Table 7.1)

Dinosaurs	Total divarication digits II–IV (degrees)	Interdigital angles II–III III–IV (degrees)	
Coelurosaurs (N = 15)	49 ± 23	24 ± 11	23 ± 9
Carnosaurs (N = 14)	53 ± 12	26 ± 8	27 ± 8
Small ornithopods (N = 15)	62 ± 9	29 ± 11	29 ± 12
Large ornithopods (N = 11)	60 ± 19	30 ± 11	31 ± 10

medially or straight backwards (e.g. Figure 6.4e–g), whereas that of ornithopods extended antero-medially or straight inwards, towards the midline of the trackway (e.g. Figure 6.27g). There is one possible exception: some early coelurosaurs, such as *Syntarsus*, may have had the hallux directed forwards, alongside digit II, thus resembling the smaller ornithopods (Raath 1969: 19).

Total divarication of digits II–IV

Some researchers have assumed that footprints of the theropod *Megalosaurus* had more widely divergent digits than those of the contemporary ornithopod *Iguanodon* (e.g. Blows 1978: 57). Other investigators have assumed exactly the opposite (e.g. Madeira and Dias 1983: 156). To some extent these conflicting opinions may reflect differences in methods used to measure the total divarication of digits. The figures in Table 7.2 seem to confirm that theropods have less widely divergent digits than ornithopods, though there is obviously a very wide range of variation.

Interdigital angles II–III and III–IV

Some ornithopod footprints have digit II more widely divergent than digit IV (Thulborn and Wade 1984), whereas the reverse may be true in some theropod footprints (Farlow 1987: 11). However, it is doubtful that these distinctions can be extended into a general rule because the interdigital angles are so variable and so difficult to measure in consistent fashion. A brief analysis (Table 7.2) seems to indicate that interdigital angles II–III and III–IV are roughly equal in theropod footprints as well as in ornithopod footprints.

Rear margin of the footprint

In many instances the rear half of a theropod footprint is angular or V-shaped in outline (e.g. Plate 10, p. 160, top right). By contrast, the rear half of an ornithopod footprint tends to have a broader and almost U-shaped outline (e.g. Plate 9, p. 140, centre left). This difference in shape seems to be correlated with the greater divarication of digits II–IV in ornithopod footprints (Langston 1974: 95).

Interdigital webbing

Interdigital webs have been reported in various ornithopod footprints (e.g. Currie and Sarjeant 1979; Gierliński and Potemska 1987) but may also occur in some theropod footprints as well (e.g. Blows 1978: 57). It is doubtful that the presence of webbing alone is sufficient to distinguish ornithopod tracks from theropod tracks.

Footprint rotation

M.G. Lockley has suggested (1987: 117) that inwards rotation of the pes prints is typical of many ornithopod trackways. Even so, this feature cannot be regarded as diagnostic of ornithopod tracks, because it is equally typical of many theropod trackways. In fact, the direction in which the pes prints are rotated, whether positively (inwards) or negatively (outwards), seems to be correlated with the gait of a trackmaker, rather than with its taxonomic identity: the hindfeet tend to be turned inwards in bipedal dinosaurs and outwards in quadrupeds. This gait-related change in the attitude of the hindfeet is apparent in the trackway of a single dinosaur that switched from two legs on to all fours (see Figure 6.26c).

Drag-marks

At one site it has been noticed that marks made by the dragging toe-tips are more common in the footprints of ornithopods than those of theropods (Thulborn and Wade 1989). This distinction might, perhaps, be correlated with the presence of thicker, and presumably less flexible, toes in ornithopods. It is not known if comparable differences exist at other sites, and it should be noted that drag-marks do sometimes occur in undoubted theropod prints (e.g. Figure 5.16e).

THEROPOD OR BIRD?

The older literature contains scattered reports of supposed bird tracks in Mesozoic sediments. Those reports were largely influenced by Edward Hitchcock's opinions about the avian origin of the 'Ornithichnites' in the Connecticut Valley (see Plate 13, p. 306, bottom), and most of them were probably based on the tracks of small dinosaurs (Lessertisseur 1955; Brodkorb 1978). In fact, it is unlikely that the footprints of late Jurassic and Cretaceous birds could ever be distinguished with certainty from those of coelurosaurs, which shared an almost identical pattern of foot structure. However, the earliest birds do seem to have been relatively rare creatures, to judge from the scarcity of their skeletal remains, and one might suspect that they were responsible for few tracks of late Jurassic and early Cretaceous age. On the other hand, most Cretaceous birds appear to have been dwellers in or near water (Feduccia 1980), so that there would seem every likelihood of their tracks having been preserved along ancient shore-lines (e.g. Ambroggi and de Lapparent 1954b; Alonso and Marquillas 1986). In short, some Cretaceous footprints may well be those of birds, but it may be an almost impossible task to distinguish them from the footprints of small bipedal dinosaurs.

Nevertheless, there have been attempts to discriminate between the tracks of birds and those of small dinosaurs. For example, R.S. Lull (1904) suspected that the fleshy pads on the undersides of the toes were predominantly mesarthral in position among dinosaurs but arthral among most modern birds. G. Heilmann (1927) doubted the existence of such a clear-cut distinction and suggested, instead, that dinosaur footprints might have less widely divergent toes ($30°$–$45°$) than bird footprints (about $90°$). Heilmann's idea was endorsed by P.J. Currie (1981), who described a series of supposed bird tracks from the early Cretaceous sediments of British Columbia, Canada. Currie maintained that the 'divarication between digits II and IV in even the smallest dinosaurs never exceeds $100°$ on an average per trackway' (1981: 257). By contrast, the total divarication of the digits tended to exceed $100°$ in a variety of modern bird tracks (see also Johnson 1986). In addition, Currie found that bird tracks tended to be broader than long and that they usually showed a well-defined impression from the distal end of the metatarsus. However, it must be noted that these criteria are general tendencies within a large sample of tracks, and that there are exceptions. So, for instance, Currie observed that

the total divarication of digits was as little as 75° in some bird footprints and that the ratio *FW/FL* sometimes fell as low as 0.84. On the
other hand, some dinosaur tracks may have a total divarication much
greater than 90° (e.g. *Taupezia*, Delair 1963) and some are consistently
wider than long (e.g. *Wintonopus*, Thulborn and Wade 1984). In short,
the criteria described by Currie (1981) may be applicable to large
assemblages of tracks but are probably less reliable for evaluating footprints that are found singly or in small numbers.

ANOMALIES AND PITFALLS

A considerable variety of problems, both potential and actual, may
sometimes arise in the identification of genuine dinosaur tracks. The
following survey is not an exhaustive review of all problems and
anomalies but is merely a selection of cautionary tales and points to
be borne in mind.

Seemingly anomalous age

Tridactyl footprints are occasionally discovered in Palaeozoic
sediments, deposited long before the advent of dinosaurs. Such tracks
presumably originated from primitive reptiles or amphibians but may,
nonetheless, bear an extremely close resemblance to the footprints of
dinosaurs (Figure 7.4a,c). Some of them, such as those in the
ichnogenus *Ornithoidipus*, may even appear to have been produced by
bipeds (Sternberg 1933a).

Dinosaur-like tracks that occur in Tertiary and Quaternary
sediments, deposited after the extinction of the last dinosaurs, are
invariably the work of birds. Some of them, produced by giant flightless birds such as the New Zealand moas and the Australian
mihirungs, may bear an uncanny likeness to dinosaur tracks (Figure
7.4b,d). Occurrences such as these underline the need to examine not
only the morphology of fossil tracks but also their stratigraphic position, their mode of preservation, and any associated fossils.

Limulid tracks

The tracks of limulids (horseshoe crabs) are frequently mistaken for
those of small vertebrates, including amphibians, birds, mammals,
pterodactyls and dinosaurs. Particularly fine examples of these tracks

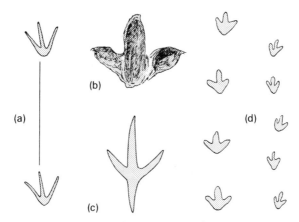

Figure 7.4 'Dinosauroid' tracks made by animals other than dinosaurs. (a) *Ornithoidipus pergracilis*, a track of problematical origin, from the Upper Carboniferous of Nova Scotia; footprints about 4 cm long. (b) Left pes print of a moa, from the Pleistocene or Holocene of New Zealand; about 37 cm long. (c) *Ornithichnites gallinoides*, a track of problematical origin, from the Upper Carboniferous of Canada; about 17.5 cm long. (d) Two trackways of moas, probably adult and juvenile, from the Pleistocene of New Zealand; footprints of the bigger animal about 20 cm long. (Adapted from Sternberg 1933a (a), H. Hill 1895 (b), H. Schmidt 1927 (c), W.L. Williams 1872 (d).)

occur in the late Jurassic limestones of Solnhofen, in Bavaria, and were responsible for practices that K.E. Caster has described as 'completely incomprehensible' (1957: 1026). Up until 1935 those Soln-hofen tracks terminating at the body remains of limulids were unhesitatingly attributed to those animals. . . whereas identical tracks without a fossil limulid at the end were unfailingly attributed to vertebrates.

Limulid tracks are distinguishable from dinosaur trackways on account of:

1. their small size;
2. the common occurrence of a median groove made by the telson or tail-spine;
3. the indications of three or four pairs of walking legs, in addition to the 'footprint-like' impressions (see Figures 7.5, 7.6);

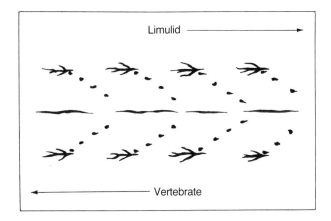

Figure 7.5 Diagram of the trackway made by an adult limulid or horseshoe crab. Arrow at top indicates direction in which the track-maker actually moved. Such tracks have often been mistaken for those of small vertebrates, including dinosaurs, that moved in the opposite direction. (Adapted from Caster 1944.)

4. the fact that the left and right impressions occur in pairs, rather than in an alternating pattern.

Despite these, and other, distinctive features (Caster 1939, 1944; Nielsen 1949) limulid traces continue to be mistaken for vertebrate tracks. Authoritative surveys of fossil reptile tracks (Haubold 1971, 1984) still include the ichnogenus *Crucipes*, which is almost certainly a limulid track of late Carboniferous age, and a popular report of the 'smallest footprint of a dinosaur found anywhere in the world' (Anonymous 1986a) might well be based on the work of a limulid. The same may be true for tiny 'coelurosaur' tracks, some no more than 5 mm long, which came to light in borehole cores from the Lower Triassic of the English Midlands (Wills and Sarjeant 1970).

Worm burrows

Some supposed dinosaur footprints, discovered in Jurassic sediments on the coast of Yorkshire, England (M. Schmidt 1911), were subsequently identified as weathered examples of fossil worm burrows (Bather 1926, 1927). On weathering, each U-shaped burrow had developed into a slot-like cavity resembling a dinosaur's toe-print, and

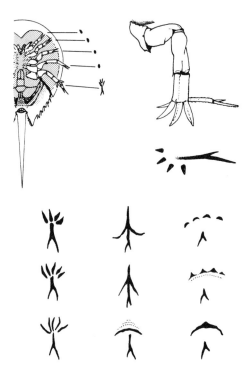

Figure 7.6 A variety of limulid tracks, often mistaken for those of small vertebrates. Top left: underside of an immature limulid (left side only), showing jointed appendages and the general pattern of markings they leave on the substrate; note that the four walking legs make simple impressions whereas the large 'pusher' (flabellum) at rear produces a bigger and more complex impression. Top right: closer view of 'pusher' and the trace it may produce. Below: various 'pusher' traces left by different growth stages of limulids on different substrates; in all cases the animals were headed up the page. Traces similar to those at centre and top centre have frequently been attributed to small dinosaurs, birds and pterodactyls, all assumed to be heading down the page. (Adapted from Caster 1939, 1944.)

in some places these cavities were so abundant that the bedding planes looked as if they had been thoroughly trampled by dinosaurs.

Stump holes

In 1963, D. Baird documented one case where tree trunks surrounded by sediment had rotted away to leave cylindrical cavities that were mistaken for the tracks of a giant amphibian (Martin 1922a,b). The supposed footprints were quite regularly spaced, because the distance between one tree and the next was dictated by the spread of the branches. Some examples also possessed marginal notches, resembling irregular digits, that were formed by the buttress-like bases of roots. And, to complete the resemblance to genuine footprints, one tree sometimes grew inside the cavity left by the decay of another, thus giving the impression of a pes print superimposed on a manus print.

Man-tracks

Human footprints do occur in the fossil record, in sediments deposited long after the extinction of dinosaurs, but they are unlikely to be mistaken for dinosaur tracks. On the other hand, some unquestionable dinosaur tracks have been interpreted, either accidentally or intentionally, as the footprints of humans. The most notorious examples are from the Cretaceous limestones of the Paluxy River, in Texas, USA, and were reviewed in some detail by J.D. Morris (1980). Painstaking investigations have established that none of these so-called 'man-tracks' originated from humans (Milne and Schafersman 1983; Cole *et al.* 1985; Farlow 1985a: 215–16; Godfrey and Cole 1986; Hastings 1987). Some are carvings or erosion features, whereas others are mistaken interpretations of dinosaur footprints. It seems that dinosaur tracks of at least two types have been misinterpreted in different ways.

First, there are some large 'sandal-like' impressions, which appear to be nothing more than the middle digit in tridactyl dinosaur tracks. This digit bore much of the dinosaur's weight and, for that reason, tended to sink deeply into the substrate. When the footprint is weathered, the shallower toe-prints at the sides may be obliterated, leaving only a vague impression of the middle one (Figure 7.7a,b). The resemblance to a human footprint is sometimes increased by a natural drag-mark, looking somewhat like the imprint of the big toe.

A second type of man-track is based on the misinterpretation of a whole dinosaur footprint (Figure 7.7d–f). Here the footprint has a long heel-like impression of the metatarsus, presumably indicating that the track-maker had adopted a flat-footed gait (Kuban 1986, 1989a; Farlow 1987, fig. 39). Footprints of this type are not confined to the Paluxy River of Texas: they have also been reported in Morocco (Ambroggi and de Lapparent 1954b), Spain (Brancas *et al.* 1979) and

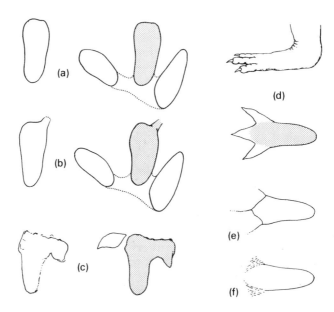

Figure 7.7 Misinterpretations of dinosaur tracks. (a) Typical example of an ornithopod footprint is shown at right; if digits II and IV are eroded, concealed or overlooked, digit III (shaded) may be interpreted as a human-like footprint (left). (b) Similar example with drag-mark from digit III resembling trace of the human big toe. (c) Alleged 'bear track' from the Lower Cretaceous of the Paluxy River, Texas, USA (left), for comparison with digits III and IV (shaded) of an ornithopod footprint from the mid-Cretaceous of Queensland, Australia (right). (d–f) Origin of elongated dinosaur footprints, sometimes mistaken for human-like tracks, in the Lower Cretaceous of the Paluxy River, Texas. (d) Dinosaur foot in plantigrade posture (above), and corresponding footprint (below); if toe-prints are disregarded or destroyed by erosion, the remainder of the print (shaded) resembles a human-like track. Similar misinterpretations are possible when toe-prints are obliterated by slumping of substrate (e) or fail to print on exceptionally hard substrates (f). (Adapted from Thulborn and Wade 1984 (dinosaur footprints, *Wintonopus*, in (a–c)), J.D. Morris 1980 ('bear track' in (c)), Farlow 1987 (d–f).)

Australia (Thulborn and Wade 1984; Plate 6, p. 94, top centre and right). Such footprints are unlikely to be mistaken for human tracks when they are well preserved, with sharp imprints of the toes, but in slushy sediments the narrow toe-prints sometimes collapsed on withdrawal of the track-maker's foot, so that little remained beyond a vague impression of the metatarsus. This long heel-like impression bears superficial resemblance to a human footprint, particularly when weathered.

Other sedimentary structures of organic origin

Aside from limulid tracks, worm burrows and stump holes there are several other organic traces that may be misidentified as dinosaur tracks. Potentially the most confusing are dish-like features, which might easily be taken for the poorly preserved or weathered tracks of sauropods (e.g. those illustrated by Monbaron 1983, pl. 8, and by Langston 1974, pl. 4, fig. 1). Such basin-like structures are common in some ancient sediments and probably originated in a variety of ways. In modern sediments they result from the activities of animals as diverse as fishes (Howard *et al.* 1977), birds (von Sivers 1929) and whales (Nelson *et al.* 1987). The origin of one such basin-like structure, which 'would surely be puzzling if found fossilized', was vividly described by W.H. Bradley (1957: 660):

> Foraging gulls wade around in the shoal water of an ebbing tide looking for anything that moves and that they can catch. Occasionally one gets a firm grip on the siphon of a large clam. Then the gull tramps round and round, pushing his webbed feet down vigorously, thereby digging out a crater around the clam by pushing the sand aside and heaping it up to make the crater wall. After considerable hard work, he dislodges the clam and, supposedly, feels a justifiable satisfaction with his feat.

In addition, ancient sediments are sometimes marked with smaller cavities and depressions that might be mistaken for the toe-prints of dinosaurs. Typical examples are the numerous 'gouges', presumably made by fishes, which were illustrated by S. Dzulynski and J. Kinle (1957, pl. 26).

Sedimentary structures of inorganic origin

There is an enormous range of natural sedimentary structures that might be misidentified as dinosaur tracks. These include various flutes, scour-marks and prod-marks (e.g. Enos 1969, figs 5d, 7a, 12a; Boyd 1975, fig. 5.1), many of which resemble the traces of swimming animals (e.g. Peabody 1956b; Boyd and Loope 1984). Objects carried by water currents sometimes went bouncing and spinning along the substrate, leaving a regular series of alternating skip-casts that might easily be mistaken for a trackway (e.g. Dzulynski *et al.* 1959, fig. 12). Current crescents, which resulted from the scouring effect of water eddying around pebbles, have on occasion been identified as the fossil 'heel marks' of reptiles (Peabody 1947).

An entire range of track-like features is known to result from the mobilization and settling of unconsolidated sediments. These include load casts (Kuenen 1957), dish structures (Rautman and Dott 1977), sand vólcanoes (J.R.L. Allen 1982: 137–41), basin casts and air-heave structures (Prentice 1962). Basin-like cavities may also result from the collapse of sediments following the solution of underlying salt deposits (Hunt 1975, fig. 38). Shrinkage cracks developed in muddy sediments sometimes resemble small footprints (see especially the 'incomplete' types illustrated by J.R.L. Allen 1982, fig. 13.24), and certain synaeresis cracks have a 'characteristic bird's foot shape' (Clemmey 1978, fig. 3). The likelihood of confusing such sedimentary structures with footprints is often increased by the effects of erosion (e.g. J.D. Morris 1980: 103) and weathering (e.g. Stokes 1986).

Artefacts

Rock engravings (petroglyphs) and carvings are unlikely to be confused with genuine footprints, provided that they are examined with care. For a start, they can be found in igneous or high-grade metamorphic rocks, and they are sometimes accompanied by more obvious artefacts such as incised geometric designs or human and animal figures. Chisel-marks or bruising may be apparent, and the 'footprints' are sometimes engraved on fracture planes or erosional surfaces that cut obliquely across the bedding. More significantly, anatomical details such as the digital nodes tend to be portrayed diagrammatically, if at all, whereas normal preservational features such as transmitted impressions and drag-marks are conspicuous by their absence. Even so, carved footprints have on occasion been mistaken for genuine fossil tracks (see Monroe 1987).

Holes dug for fence-posts represent another pitfall (Baird 1963): they are consistent in their size and shape and are usually spaced at regular intervals in a linear series, rather like the footprints in a genuine trackway.

Altered tracks

Finally, and potentially the most misleading of all, are those genuine tracks that have been deliberately altered by human handiwork. Such tracks may turn up in museum collections, where at some stage they have been 'improved', 'repaired' or 'enhanced' for public display. Shallow prints are sometimes found to have been artificially deepened, while those with vague outlines may have been sharpened up with a hammer and chisel (e.g. Lockley 1986b: 26). In other instances, poorly defined footprints have been outlined with paint or ink (e.g. Sarjeant 1975: 285) and imperfect specimens have been restored with plaster or concrete (e.g. J.D. Morris 1980: 123-4). Such alterations can usually be detected without too much difficulty, though there is always some risk that they might be mistaken for original features of fossil footprints.

8

Estimating the size of a track-maker

> What scale should the biologist use to measure the size of an organism? Two fundamental quantities can be measured with relative ease: mass and linear dimension. Mass is usually much to be preferred, but in some circumstances a linear measurement may be more convenient, more meaningful, and even more revealing.
> K. Schmidt-Nielsen, *Scaling: Why is Animal Size so Important?* (1984)

A glance at a dinosaur's tracks will usually allow an educated guess at the overall size of the animal, be it 'as small as a chicken' or 'as big as an elephant'. Such rough estimates certainly help to visualize a track-maker but are too vague to be of use in assessing its speed and gait. Theoretically, it should be possible to predict the size of a dinosaur from the size of its footprints, but in practice there are some awkward problems.

SIZE OF FOOTPRINTS AND TRACKWAYS

An ideal measure of size for dinosaur tracks should be both reliable and convenient: the most reliable measurement would show little or no variation along a trackway, while the most convenient measurement would be easy to make and would allow accurate prediction of the dinosaur's body size. Unfortunately, there is no single measurement that satisfies both criteria.

Footprint size

The size of dinosaur footprints may be quantified in several ways – as footprint length (FL), footprint width (FW) or footprint area. It is also possible to use an **index of footprint size** (SI), such as the following:

$$SI = (FL \times FW)^{0.5} \tag{8.1}$$

This index is expressed in the same units of measurement as FL and FW.

Table 8.1 Analysis of variance for measurements of ornithopod trackways (*Wintonopus latomorum*) in the Cretaceous of Queensland, Australia

Variable	Variation within trackways (%)	Variation among trackways (%)
Footprint length (*FL*)	15.1	84.9
Footprint width (*FW*)	6.0	94.0
Footprint size index (*SI*)	0.8	99.2
Pace length (*PL*)	7.5	92.5
Stride length (*SL*)	2.4	97.6
Pace angulation (*ANG*)	77.9	22.1

Note: The analysis is based on a sample of 287 footprints in 57 trackways (from Thulborn and Wade 1984).

Which of these various measures is the most reliable and the most convenient? Analysis of dinosaur trackways reveals that *FL* is more variable than *FW* or *SI* (e.g. Figure 5.1) and that it is, in fact, so inconsistent that it is probably the least reliable indicator of footprint size. Footprint length is probably so variable because it is easily affected by the angle at which the foot enters and leaves the substrate and by the development of extramorphological features such as drag-marks (see Figures 5.11–5.16). From the analysis in Table 8.1 it appears that the footprints in a dinosaur trackway tend to show least variation in *SI* and most variation in *FL*.

Despite its variability *FL* is often used to predict a track-maker's body size and, thereafter, its gait and speed. Clearly, this dimension needs to be measured with some care if it is not to generate misleading conclusions about dinosaur locomotion. One obvious precaution is to measure *FL* on the best-preserved and least-disfigured footprint in each trackway. The alternative is to use the mean figure for *FL*, based on all the well-preserved footprints in a trackway.

In summary, the various measures of footprint size have different advantages and disadvantages. The less variable ones, such as *SI* and *FW*, would be most appropriate for defining and describing ichnotaxa. Footprint area (or its square root) might prove to be equally useful but is difficult and time-consuming to measure. Footprint length is a less

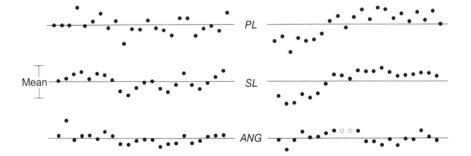

Figure 8.1 Variation in pace length (*PL*), stride length (*SL*) and pace angulation (*ANG*) along two dinosaur trackways. Example at left is a trackway of 24 footprints made by a running coelurosaur, from the mid-Cretaceous of Queensland, Australia; example at right is a trackway of 23 footprints made by a walking iguanodont, from the Upper Jurassic of Dorset, England. Scale bar at centre left indicates 20% variation above and below the mean (horizontal line) and applies to all six diagrams. Open circles represent missing data points and are interpolated purely for the sake of visual continuity.

reliable measure of size but is frequently used to estimate a track-maker's body size. By comparison there have been few attempts to estimate the size of a track-maker from *FW* or *SI* (e.g. Lockley *et al.* 1986: 1171; Thulborn and Wade 1984: 437).

Trackway size

The size of dinosaur trackways might also be expressed in several ways – by means of stride length (*SL*), pace length (*PL*), trackway width (*TW*) or pace angulation (*ANG*). None of these measures really satisfies the two criteria of reliability and convenience.

First, it is clear that *SL*, *PL* and *ANG* (and, hence, *TW*) may show considerable variation within any one dinosaur trackway (Figure 8.1). None of these measures gives a very reliable indication to the overall size of a trackway, nor to the size of the animal that made it. It is true, in general terms, that big dinosaurs would leave big trackways whereas smaller dinosaurs would produce correspondingly smaller trackways. Even so, *SL* and *PL* are known to vary with the gait and speed of a track-maker, so that a small running dinosaur might have taken longer strides than a large dinosaur that was merely walking. In other words, SL and PL do not correspond in a straightforward manner to the size of a track-maker: instead, these dimensions are regulated by a complex interaction between a track-maker's size, gait and speed.

Pace angulation also varies with the gait and speed of a track-maker, as does TW. So, for example, ANG is seen to decrease markedly where an ornithopod dinosaur shifted from a quadrupedal gait to a bipedal gait (see Figure 9.14). Pace angulation tends to increase as an animal accelerates and takes longer strides (Peabody 1959, table 1), and it may change dramatically when a track-maker deviates from a perfectly straight course.

In short, measures such as SL and ANG give no reliable indication to a track-maker's body size, though they may provide important information about its posture and gait.

THE SIZE OF DINOSAURS

What is the best way to measure the body size of a dinosaur? The word 'best' implies that any such measure should meet certain important criteria: it should be applicable to dinosaurs of different shapes, both bipeds and quadrupeds; it should not depend on assumptions about the 'normal' posture of dinosaurs; it should be easily derived from the evidence of footprints; it should form a convenient starting-point for investigating the gaits and speeds of dinosaurs; and it should allow meaningful comparisons of size, gait and speed among dinosaurs, and between dinosaurs and other animals. Several measures of dinosaurian body size are available, some of them more useful than others.

Body length

It is impractical to use total body length, from snout to tail-tip, as a measure of dinosaur size. Few dinosaur skeletons are complete enough to provide this measurement, which, in any event, may bear little relationship to other measures of body size, such as height and weight. For example, the sauropod *Diplodocus* had a length of about 25.5 m (McGinnis 1982) and is estimated to have weighed slightly less than 11 t (Colbert 1962). One specimen of the contemporary sauropod *Apatosaurus* is slightly shorter (23 m) but is estimated to have weighed about three times as much (28–32 t). A third sauropod, *Brachiosaurus*, was also slightly shorter than *Diplodocus* but, according to some estimates, might have weighed seven times as much – a staggering 78 t. Evidently, a dinosaur's length gives little indication to its overall bulk or weight.

While it would seem impossible to predict the total length of a dinosaur from the size of its footprints, P. Ellenberger (1972: 60) did attempt to do so for one track-maker:

> For *Apatosaurus* (*Brontosaurus*) or *Iguanodon*, the body length is equivalent to some 18 times the functional length of the foot. If this rule of size is trustworthy, [the maker of the track] *Pseudotetrasauropus mekalingensis* must have exceeded a good ten metres [translation].

The idea that *FL* represents about one-eighteenth of body length can be regarded as nothing more than a very rough generalization.

Snout-vent length

Many studies of living reptiles use snout–vent length as a standard measure of body size. This measure avoids complications caused by the reptilian tail, which may be extremely variable in length and is sometimes incomplete or damaged. Snout–vent length cannot be measured very accurately on a dinosaur skeleton because there is no exact indication of where the cloaca was located. Moreover, the measurement may be affected by the posture in which the skeleton is mounted. Nor can snout–vent length be measured on dinosaur trackways, even though a few of them do include the impression of a swelling or callosity in the region of the cloaca (see Figure 4.14).

Gleno-acetabular distance

The distance between shoulder joint (glenoid) and hip joint (acetabulum) may be estimated from the trackway of a quadrupedal dinosaur. This gleno-acetabular distance is most easily measured from the midpoint between a pair of hindfoot impressions to the midpoint between the *next* pair of forefoot impressions (Figure 8.2). It is assumed that the track-maker used an alternating gait, with its contralateral forelimb and hindlimb moving almost synchronously. D. Baird pointed out that this assumption is valid for crocodilians (1954: 178), concluding that it 'cannot be far wrong for the trackway of a quadrupedal dinosaur' (1980: 224).

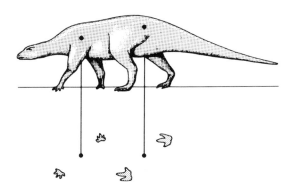

Figure 8.2 Gleno-acetabular distance. Points midway between paired manus prints and paired pes prints correspond approximately to location of the track-maker's shoulder joint and hip joint. The track-maker is shown using an alternating gait: the left manus has just touched the ground, and the right pes is being lifted.

Sometimes, slightly different methods have been used to determine gleno-acetabular distance. For instance, W. Soergel, in his study of *Chirotherium* (1925: 57), suspected that the track-maker's gait was not exactly alternating: when both hindfeet were planted on the ground, one of the forefeet would have been reaching forwards at the start of its stride. Consequently, Soergel recommended that gleno-acetabular distance should be measured along the midline of the trackway, from a point midway between the hindfoot prints to the *foremost* of the next two handprints.

Despite these minor complications gleno-acetabular distance is an excellent indication of a track-maker's body size. Unfortunately, it is of limited value in the study of dinosaurs because so many of these animals moved bipedally, leaving no handprints.

Height in standing pose

Popular works sometimes refer to the height of dinosaurs. The references may be rather imprecise (e.g. the sauropod *Brachiosaurus* could peer over the top of a three-storey building) or fairly exact (e.g. *Tyrannosaurus* stood about 18 feet high). Such descriptions are of doubtful value because they depend entirely on the pose in which the skeleton has been mounted. *Tyrannosaurus* is a case in point. In its classic standing pose, in the American Museum of Natural History,

New York, this dinosaur stands 5.6 m high. Its backbone is aligned at an angle of about 45° to the ground, so that the animal looks as tall and as imposing as possible. In 1969, a *Tyrannosaurus* skeleton was installed in the British Museum (Natural History), London. Here the skeleton was mounted with its backbone horizontal, so that the head is less than 3.5 m above the ground (Newman 1970). Still more recently, it was suggested that the backbone of *Tyrannosaurus* was neither horizontal nor at 45°, but somewhere in between (Tarsitano 1983). All three poses are quite life-like and realistic, yet each of them provides a different estimate of total height.

Similar problems are evident in reconstructions of sauropod dinosaurs. Sometimes it is assumed that these animals carried the neck horizontally, yet on other occasions the neck has been positioned almost vertically. In other words the maximum height of a sauropod may be varied tremendously, depending on assumptions about the animal's 'normal' posture (see Strunz 1936). Clearly, then, total height is one of the least useful measures of dinosaurian body size.

Even so, there have been a few attempts to estimate the height of dinosaurs from the evidence of their tracks (e.g. Anonymous 1909: 422). A good example is the interpretation by B. Brown (1938: 202) of some exceptionally large footprints from Colorado:

> The tracks [= footprints] measured 34 inches from the heel to the end of the middle toe and 34 inches across the side toes, showing the right and left feet in normal stride where the giant had stepped 15 feet. Compared with the skeleton of *Tyrannosaurus*, 18 feet tall, which in life could step nine feet, these tracks indicate the mystery dinosaur to have been a beast that towered to a height of approximately 35 feet.

A size-estimate such as this must be regarded as little better than a rough guess.

Height at the hip

Height at the hip (*h*) is one of the most useful measures of dinosaur body size, as it meets nearly all of the requirements that were specified earlier: it is readily applicable to dinosaurs of different shapes, including both bipeds and quadrupeds; it is not seriously affected by changing assumptions about the 'normal' posture of a dinosaur; it has

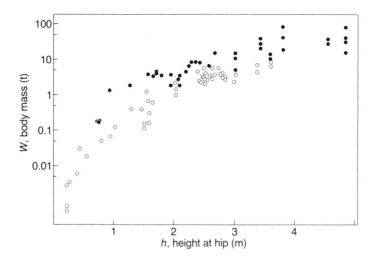

Figure 8.3 Estimates of body mass (*W*, in tonnes, on logarithmic scale) related to height at the hip (*h*, in metres) for various dinosaurs. Sample of bipeds (open circles) includes both theropods and ornithopods; sample of quadrupeds (solid circles) comprises stegosaurs, ankylosaurs, ceratopsians and sauropods. Several different estimates of body mass are shown for some dinosaurs (e.g. four estimates for the sauropod *Brachiosaurus*, at top right). (Data compiled from several sources mentioned in text.)

frequently been used to investigate the gaits and speeds of dinosaurs (see Chapters 9 and 10); it facilitates comparisons among dinosaurs in terms of size, gait and speed; it allows those comparisons to be extended to other animals, such as mammals and ground-dwelling birds; and it is a measure of size that can be derived fairly readily from the dimensions of footprints. In addition, it is reassuring to discover that there may be a reasonably good correlation between height at the hip and body mass (Figure 8.3). The dimension *h* will be discussed more fully at a later point, along with the several methods that have been used to estimate it from the evidence of trackways.

Body mass

Zoologists often select body mass (*W*) as a standard measure of size for living animals. This measure is intuitively attractive because it expresses the 'bulk' of an animal and is convenient because its cube root is a linear dimension readily incorporated in statistical analyses.

Moreover, there are many physiological variables that may be expressed as functions of body mass (Calder 1984; Schmidt-Nielsen 1984). Those well-known relationships between body mass and physiology allowed D.A. Russell (1980) to speculate about the natural history of an 'average' sauropod dinosaur weighing 20 t. According to Russell, such a dinosaur might have produced between 24 and 70 eggs, each weighing about 8.5 kg, and laid in clutches of four. Hatchlings weighing about 7.7 kg would have emerged after an incubation period of about 70 days and would have grown at a rate of about 0.366 kg per day. Eventually, they would have achieved sexual maturity at an age of 45 years and a weight of 6 t. Russell was able to make these predictions, and others, simply by extrapolating from the general relationships of body mass to body functions in living animals.

There is one major shortcoming to comparisons based on W: they take no account of differences in the shape of animals. Body mass is a valuable standard of comparison for animals that are generally similar in shape (e.g. a series of antelopes or a selection of passerine birds) but it is far less useful for comparing animals of radically different shape (e.g. a crocodile, an ostrich and a cow).

The locomotor abilities of living animals have often been judged on the basis of W, and it would be interesting to extend those comparisons to dinosaurs. To achieve that aim it is necessary to obtain reliable estimates of W for dinosaurs that may be known from nothing more than skeletal fragments or footprints. In fact, there have been several attempts to estimate the live body weights of dinosaurs, usually from the evidence of the best-preserved skeletons that are available.

In 1905, W.K. Gregory proposed an elegant method to calculate the live body weights of dinosaurs. This method, which was subsequently refined and extended by E.H. Colbert (1962), is quite straightforward and seems to provide fairly realistic estimates. The first step is to construct a scale model of a dinosaur as it would have appeared in life. It is important that this three-dimensional model should be as accurate as possible. The volume of the model is ascertained from its displacement of water (or sand; Colbert 1962) and is then scaled up to estimate the volume of the living dinosaur. Then, on the assumption that the specific gravity of the dinosaurian body was roughly equal to that of water, it is possible to estimate the weight of the dinosaur.

There are two potential sources of error. The first is the accuracy

Table 8.2 Estimates of live body weights for
dinosaurs (from Colbert 1962)

Dinosaur	Estimated weight (kg)
Sauropod *Brachiosaurus*	78 258
Sauropod *Apatosaurus**	32 418
Sauropod *Apatosaurus**	27 869
Sauropod *Diplodocus*	10 562
Ceratopsian *Triceratops*	8 478
Carnosaur *Tyrannosaurus*	6 895
Iguanodont *Iguanodon*	4 514
Hadrosaur *Corythosaurus*	3 820
Ceratopsian *Styracosaurus*	3 686
Ankylosaur *Palaeoscincus*	3 472
Hadrosaur *Anatosaurus*	3 071
Carnosaur *Allosaurus*	2 087
Stegosaur *Stegosaurus*	1 777
Iguanodont *Camptosaurus*	383
Ceratopsian *Protoceratops*	177

* Two different models were used for *Apatosaurus*
(*Brontosaurus*).

of the scale model; even slight inaccuracies may introduce large
volumetric errors in scaling-up to the size of the living dinosaur.
Second, the specific gravity of the dinosaur body is unknown, though
it is unlikely to have been much greater or less than 1.0. Gregory
(1905) assumed that the specific gravity of *Brontosaurus* (*Apatosaurus*)
was slightly greater than 1.0, so that the animal would have been
sufficiently ballasted to walk on the floors of lakes and watercourses.
On the other hand, Colbert (1962) took the specific gravity to be
about 0.9 (the figure he obtained from a juvenile alligator), so that
dinosaurs would have floated in water. Most recently, R.McN. Alexander (1985) has assumed that dinosaurs had the same density as
water; even so, his estimates of body mass for various dinosaurs
(including *Tyrannosaurus*, *Diplodocus* and *Iguanodon*) are in fair agreement with those obtained earlier by Colbert (Table 8.2).

These volumetric methods have been applied almost exclusively to
large dinosaurs, the smallest being a specimen of *Protoceratops* with a

skeletal hip height of 75 cm (listed in Table 8.2). The body weights of smaller dinosaurs have been estimated by different methods or remain unknown.

A second approach to estimating the weight of a dinosaur depends on the obvious fact that the major limb bones must have been strong enough to have supported the animal. That is, we should expect to find some correlation between the thickness of the leg bones and the weight of the animal. This approach has been followed in studies by R.T. Bakker (1972), D.A. Russell *et al.* (1980), J.F. Anderson *et al.* (1985) and by D.W. Yalden (1984) – though the latter was more interested in the weight of the primitive bird *Archaeopteryx* and mentioned the dinosaur *Compsognathus* only in passing.

The first three of these studies predicted relationships between the weight of a dinosaur and the cross-sectional dimensions of its major limb bones. The value of such predictions is uncertain because they assume that safety factors were about the same for dinosaur leg bones as for the leg bones of modern animals. Alexander (1981) pointed out that different animals, with varied lifestyles, may have quite different safety factors. Some animals may have exceptionally large safety factors, reflected in unusually thick limb bones, whereas other animals of similar weight may have narrower bones and a reduced margin of safety. Consequently, there might be no very straightforward relationship between the weight of an animal and the dimensions of its limb bones. Alexander (1985) checked this point by comparing the estimated weights and bone dimensions of dinosaurs to those of mammals in general, and to those of bovids (cattle) in particular. He found that, on a weight-for-weight basis, dinosaurs had femora shorter than those of mammals in general but longer than those of bovids. The same was true for the humeri of quadrupedal dinosaurs, though the tibiae of dinosaurs in general had roughly the lengths predicted from bovids.

The relationship of body mass to skeletal dimensions is further complicated by the fact that

> . . . the ability to support static loads is not the ultimate demand on the skeleton. . . Indeed, the support of static loads is probably irrelevant, for the stresses on the bones are much greater during locomotion when forces of acceleration and deceleration dominate and far exceed the forces of static loads. (Schmidt-Nielsen 1984: 6)

That is, skeletal dimensions will to some extent reflect the behaviour and gait of an animal, and not merely its body mass.

Yalden's study (1984) examined the relationship of body mass to bone dimensions in a great variety of animals, including ground-dwelling birds, flying birds, quadrupedal and bipedal mammals. From the relationships in cursorial birds he estimated that the small theropod dinosaur *Compsognathus* may have weighed between 532 and 638 g. Previously, J.H. Ostrom (1978) had suspected its weight to be somewhere between 3.0 and 3.5 kg.

In some instances, two or more methods have been used to estimate the weight of a single dinosaur. The enormous sauropod *Brachiosaurus* has attracted particular attention: four different estimates of weight for this dinosaur are 78 t (Colbert 1962), 40 t (Bakker 1975b), 29 t (Anderson *et al.* 1985) and 15 t (Russell *et al.* 1980). Different estimates of W are not so widely divergent for other dinosaurs. For example, Colbert (1962) estimated the weight of *Tyrannosaurus* at slightly less than 7 t, whereas Alexander (1985) used different assumptions to estimate it at about 7.4 t. Despite such differences it is possible to assemble a reasonably consistent set of weight estimates for dinosaurs in general (Figure 8.3). Again, it should be apparent that most efforts have been directed to predicting W in large dinosaurs. Few estimates of W are available for small dinosaurs, and some of them are nothing more than rough guesses.

Most estimates of body weight for dinosaurs have been derived directly or indirectly from skeletal remains, and indeed it might seem impossible to estimate a dinosaur's weight directly from the evidence of its trace fossils. Nevertheless, M.A. Raath (1974) did make one ingenious attempt to predict the live body weight of a prosauropod dinosaur from the evidence of the gastroliths preserved within its ribcage (Plate 2, p. 48, top left). This attempt was based on H.B. Cott's finding (1961) that full-grown Nile crocodiles carry a 'standard load' of pebbles representing about 1 % of total body weight. As the total weight of gastroliths carried by Raath's prosauropod was slightly less than 0.5 kg, the live body weight of the animal was suggested to have been roughly 50 kg. However, this extrapolation from crocodiles to dinosaurs may not be legitimate. Gastroliths may represent a 'standard load' in crocodiles because they function as ballast, but this was not necessarily the case in terrestrial dinosaurs such as prosauropods. Here, the gastroliths may have served to grind up food, and they need not have formed a consistent fraction of total body weight. On the other hand, M. Wade (1989) has suggested that gastroliths 'may have been essential as ballast' in prosauropods – a view that would certainly seem to endorse Raath's approach.

Theoretically, the weight of a dinosaurian track-maker should be related to the surface area and the depth of its footprints. But, in practice, those relationships involve so many imponderables, such as the behaviour of the track-maker and the physical properties of the substrate, that they cannot be used to predict the track-maker's weight. Even so, it is possible to investigate the general relationship between the weight of a dinosaur and the area and depth of its footprints.

Such an investigation was carried out by R. McN. Alexander (1985), who calculated the pressures exerted on the ground by the feet of certain dinosaurs. In the case of a trackway made by a sauropod similar to *Apatosaurus* the total sole area (all four feet) was 1.2 m^2 and the animal's weight was estimated at 34 t (using a slightly modified version of Colbert's (1962) method). From these estimates it was calculated that the average standing pressure was about 280 kPa. Similar calculations for *Tyrannosaurus* and *Iguanodon* indicated standing pressures of about 130 kPa and 120 kPa, respectively. It seems that large bipedal dinosaurs exerted about the same ground pressures as living cattle (100–150 kPa) but that the sauropods exerted rather greater pressures. Alexander inferred that bipedal dinosaurs could probably have walked on soft ground as easily as cattle but that sauropods would have faced a greater risk of becoming mired. Some sauropods and ceratopsians did indeed perish by becoming stuck fast in boggy ground, and even the lightweight bipedal dinosaurs might not have been completely immune from such danger, as mentioned in Chapter 2. Alexander's (1985) estimates of ground forces apply to standing dinosaurs, and they should be approximately doubled to obtain estimates of peak pressures during walking. This implies that a walking sauropod might have generated ground pressures greater than 500 kPa.

It is worth noting that predictions of body mass (and also of speed; see Chapter 10) are not seriously affected by the possibility that the intensity of the gravitational force, g, has decreased through time. The rate of change in G, the universal gravitational constant, has been the subject of much speculation by physicists but has not yet been determined with any precision. Even so, the decline in G is unlikely to introduce major errors into the various calculations and predictions mentioned in this book. For example, if one adopts the working assumption that G might have decreased by 1 part in 10^{10} per year (following Stewart 1970: 414), this would indicate that g was about 1.04 times the current value towards the close of the Triassic period, when dinosaurs first began to appear in abundance. Such a difference would have negligible effects on estimates of weight and speed for even the earliest dinosaurs (see also Economos 1981: 214–15). In fact, some of the latest estimates, using data from Viking landers on Mars (Hellings *et al.* 1983), indicate that the possible range of variation in G is much narrower than was previously suspected. Somewhat surprisingly, the study of dinosaurs and their footprints might help to shed some light on the rate of decline in G. In 1970, A.D. Stewart compiled a list of geological and palaeontological 'gravity sensitive systems' that might be expected to betray any substantial variations in the Earth's gravitational field during the remote past. Among those gravity sensitive systems were included the dimensions and weight-supporting potential of the limb bones in dinosaurs, the ground pressures exerted by the feet of dinosaurs, and variation in the depth of their footprints!

DEFINITION AND CALCULATION OF HEIGHT AT HIP (h)

It was mentioned earlier that the dimension h is probably the most convenient measure of body size in dinosaurs. At this point it will be useful to specify the exact meaning of this dimension and to explain how it may be predicted from the evidence of footprints and trackways.

Definition of the dimension h

In 1976, R. McN. Alexander defined the dimension h as the height of the hip from the ground. Subsequently, h has often been termed 'hip height' or 'height at the hip'. A rather similar dimension is 'skeletal hip height' (Thulborn and Wade 1984), which denotes the height above the ground of the hip joint in a dinosaur skeleton. A third dimension, distinguished as H ('height of the hindlimb, from the summit of the femur to the sole of the foot'), has been defined as the

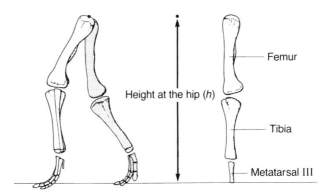

Figure 8.4 For dinosaurs represented by skeletal remains, height at the hip (*h*) may be estimated fairly accurately by summing the lengths of the principal hindlimb bones.

combined lengths of femur, tibia and longest metatarsal, plus an increment of 9 % to account for the ankle bones and for soft tissues at the knee, ankle and sole (Thulborn 1982: 228). The figure of 9 % was derived from the apparently natural spacing of bones in an undisturbed hindlimb skeleton from the ornithopod dinosaur *Thescelosaurus* (Gilmore 1915, fig. 14).

For practical purposes these various dimensions are assumed to be roughly equivalent; they will be greatly different only in the largest dinosaurs. Moreover, all these measurements share the assumption that height at the hip should be measured from about the hip joint to the ground. Consequently, all these dimensions can be (and have been) estimated by summing the lengths of the major hindlimb bones (femur, tibia and longest metatarsal). In other words, the height at the hip will be roughly equal to the length of the hindlimb skeleton, excluding the phalanges of the toes (Figure 8.4).

The dimension *h* may be measured directly on dinosaur skeletons or estimated by various methods from the evidence of trackways. It may also be measured in animals other than dinosaurs, thus allowing comparisons between dinosaurs, humans, ungulates, ground-dwelling birds, and reptiles such as lizards and crocodiles. A corresponding dimension *s*, or height at the shoulder, may be calculated by summing the lengths of the major forelimb bones (humerus, radius and longest metacarpal). The dimension *s* is useful for investigating the locomotion of quadrupedal dinosaurs (see Chapter 9).

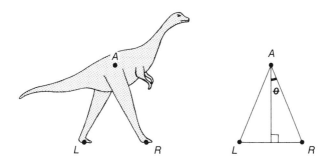

Figure 8.5 Trigonometric method for predicting height at the hip (*h*) in a dinosaurian track-maker. With a measurement of pace length (*L−R*) and an assumption about the angle of gait (2Ø), the height of the hip joint (*A*) above the ground may be estimated by simple trigonometry. (Dinosaur's body outline adapted from a sketch by Avnimelech 1966.)

Methods to predict the dimension h

Geometric methods

In the simplest of these methods pace length, measured directly from the trackway, is taken to represent the base of an isosceles triangle (L–R in Figure 8.5). The two equal sides of the triangle represent the dinosaur's hindlimbs, one extended backwards and the other forwards (A-L and A-R), which enclose the angle of gait (2σ; also known as the 'angle of step' and 'walking angle'). A vertical line bisecting this angle determines the height of the hip joint (A) above the ground. The height of the hip and the length of the extended hindlimb may be calculated by simple trigonometry, providing that one knows the angle of gait. The problem here, of course, is that we do not know this angle in dinosaurs. Exactly the same problem arises in more elaborate geometric methods that take into account lateral displacement of the hindfeet (e.g. Demathieu 1970: 30) or that treat the hindlimb as a pendulum (e.g. Demathieu 1984: 441). In living animals the angle of gait varies considerably; it may be as little as 25° in a horse and as much as 90° in a lizard (Demathieu 1970). Moreover, this angle is known to increase as an animal accelerates through its repertoire of gaits: it may increase from 25° to 55° or even 60° when a horse shifts from a walk into a faster gait (see also Cracraft 1971; Hildebrand 1984).

Such geometric methods require untestable assumptions about the angle of gait in dinosaurs. So, for example, Demathieu (1984) assumed that the angle was about 40° in a variety of dinosaurs, including the ornithopod *Iguanodon*, the carnosaur *Gorgosaurus* (*Albertosaurus*) and the ostrich dinosaur *Struthiomimus*. By contrast, Avnimelech (1966, fig. 1) assumed that the angle of gait was about 33° in a dinosaur resembling *Struthiomimus*. The validity of such assumptions remains unknown, and the results provided by these geometric methods should be treated with appropriate caution.

Occasionally, the length of a track-maker's legs has been estimated from the dimensions of pace or stride, without any explicit assumption about the angle of gait. This approach was adopted in a study of bipedal dinosaur tracks by S.M. Anderson (1939: 363), who assumed that 'a stride of 3 feet would probably be made by legs no more than 5 feet long'. From Anderson's chart of the tracks it is clear that he used the term 'stride' to signify a pace. This means that the angle of gait would have been about 35°. In fact, Anderson's generalization, that pace length (PL) $\simeq 0.6h$, may not be too far from the truth: more recent findings suggest that bipedal dinosaurs tended to walk, on average, with strides about 1.3 times their height at the hip (Thulborn 1984), implying that PL represented about $0.65h$ and that the angle of gait was about 38°. Even so, PL tends to be so variable that it is unlikely to furnish reliable estimates of h (see Figure 8.1).

Morphometric ratios

A good example of this approach is provided by M.A. Avnimelech's (1966) study of dinosaur tracks in the Cretaceous of Israel. Avnimelech observed that in the skeletons of three-toed bipedal dinosaurs the femur, tibia, metatarsus and digit III had lengths in the ratios 11/11/6/5. That is, the length of digit III represented about 5/28 (or 18 %) of a dimension equivalent to h (combined lengths of femur, tibia and metatarsus). As the footprints studied by Avnimelech had digit III about 25 cm long he deduced that the track-maker was about 140 cm high at the hip.

Elsewhere, Alexander (1976) suggested that h might be calculated as approximately four times footprint length in a variety of dinosaurs, both bipeds and quadrupeds, and his suggestion has been rather widely adopted. However, Alexander also mentioned that FL could represent anything from $0.23h$ to $0.28h$ in the dinosaurs that he examined, implying that an estimate of h might be expected to lie anywhere between $3.6FL$ and $4.3FL$. M.G. Lockley *et al.* (1983) noted that *h* was between five and six times the length of the foot in a skeleton of the hadrosaur *Anatosaurus*, and J.L. Sanz *et al.* (1985) found that the ratio *h/FL* ranged from about 3.4 to 5.2 in a variety of iguanodontids. Alexander's observation that $h \simeq 4FL$ is certainly easy to remember and easy to use, but it is best regarded as a rule of thumb rather than an infallible guide.

The use of any such ratio assumes that the size of a dinosaur's foot, and hence of its footprint, bears a constant relationship to *h*. This assumption is likely to be wrong for at least two reasons. First, the ratio *h/FL* probably varies in a systematic fashion between dinosaur taxa. It is unlikely, for example, that the average ratio in coelurosaurs would be identical to that in hadrosaurs. Second, the ratio *h/FL* is certain to have changed throughout the life of any one dinosaur, on account of the allometric growth that prevails in terrestrial vertebrates. Juvenile tetrapods often have relatively large feet, but as they grow to maturity their feet grow less rapidly than other parts of the hindlimb, so that the adults are relatively small-footed by comparison with juveniles. This means that an average ratio *h/FL* might generate overestimates of *h* for juveniles and underestimates for adults, even within a single species.

Nevertheless, the use of ratios is convenient for the preliminary analysis of trackway data, especially in the field. In such circumstances it is preferable to use a separate ratio for each taxonomic group of dinosaurs, as follows:

$$\text{Small theropods } (FL < 25 \text{ cm}): h \simeq 4.5FL \qquad (8.2)$$
$$\text{Large theropods } (FL > 25 \text{ cm}): h \simeq 4.9FL \qquad (8.3)$$
$$\text{Small ornithopods } (FL < 25 \text{ cm}): h \simeq 4.8FL \qquad (8.4)$$
$$\text{Large ornithopods } (FL > 25 \text{ cm}): h \simeq 5.9FL \qquad (8.5)$$
$$\text{Small bipedal dinosaurs in general } (FL < 25 \text{ cm}): h \simeq 4.6FL \qquad (8.6)$$
$$\text{Large bipedal dinosaurs in general } (FL > 25 \text{ cm}): h \simeq 5.7FL \qquad (8.7)$$

These ratios were derived from cursory analysis of osteometric data (Thulborn 1989) and might well be improved by more detailed study. Roughly speaking, they are about as reliable as the ratios used by trackers and hunters, who judge the stature of a track-maker in multiples of footprint size. For instance, the height of a human may be estimated as approximately 6.6 times maximum footprint length (Napier 1972: 119), and in elephants the height at the shoulder is between 2 and 2.5 times the circumference of the forefoot (Baze 1955: 23, on Indian elephants; Sikes 1971: 44, on African elephants).

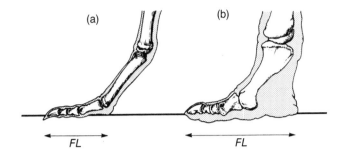

Figure 8.6 For small and lightweight dinosaurs (a) the length of the metatarsus (*MT*) may be roughly equivalent to footprint length (*FL*); metatarsus length may then be used to predict height at the hip (see Figure 8.7). In bigger and heavier dinosaurs (b), such as hadrosaurs and sauropods, footprint length was exaggerated by a substantial cushion of soft tissues in the 'heel' region; consequently, it is more difficult to predict any skeletal dimension on the basis of footprint length. (Adapted from Thulborn and Wade 1984 (a), Langston 1960 (b).)

No ratios are yet available for quadrupedal dinosaurs, though it has been suggested that the commonest of these – the sauropods – resembled the bigger ornithopods in having $h \simeq 5.9FL$ (Thulborn 1989). At first glance this would seem to provide excessively large estimates, compared to the assumption that h was about $4FL$ (Alexander 1976). However, in four sauropod hindlimbs measured by Coombs (1978a: 416) the length of the metatarsus, MT, represents some 7.8 % of skeletal h. Consequently, h might be estimated at about $12.8MT$. The problem now, of course, is to relate the skeletal dimension MT to footprints, bearing in mind that the sauropod foot probably incorporated a substantial wedge of soft tissues in the 'heel' region (Figure 8.6). The assumption that $h \simeq 5.9FL$ would mean that MT represented some 46 % of total FL; by comparison, Alexander's (1976) assumption that $h \simeq 4FL$ implies that MT was about 31 % of FL. At present, it is impossible to choose between these different assumptions without undertaking further research into the anatomy and posture of the sauropod foot.

Most of the ratios mentioned above predict h in multiples of FL. Less commonly, h has been predicted as a multiple of some other footprint dimension. For instance, Lockley *et al.* (1986) estimated h as

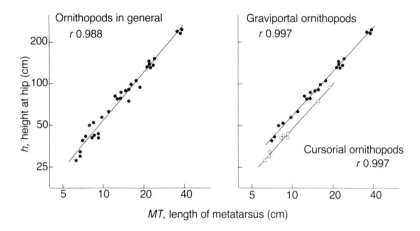

Figure 8.7 Relationship between height at the hip (*h*) and metatarsus length (*MT*) in the skeletons of ornithopod dinosaurs. (a) A heterogeneous sample of 32 ornithopod skeletons; least-squares regression line expresses equation 8.14 in text. (b) The same sample, but with graviportal ornithopods (solid symbols) separated from cursorial ornithopods (open symbols). Graviportal ornithopods were distinguished by having the tibia shorter than the femur whereas cursorial ornithopods have the tibia longer than the femur; *h* represents the sum of the lengths of the femur, tibia and longest metatarsal (as in Figure 8.4). The least-squares regression line for cursorial ornithopods expresses equation 8.12 in text; the regression line for graviportal ornithopods expresses equation 8.13. Equally good correlations exist in other groups of dinosaurs, making it possible to predict *h* on the basis of *MT*. (Adapted from Thulborn and Wade 1984.)

4FW in a series of sauropod trackways where it was difficult to obtain reliable measurements of *FL*.

Allometric equations
In each major group of dinosaurs the dimension *MT*, the length of the metatarsus, is strongly correlated with the sum of the lengths of femur, tibia and metatarsus (Figure 8.7). Consequently, this latter dimension, which is roughly equivalent to *h*, may be predicted with a fair degree of accuracy from an estimate or measurement of *MT*. Such a prediction involves an allometric regression equation of the general form

$$h = aMT^b \qquad (8.8)$$

where the exponent *b* (allometric coefficient) defines the slope of the regression line, and the intercept *a* (termed the 'proportionality coefficient' by Schmidt-Nielsen 1984) expresses size differences between regression lines of similar slope. An equation of this form acknowledges that dinosaurs of different size or age are likely to differ in their hindlimb proportions, whereas the use of a simple ratio MT/h would assume hindlimb proportions to remain constant regardless of size or age.

A series of appropriate predictive equations has been derived from measurements of dinosaur skeletons (Thulborn 1984; Thulborn and Wade 1984) and is listed below. To apply such an equation to a dinosaur trackway it is necessary first of all to obtain an estimate of MT for the track-maker. In most bipedal dinosaurs the dimension MT is roughly equal to the summed lengths of the phalanges in digit III (see illustrations of foot skeletons in Chapter 6). The length of digit III will in many cases correspond rather closely to total footprint length, which may be measured directly on the footprint. In other words, a measurement of FL may be substituted for MT, allowing the use of the following predictive equations:

$$\text{Small theropods } (FL < 25 \text{ cm}): h \simeq 3.06FL^{1.14} \tag{8.9}$$
$$\text{Large theropods } (FL > 25 \text{ cm}): h \simeq 8.60FL^{0.85} \tag{8.10}$$
$$\text{Theropods in general: } h \simeq 3.14FL^{1.14} \tag{8.11}$$
$$\text{Small ornithopods } (FL < 25 \text{ cm}): h \simeq 3.97FL^{1.08} \tag{8.12}$$
$$\text{Large ornithopods } (FL > 25 \text{ cm}): h \simeq 5.06FL^{1.07} \tag{8.13}$$
$$\text{Ornithopods in general: } h \simeq 3.76FL^{1.16} \tag{8.14}$$

In all these equations FL and h are expressed in centimetres. Thus, for a small theropod trackway with FL measured at 10 cm, equation 8.9 predicts h to be $3.06 \times (10)^{1.14}$ cm, or 42.2 cm. It scarcely needs pointing out that FL should be measured as carefully as possible in the first instance.

There are a few complications. First of all, a trackway may be so poorly preserved that its maker can be identified no more precisely than 'a bipedal dinosaur'. In such circumstances it is advisable to select the two most appropriate equations from the list above and to calculate a mean estimate for h. In the case of a 'small bipedal dinosaur', for instance, the mean estimate of h should be obtained with equations 8.9 and 8.12.

Next there is the case of the ornithomimids. These ostrich dinosaurs had unusual foot proportions, with an exceptionally long metatarsus

Table 8.3 Predictions of height at the hip (*h*) in dinosaur track-makers: comparison of various methods

	FL (cm)	Methods* A	B	C	D	E	Mean
Small theropod	10	56	40	50	45	42	47
Medium theropod	30	168	120	150	147	155	148
Large theropod	50	280	200	250	245	239	243
Small ornithopod	10	56	40	50	48	48	48
Medium ornithopod	30	168	120	150	177	194	162
Large ornithopod	50	280	200	250	295	333	272

Note: Track-makers are hypothetical; all estimates are in centimetres.

* A, length digit III = 18 % *h* (after Avnimelech 1966); B, *h* = 4*FL* (after Alexander 1976); C, *h* = 5*FL* (after Lockley *et al.* 1983); D, separate ratio *h*/*FL* for each dinosaur group (equations 8.2–8.7); E, allometric equations (equations 8.9–8.14).

and relatively short toes (see Figure 6.11c,d). On average, it seems that MT was about 1.5 times *FL* (Thulborn and Wade 1984: 441), so that *h* may be predicted as follows:

$$h \simeq 3.49(1.5FL)^{1.02} \tag{8.15}$$

As before, *h* and *FL* are expressed in centimetres.

The preceding equations apply to the relatively common tracks of bipedal dinosaurs. No such equations are available for the semibipedal prosauropods nor for any of the quadrupedal dinosaurs. Genuine prosauropod tracks are rare, as are those of habitually quadrupedal ornithischians (stegosaurs, ankylosaurs and ceratopsians), and until they become better known it will probably suffice to use rough estimates of *h* derived from ratios. The tracks of sauropods are more common, but it is impossible to provide a trustworthy equation predicting *h*. This is because there are few complete examples of the sauropod skeleton and because there is some uncertainty about the anatomy and posture of the sauropod foot (and, hence, about the exact relationship of footprint dimensions to *h*). Pending further research, *h* might best be predicted for sauropod track-makers by means of the ratios mentioned earlier.

In summary, there is little doubt that the linear dimension *h* is the most convenient and useful measure of dinosaurian body size. It is easy to envisage and may be applied to all dinosaurs, regardless of their shape or posture. Moreover, the dimension h seems to be reasonably well correlated with estimates of dinosaurian body mass, and it may be predicted from the evidence of trackways by several methods (Table 8.3).

The various predictions shown in Table 8.3. are in fair agreement, especially for the smaller dinosaurs and for theropods in general. The most divergent predictions are those for big ornithopods. Here it seems that the larger estimates are probably more realistic, since Lockley *et al.* (1983) found that FL could represent as little as one-fifth or one-sixth of skeletal hip height in hadrosaurs. Alexander's (1976) assumption that FL represented about 25 % of skeletal hip height would probably provide underestimates in the case of these larger ornithopods.

Finally, it should be noted that the choice of method to predict h may affect conclusions about the gait and speed of a track-maker. This important point is discussed in the next chapter, where it is suggested that the most enlightening estimates of h are obtained by means of allometric equations.

9

Gaits of dinosaurs

If (as I hope you will) you observe the movements of animals for yourself, you must not be surprised or disappointed if you find exceptions to the rules. . . you must be prepared to find that some animals, and even individual animals, have their own peculiar gaits.

James Gray, *How Animals Move* (1953)

CRITERIA FOR DEFINING AND DESCRIBING GAITS

The various gaits of living animals are distinguished by differences in the sequence and duration of footfalls – by the order in which the limbs move and by the amount of time that each foot remains on the ground. These criteria have been used to describe numerous gaits, and variations of gaits, among living mammals (e.g. Muybridge 1899; Hildebrand 1965; Brown and Yalden 1973), but they cannot be applied very easily to extinct animals such as dinosaurs because there is insufficient evidence about the sequence and duration of footfalls. The pattern of limb movements in living animals may be studied frame by frame in movie films, but this technique is obviously not applicable to dinosaurs. Even so, it is possible to ascertain the general pattern of limb movements in these animals.

Sequence of footfalls

The sequence of footfalls is readily apparent in the trackways of bipedal dinosaurs, where the regular alternation of left and right footprints reveals that these animals walked and ran like humans or ostriches. In some trackways there seems to be an alternation of long and short paces (e.g. Figure 8.1; Plate 11, p. 208, bottom left) which, in one case, led W.P. Coombs (1980b) to suggest that a theropod dinosaur might have been moving with a 'gallop' rhythm. The notion that bipedal dinosaurs might have used a ricochetal (hopping) gait has not been substantiated by the discovery of appropriate trackway patterns.

It is more difficult to determine the sequence of footfalls in the trackways of quadrupedal dinosaurs. However, it is noticeable that the left and right paces, whether for forefeet or for hindfeet, are fairly consistent in their length (e.g. Figures 6.22, 6.41), which implies that the track-maker used a regular or symmetrical gait, with each left foot half a stride out of phase with its right counterpart. Thus, the two hindlegs of a quadrupedal dinosaur would have moved like the legs of a walking human; the same would have been true for the forelegs, so that a quadrupedal dinosaur might be likened to two humans walking in tandem. There remains the question of when the forelimbs were moved in relation to the hindlimbs. This question concerns **relative phase**, which may be defined as the time at which a foot is set down, expressed as a fraction or percentage of the stride that has elapsed since the setting down of an arbitrarily chosen reference foot (Alexander and Jayes 1983; Alexander 1985). In the case of the sauropod trackways shown in Figure 6.15 the relative phases of the four feet may be expressed as follows (as fractions of stride elapsed, and with the left forefoot selected as the reference foot):

Left forefoot: 0   Right forefoot: 0.5
Left hindfoot: p Right hindfoot: $p + 0.5$

Unfortunately, the quantity p, which is some fraction between 0 and 1, cannot be measured on a trackway or on a dinosaur skeleton; at best, it may be estimated very roughly from restorations or scale models of dinosaurs in what are assumed to be their most life-like postures. R.McN. Alexander (1985) attempted to estimate relative phase in a sauropod track-maker, but his results were inconclusive in that he obtained two very different values for p. The first of these estimates ($p = 0.73$) implied that the track-maker walked with a standard alternating gait. The second estimate ($p = 0.94$) implied that the ipsilateral limbs moved almost synchronously in the rolling and shambling gait ('walking pace' or 'amble') that is sometimes used by elephants, bears and long-legged animals such as camels. At present there is insufficient evidence to allow a choice between these two possibilities.

Duration of footfalls

The amount of time that a foot remains on the ground is termed its **duty factor**, which is usually expressed as a fraction or percentage of total stride duration. Duty factor decreases as animals accelerate (Brown and Yalden 1973) and is known to be roughly equal for the forefeet and hindfeet in quadrupedal mammals (Alexander and Jayes 1983). Similar generalizations doubtless applied to dinosaurs, though it is impossible to calculate duty factor directly from the evidence of their trackways.

Relative stride length

Dinosaur tracks provide limited information about the sequence of footfalls and no information at all about duty factor. Consequently, the gaits of dinosaurs must be defined and described by some other method that does not rely on these criteria.

Most terrestrial vertebrates use relatively few gaits. They often have a slow gait (walking), an intermediate gait (such as trotting) and a fast gait (such as galloping). Also, it is well known that animals take relatively short strides while walking and that they take increasingly longer strides as they accelerate through their faster gaits. These generalizations apply to all land vertebrates, including dinosaurs, and they allow the definition of three gaits – walk, trot and run – each characterized by successively longer strides. In practice, the criterion of stride length must be modified to take into account the differing sizes of animals. This scaling is necessary because the strides of a small running animal (e.g. a mouse) may be far shorter, in absolute terms, than the strides of a big walking animal (e.g. an elephant). Stride length scaled in accordance with the size of an animal is termed **relative stride length** (or, sometimes, 'standardized stride length').

Relative stride length is conveniently defined as SL/h, where SL is the length of the animal's stride and h is the animal's height at the hip. Stride length may be measured directly on a dinosaur's trackway (see Figure 4.10), while h may be estimated from the dimensions or spacing of the track-maker's footprints, using methods explained in the previous chapter.

Alexander (1976) found that living terrestrial vertebrates change from a walking gait to a trotting or running gait when SL/h reaches a value of about 2.0, and he suggested that the same was probably true for dinosaurs. In later studies of dinosaur locomotion (Thulborn 1982; Thulborn and Wade 1984) Alexander's observations on the gaits of living vertebrates (1976, 1977) were extended to define three dinosaurian gaits:

Walk: SL/h less than 2.0
Trot: SL/h between 2.0 and 2.9
Run: SL/h greater than 2.9

Relative stride length is not only an indicator of gait. It can also be used to estimate the absolute speed of the track-maker, and it seems to be one of the best available criteria for appraising and comparing the locomotor abilities of dinosaurs. In short, estimates of SL/h can yield valuable insights into the locomotion, behaviour and, perhaps, even the physiology of dinosaurs. The value of such estimates will depend, of course, on their accuracy. In practice, there is no difficulty in measuring SL on a dinosaur's trackway; inaccuracies are more likely to be introduced in attempting to estimate the track-maker's height at the hip (h). The several methods that are available to predict h were discussed in the previous chapter and they will be mentioned again in this chapter because they have an important bearing on the interpretation of relative stride length.

GAITS OF BIPEDAL DINOSAURS

From the evidence of trackways it is possible to glean some interesting details about the movements of the body and the limbs in bipedal dinosaurs. First, it is clear that the tail was carried clear of the ground, since it so rarely left any trace on the substrate. The tail probably extended more or less horizontally, serving to counterbalance those parts of the body in front of the hips. Older restorations of bipedal dinosaurs sometimes portray these animals in an almost kangaroo-like posture, with the backbone inclined at about 45°, but these are probably incorrect. It seems more likely that bipedal dinosaurs would have travelled with the backbone almost horizontal (Galton 1970; Newman 1970).

The trackways of bipedal dinosaurs are often surprisingly narrow. Pace angulation is commonly in the range 160°–170°, and the left and right footprints often seem to fall in a single line rather than in a zig-zag pattern. Evidently, each foot was planted, in turn, under the midline of the body, thus providing the most stable support during the stride (Figure 9.1). This need for stability would have been particularly important in the heavier bipeds such as iguanodonts, hadrosaurs and carnosaurs. For this reason the bigger bipedal dinosaurs may have had a 'knock-kneed' appearance, with their knee

Figure 9.1 The Jurassic carnosaur *Allosaurus* in bipedal walking gait. Note that the feet are placed under the midline of the body, leaving a narrow trackway.

joints closer together than their hip joints. To stand on one foot these animals would merely have leaned sideways, thus bringing the centre of gravity directly over the planted foot. In walking, they would have rolled the body from side to side as each foot, in turn, supported the entire body weight.

During the stride each foot described an outwards sweeping arc, rather than being swung straight ahead. This outswinging of the foot is often betrayed by scrape-marks extending antero-laterally from the footprints (e.g. Plate 7, p. 118, centre right), and to some extent it would have been automatic, as the dinosaur's body would have been rolling over to the opposite side in order to bring the centre of gravity directly over the planted foot. This outwards excursion of the moving foot ensured that it did not collide with the supporting foot, despite the narrowness of the trackway. Towards the end of its step the moving foot was swung inwards again, under the body, and would tend to be planted on the ground with positive (inwards) rotation.

In addition to rolling its body from side to side, a moving bipedal dinosaur probably rotated its hip region round a vertical axis. Similar

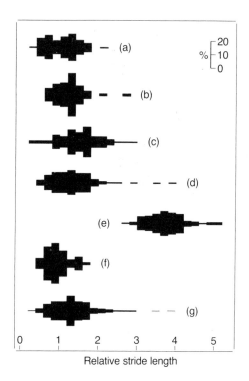

Figure 9.2 Frequency distributions for estimates of relative stride length (SL/h) in various samples of dinosaur trackways. Scale of percentage frequency (top right) applies to all diagrams. (a) Trackways of bipedal dinosaurs (N = 85) from the Lower Cretaceous of Canada (mean SL/h = 1.05). (b) Trackways of bipedal dinosaurs (N = 60) from the Lower Cretaceous of Texas (mean SL/h = 1.22). (c) Trackways of bipedal dinosaurs (N = 120) from the Upper Triassic and Lower Jurassic of southern Africa (mean SL/h = 1.49). (d) Trackways of bipedal dinosaurs (N = 175), Upper Triassic through Lower Cretaceous, worldwide (mean SL/h = 1.30). (e) Trackways of bipedal dinosaurs (N = 92) from the mid-Cretaceous of Queensland, Australia (mean SL/h = 3.73). (f) Trackways of quadrupedal dinosaurs (N = 49), Upper Triassic through Lower Cretaceous, worldwide (mean SL/h = 0.95). (g) Samples (a), (b), (c) and (d) combined (440 trackways of bipedal dinosaurs, mean SL/h = 1.29). For the quadrupedal dinosaurs in sample (f), h was estimated as four times footprint length; in all other samples h was estimated by means of allometric equations listed in Chapter 8. (Based on data from Currie 1983 (a), J.O. Farlow personal communication (b), P. Ellenberger 1972, 1974 (c), Thulborn 1984 (d), Thulborn and Wade 1984 (e). From Thulborn 1989.)

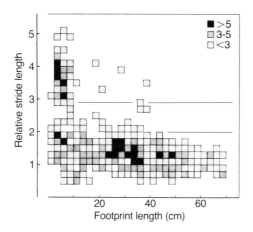

Figure 9.3 Relationship between estimated relative stride length and footprint length for bipedal dinosaurs. Based on data from 532 trackways of bipedal dinosaurs (samples (a) to (e) in Figure 9.2). Horizontal lines at relative stride length 2.0 and 2.9 indicate walk-trot transition and trot-run transition respectively. (From Thulborn 1989.)

rotation is evident in lizards (Brinkman 1981: 93) and also in humans, where it serves to extend step length and ensures a smoother ride by reducing vertical displacements of the pelvis (Saunders *et al.* 1953). Rotation at the hips might have furnished bipedal dinosaurs with a small increase in stride length, a slight improvement in the efficiency of the major hindlimb muscles and a minor advantage in manoeuvrability (Thulborn 1982: 232–3).

The leg movements of theropod dinosaurs were essentially identical to those of existing ratites (Padian and Olsen 1989), and the same pattern of movements probably occurred in ornithopods as well, to judge from the generally similar layout of their trackways. In one respect, however, a moving bipedal dinosaur would have looked quite different from a flightless bird: it possessed a large mobile tail. The tail probably oscillated from side to side, in synchronism with the striding of the hindlimbs (Hamley in press), and it would have performed its role of a dynamic counterbalance by being moved up and down.

Preferred gaits

Measurements gathered from several hundred trackways reveal that

bipedal dinosaurs favoured a walking gait, and that the tracks of trotting or running animals are uncommon (Figure 9.2). This preference
for a walking gait is is evident in dinosaurs of all sizes, even though
those of any particular size may show considerable variation in SL/h
(Figure 9.3). Despite this variation it has been suggested that SL/h
about 1.3 defines an average walking gait for bipedal dinosaurs in
general (Thulborn 1984).

Trackways made by fast-running bipedal dinosaurs are rare. Most of
the world's examples were discovered at a single site in Queensland,
Australia, where it seems that an aggregation of small ornithopods
and coelurosaurs was startled into a stampede by the approach of a
carnosaur (Thulborn and Wade 1979, 1984). Among these small running dinosaurs the mean value for SL/h was about 3.7. Several other
trackways of fast-moving bipedal dinosaurs were reported by J.O.
Farlow (1981) from the Lower Cretaceous of Texas, USA. Overall, it
appears that bipedal dinosaurs normally used a walking gait and that
they resorted to running only on rare occasions or in unusual circumstances.

Few bipedal dinosaurs seem to have used a trotting gait, with SL/h
between 2.0 and 2.9 (Figures 9.2, 9.3). The rarity of tracks made by
trotting animals might be fortuitous or it might be an indication to
the gait preferences of bipedal dinosaurs. If these animals did avoid
the trotting gait they might have done so for any of three reasons –
anatomical, behavioural or physiological.

First, the gait preferences of bipedal dinosaurs might be explained
in terms of locomotor anatomy. C.J. Pennycuick (1975) pointed out
a definite correlation between anatomy and gait preferences in living
African ungulates: those with horizontal backbones (e.g. zebra and
Thomson's gazelle) frequently use a trotting gait whereas those with
sloping backbones (e.g. gnu and giraffe) rarely trot and usually shift
from a walk straight into a running gait (see also Dagg 1973). There
might have been some similar correlation between anatomy and gait
preferences in bipedal dinosaurs, though it seems impossible to identify an appropriate anatomical feature (or set of features). However, it
should be remembered that the dinosaurian 'trot' is an arbitrarily
defined gait that is not marked off from the walk and the run by any
change in the sequence of limb movements. If a bipedal dinosaur
could walk with short strides and run with long strides there seems
no reason why it could not have trotted with strides of intermediate
length. It is difficult to imagine any anatomical feature that would

allow an animal to take short strides and long strides but would prohibit strides of intermediate length.

Secondly, it is possible that the gait preferences of bipedal dinosaurs had a behavioural basis. The trot is a characteristically mammalian gait that is commonly used by herbivores moving in herds or on migrations; it is also used by those mammals that show powers of 'mental abstraction' or 'foresight' - notably, humans and those carnivores that hunt in packs or lie in ambush for unseen prey. In other words, the trot is a gait that seems to be related to distinctively mammalian patterns of behaviour. By contrast, lizards and crocodilians seem to use only slow and fast gaits, without any consistent intermediate gait that could be closely compared to mammalian trotting. The gait preferences of bipedal dinosaurs may indicate that these animals had patterns of behaviour rather more like those of living reptiles than those of living mammals. Nevertheless, there is good fossil evidence to indicate that some types of dinosaurs, at least, were gregarious (see Chapter 11). If dinosaurs did gather in herds, or packs, it is only reasonable to suppose that some individuals would have used a trotting gait from time to time, to prevent their falling behind the group, and that juveniles would have trotted alongside their walking parents. What little is known of dinosaur behaviour does not rule out the possibility that some animals trotted on some occasions.

Finally, it is possible that physiological reasons underlie the gait preferences of bipedal dinosaurs. Some living mammals, including humans and horses, are known to change gaits in such a way as to minimize energy consumption: that is, the energetic cost of a very fast walk may be greater than that of running at the same speed, and mammals change gait to keep energetic cost to a minimum. This was demonstrated very clearly for horses by D.F. Hoyt and C.R. Taylor (1981): they showed that a free-moving horse selected particular speeds and gaits so as to minimize energy consumption, and that there were ranges of speeds (coinciding with gait transitions) that the animal never used for any sustained period. Within each of its gaits the horse selected those speeds that represented energetic optima. Migrating African ungulates seem to select their gaits and speeds in similar fashion (Pennycuick 1975), and Hoyt and Taylor suggested that their findings for horses (1981) might apply to terrestrial animals in general. Similar physiological factors might well have controlled the gait preferences of bipedal dinosaurs: the speeds and gaits defined by SL/h 1.3 (walk) and 3.7 (run) might represent energetic optima whereas the trot (SL/h 2.0–2.9) might have been a transitional gait of high energetic cost (Thulborn 1984). It is reasonable to suppose that dinosaurs, like other animals, should have selected energetically optimal gaits.

Hopping dinosaurs?

Many dinosaurs had forelimbs that were considerably shorter than their hindlimbs. This fact led some nineteenth-century palaeontologists to conceive of dinosaurs as reptilian kangaroos: they envisaged plant-eating dinosaurs resting on hindlegs and tail, in tripod-fashion (see Figure 3.1), and imagined that predatory dinosaurs might have pounced on to their victims by leaping and hopping. Such imaginative ideas gained even wider currency when they appeared in popular works of fiction:

> There was movement among the bushes at the far end of the clearing which I had just traversed. A great dark shadow disengaged itself and hopped out into the clear moonlight. I say 'hopped' advisedly, for the beast moved like a kangaroo, springing along in an erect position upon its powerful hindlegs, while its front ones were held bent in front of it.
>   (Arthur Conan Doyle, The Lost World, 1912)

In scientific literature the notion of hopping dinosaurs was largely dispelled by the start of the 20th century, by which time it was realized that 'Ornithoidichnites' and similar markings were the tracks of dinosaurs and not of antediluvian birds. The tracks showed conclusively that bipedal dinosaurs moved their hindfeet alternately, like humans and ostriches. Nevertheless, the notion of hopping dinosaurs is far from extinct.

In 1972, for example, M.A. Raath described three small footprints, from the early Jurassic rocks of Rhodesia (now Zimbabwe), which were clearly those of a small theropod dinosaur, possibly the coelurosaur Syntarsus. Two of the prints were preserved side by side, leading Raath to suspect that the track-maker might have been hopping rather than walking. However, they might equally well have been produced by a dinosaur that was squatting or standing still, and the only way to settle the question, as Raath explained, would be to discover a trackway rather than an isolated pair of footprints.

The notion of hopping dinosaurs was revived most recently by

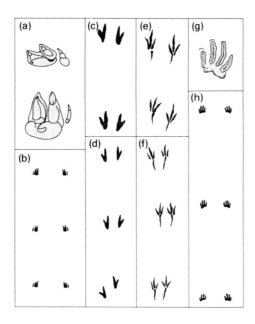

Figure 9.4 Tracks attributed to hopping animals. (a) Two right pes prints of *Saltosauropus latus*, from the Upper Jurassic of France; the upper example is a foreshortened footprint, about 18 cm wide; the lower example is a more complete print, about 20 cm wide. (b) An average trackway of *Saltosauropus* (composite). (c), (d) Tracks of wallabies, both with individual footprints about 11 cm long. (e) Track of hopping sparrow, in snow; footprints about 3.6 cm long. (f) Track of unidentified hopping bird, in dry sand; footprints about 4.5 cm long. (g) *Molapopentapodiscus supersaltator*, a right pes print from the Lower Jurassic of southern Africa; about 2.5 cm long. (h) An average trackway of *Molapopentapodiscus* (composite) – apparently made by a hopping or bounding animal, but probably not by a dinosaur. (Adapted from Bernier *et al.* 1984 (a,b), Thulborn 1989 (c–f), P. Ellenberger 1972 (g,h).)

P. Bernier and his colleagues (1984), to account for some unusual trackways in the Upper Jurassic of southeastern France (Figure 9.4a,b). Bernier (1984) has also published an appropriate restoration of such a hopping dinosaur (Figure 9.5a), while G. Demathieu (1984) has pondered the locomotor mechanics of such a creature and has even attempted to estimate the speed at which it moved. The trackways, named *Saltosauropus latus*, were attributed to theropod dinosaurs of

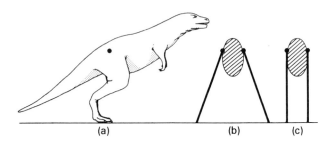

Figure 9.5 Conjectures about the posture of a hopping dinosaur, based on *Saltosauropus latus*, from the Upper Jurassic of France. (a) Outline restoration of the *Saltosauropus* track-maker envisaged by P. Bernier (1984); the solid circle indicates the approximate location of the hip joint. (b) Diagrammatic cross-section through the hip region of such a track-maker, revealing the splayed-out attitude of the hindlimbs; proportions of this diagram are based on mean trackway data for *Saltosauropus*. (c) Schematic cross-section through the hip region of a dinosaur, to show the erect posture of the hindlimbs. (Adapted from Thulborn 1989.)

moderate size, either large coelurosaurs or small carnosaurs. Their individual footprints are short and broad, with indications of three stubby digits terminating in stout claws, and their arrangement in left-right pairs certainly implies that they might have been produced by a hopping biped (Figure 9.4b).

However, the *Saltosauropus* tracks differ from those of modern hopping animals in several respects (compare Figure 9.4c–f). First, the footprints are broader than long, whereas those of modern hoppers are longer than broad. The relatively long foot is important to a hopper: it provides a stable base for take-off and, especially, for landing, and it affords improved leverage for the hindlimb muscles. Secondly, the *Saltosauropus* footprints show positive (inwards) rotation or have digital imprints that are distinctly curved (convex to the exterior). These features imply that the track-maker's foot applied lateral, as well as downwards and backwards, forces to the substrate. By comparison, the footprints of modern hoppers have rather straight digital imprints and show little or no rotation. Evidently, the feet of modern hopping animals thrust straight down and backwards, with little or no force being exerted sideways. A third and most important difference concerns the spacing of the footprints. Modern animals hop with the left and right feet close together whereas the *Saltosauropus* track-maker had its feet widely separated. In *Saltosauropus* the interpes distance (between left and right footprints) ranges between 40 % and 100 % of stride length; in the tracks of wallabies and hopping birds the interpes distance is less than 20 % of stride length. If the body proportions of the *Saltosauropus* track-maker resembled those of any known dinosaur then the animal must have hopped with its legs splayed out to an extent that seems remarkably inefficient (Figure 9.4b). Here it is worth recalling that dinosaurs are characterized by their *erect* posture, with movements of the hindlimb largely restricted to a parasagittal plane (see Figures 7.1, 9.1). The same is true for modern hoppers, such as birds and kangaroos, where the major limb joints are adapted to resist any sideways flexures. Such comparisons cast serious doubt on the idea that *Saltosauropus* is the track of a hopping dinosaur. Elsewhere, it has been suggested that *Saltosauropus* might be the track of a broad-bodied animal, possibly a turtle that was swimming in shallow water and touching down almost synchronously with its clawed flippers (Thulborn 1989). Turtle tracks are known from the same sediments, and some of them do bear a close resemblance to *Saltosauropus* (e.g. Bernier *et al.* 1982, pl. 3).

It remains possible that some bipedal dinosaurs did use a hopping gait (Emerson 1985: 58), but there is no very convincing evidence that they actually did so. Even the best of the available trackways, such as *Saltosauropus*, must be regarded as dubious evidence for hopping dinosaurs.

Coelurosaurs on crutches?

Some curious little trackways in the late Jurassic limestones of Bavaria were interpreted by M. Wilfarth (1937) as those of coelurosaurs that had used a gait somewhat similar to the so-called 'punting' or 'crawling' of kangaroos (Frith and Calaby 1969; Windsor and Dagg 1971). In this unusual gait a slow-moving kangaroo uses its tail in prop-like fashion as a fifth appendage. Supporting itself on its forelimbs and tail, the animal swings both hindfeet forwards simultaneously and plants them in front of the forefeet. Then, while resting on its hind-feet, the kangaroo reaches forwards with both forefeet and sets them down again to restart the cycle. Wilfarths' vision of a coelurosaur performing similar manoeuvres (Figure 9.6) was splendidly described by K. Caster (1944: 77):

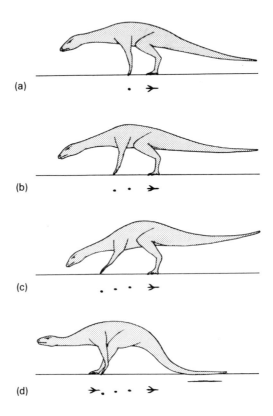

Figure 9.6 Gait envisaged by M. Wilfarth (1937) for a late Jurassic coelurosaur. Only one side of the trackway is shown below each diagram. (a) Supported on its hindfeet, the coelurosaur reaches a short way forwards to plant its forefeet. (b) The animal reaches further forwards, leaving a second manus print. (c) Reaching forwards still further, the animal plants its forefeet for the third time. (d) Supported on forefeet and tail, the track-maker swings both hindlegs forwards to restart the cycle. These conjectures were based on the late Jurassic *Kouphichnium* and similar tracks, which were almost certainly made by limulids and not by dinosaurs. (Elaborated from sketches of Wilfarth 1937.)

> He [Wilfarth] imagined that the little dinosaur hobbled along by 'trial and error'; sitting on its haunches . . . it reached forward . . . with its short dangling fore-feet and pressed its longest digits in the mud; this was apparently not far enough; it then lifted its arms, and spreading them a little further

apart, implanted them further forward . . .; still not far enough; for a third time, it reached forward, with wider spreading arms, to implant its middle digits . . .; only after this preliminary exploration did the dinosaur move forward by the wholly ingenious device of using its implanted fore-feet and longest fingers in the manner of 'crutches' by which to swing its hind-feet 'through its arm-pits,' thus to stand at last at the new position . . .; whereupon, the 'trial and error' skirmish was renewed before the next 'jump' was made!

The tentative explorations with the forelimbs would account for the fact that each pair of pes prints was preceded by several pairs of manus prints. Wilfarth's notion of coelurosaurs dragging themselves across the Jurassic mudflats was only one of many imaginative interpretations. Similar tracks from the Devonian of Pennsylvania were named *Paramphibius* by B. Willard (1935), who believed them to represent a previously unknown order of amphibians, and examples in Triassic sediments as far afield as New Jersey (Abel 1926: 35–52) and Greenland (Nielsen 1949) were, on occasion, thought to be the tracks of primitive birds, pterodactyls or leaping mammals. By comparison, their true origin, as documented by Caster (1944, 1957), is something of an anticlimax: they are merely the tracks of limulids, or horseshoe crabs (see Figure 7.5).

The idea of coelurosaurs hobbling along on hands and feet, mistaken though it is, does recall the fact that tracks of injured or diseased dinosaurs occur in the fossil record. In most instances the disabilities that are apparent, such as the loss or malformation of a toe on one foot, do not seem to have impaired the locomotor abilities of the animals (e.g. Figure 5.8). However, one trackway described by S. Ishigaki (1986) from the Jurassic of Morocco might well have been made by a limping coelurosaur: the animal's right foot had its outer two digits drawn together and consistently took shorter paces than the left foot.

Running dinosaurs

In their fastest gaits animals such as cheetahs, greyhounds and horses extend stride length by virtue of special anatomical and behavioural adaptations, many of which have also been identified in dinosaurs (Coombs 1978a; Thulborn 1982). One important stride-lengthening technique is the interpolation of an unsupported or suspended phase,

Figure 9.7 The ornithomimid *Struthiomimus* in running gait. With a live body weight estimated at about 150 kg, this long-legged theropod was certainly capable of extending its stride length by means of an unsupported phase.

when the animal lifts all feet clear of the ground and 'floats' through the air under its own inertia. The ability to use a fast running gait with unsupported phase is correlated with an animal's body mass. For example, adult giraffes, weighing about 1–1.2 t, are too heavy to use an unsupported phase while galloping, though juveniles are light enough to do so and can easily outrun their parents (Dagg and Foster 1976). Heavier mammals such as rhino (3–4 t) and elephant (4–6 t) do not attain a true galloping gait, with a long unsupported phase, but reach their maximum speeds in a fast trotting gait. Beyond the critical weight limit an unsupported phase may place dangerously high stresses on joints and limb bones when an animal lands at the end of its stride. Moreover, it requires such an input of vertical thrust that it may be energetically uneconomical for very heavy animals.

Similar constraints probably applied to dinosaurs. Small running dinosaurs clearly did exploit an unsupported phase (as in Figure 9.7), for their trackways may indicate relative stride lengths well over 3.0 and perhaps as high as 5.0. By comparison, bipedal dinosaurs with body mass greater than 1–2 t were probably too heavy to have used an unsupported interval. Such heavyweight dinosaurs may have been incapable of running and are unlikely to have attained a relative stride length greater than about 2.9 (the trot–gallop transition). This conclusion seems to be borne out by analysis of dinosaur trackways (see

Figure 9.3): none of the largest bipedal dinosaurs (with $FL > 40$ cm) is definitely known to have used a gait faster than a walk, and all the trotting or running animals were of small or moderate size ($FL < 40$ cm). Overall, it seems that large bipedal dinosaurs, including many carnosaurs, iguanodonts and hadrosaurs, were restricted to walking or trotting gaits whereas small dinosaurs could, and did, make use of a fast running gait. In other words, there is likely to have been a negative correlation between the size of a dinosaur (expressed by FL, h or W) and its maximum attainable value for SL/h. The maximum limit of SL/h for small and fast-running dinosaurs may have been about 5.0, whereas the maximum limit for the very biggest ornithopods and theropods may have been no greater than 2.0. There is no certain evidence that any bipedal dinosaurs achieved SL/h much greater than 5.0, and it is difficult to imagine that they could sustain a running gait by doing so (Thulborn and Wade 1984: 454).

Theropods versus ornithopods

The predatory theropods and plant-eating ornithopods were not very dissimilar in body build, and it seems likely that these two groups of bipedal dinosaurs would have shared roughly similar gaits. By comparing the gaits of theropods and ornithopods it is possible to demonstrate certain points that are important for the study of dinosaur gaits in general.

Figure 9.8 shows three different analyses of data from the Lark Quarry site, Queensland, Australia. The trackways selected for these comparisons were made by 34 small theropods and 35 small ornithopods, all with FL less than 7 cm. In the first analysis (Figure 9.8a) the size (h) of all the track-makers was estimated by using a single ratio ($h = 4FL$). Consequently, it appears that the theropods were roughly the same size as the ornithopods, though they seem to have used a much slower gait (mean SL/h 3.5 as opposed to 5.3). This first analysis might well lead one to assume that small ornithopods moved faster than theropods of similar size. However, the use of a single ratio to estimate h in all track-makers has ignored the fact that theropods differed from ornithopods in their body proportions. This difference is acknowledged in the second analysis (Figure 9.8b), where separate ratios were used to estimate h in each group (4.5FL in theropods and 4.8FL in ornithopods). The ornithopods now seem, on average, to have been slightly larger than the theropods, and there is a less

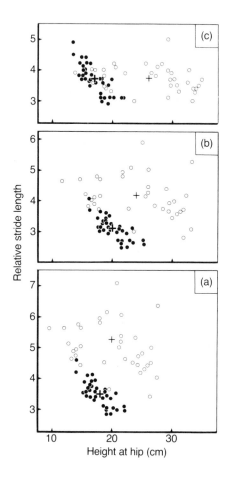

Figure 9.8 Relationship of gait (relative stride length, SL/h) to body size (height at hip, h) for small bipedal dinosaurs at the Lark Quarry trackway site, mid-Cretaceous of Queensland, Australia. Solid symbols indicate data for small theropods ($N = 34$, with the means indicated by heavy cross); open symbols indicate data for small ornithopods ($N = 35$, with the means indicated by light cross). (a) With h estimated as $4FL$ in all cases. (b) With h estimated as $4.5FL$ in theropods and as $4.8FL$ in ornithopods (equations 8.2 and 8.4 in text). (c) With h estimated by means of allometric equations (listed in text as equations 8.9 and 8.12). Analysis (a) finds the two groups to be roughly similar in size (h) but very different in their gaits (SL/h); analysis (c) finds the reverse. (From Thulborn 1989.)

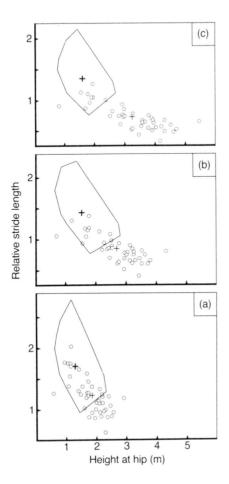

Figure 9.9 Relationship of gait (relative stride length, *SL/h*) to body size (height at hip, *h*) for bipedal dinosaurs at the Peace River trackway sites, Lower Cretaceous of British Columbia, Canada. Open symbols indicate data for ornithopods (*N* = 41, with the means indicated by light cross); extent of data for theropods is indicated by minimum convex polygon for sake of clarity (*N* = 44, with means indicated by heavy cross). (a) With *h* estimated as 4FL in all cases. (b) With *h* estimated as 4.9FL in theropods and as 5.9FL in ornithopods (equations 8.3 and 8.5 in text). (c) With *h* estimated by means of allometric equations (listed in text as equations 8.10 and 8.13). Analysis (c) reveals considerable difference in size between theropods and ornithopods. (From Thulborn 1989, based on data from Currie 1983.)

obvious difference in the gaits of the two groups (mean SL/h 3.1 in theropods as opposed to 4.2 in ornithopods). In the third analysis (Figure 9.8c) h was estimated by means of allometric equations; it is now apparent that the theropods were considerably smaller than most of the ornithopods and that these two groups shared an almost identical gait (mean SL/h 3.7 in both cases).

Figure 9.9 shows three similar analyses of trackway data from the Lower Cretaceous of Canada (44 theropod trackways versus 41 ornithopod trackways). In all three cases it seems that the ornithopods used a slower gait than the theropods, as was noted in P.J. Currie's (1983) original study of these trackways. However, the reason for this difference in gaits only becomes apparent when allometric equations are used to estimate the size (h) of the track-makers: the ornithopods used a slower gait because they were, on the whole, much bigger than the theropods (Figure 9.9c). This underlying difference in the size of ornithopods and theropods is less obvious when h is estimated by means of ratios (Figures 9.9a,b).

The foregoing comparisons demonstrate some important guidelines for the study of dinosaur gaits in general. First, it is essential that the size of the track-makers should be estimated as accurately as possible. Secondly, the most enlightening measure of a track-maker's size is h rather than FL or a multiple of FL. Thirdly, and most important, the choice of method to predict h will affect the outcome in terms of SL/h. That is, the blanket application of a single predictive equation, such as $h = 4FL$, may generate artificial differences in SL/h, simply because any set of track-makers is unlikely to have been uniform in the ratio h/FL. These misleading effects of the geometric dissimilarities among the track-makers may be suppressed to some extent by using a separate ratio to predict h in each group of dinosaurs (Figures 9.8b, 9.9b). However, those effects are best mitigated by using allometric equations to predict h (e.g. Figure 9.8c). It cannot be claimed that such equations give perfectly accurate results, but they do at least acknowledge the prevalence of allometry within and among dinosaur taxa.

GAITS OF QUADRUPEDAL DINOSAURS

The gaits of quadrupedal dinosaurs cannot be investigated in great detail because the trackways of these animals are not so well known

Figure 9.10 The Jurassic sauropod *Camarasaurus*. This dinosaur was probably restricted to walking gait on account of its great body mass, estimated at about 30 tonnes. Note that the forefeet have an erect posture whereas the multi-clawed hindfeet are relatively flat and broad-spreading.

as those of bipedal dinosaurs. Nevertheless, it is possible to reach some general and rather tentative conclusions.

Most quadrupedal dinosaurs seem to have carried the tail clear of the ground, for there are relatively few trackways that include a tail-drag. In general, the trackway tends to be broader than that of a bipedal dinosaur, though in a few instances, as in sauropods (Figure 9.10), it may be surprisingly narrow. Pace angulation is often in the range 110°–150° (for both manus and pes), so that the footprints form a zig-zag pattern rather than a linear series. In tracks of habitual quadrupeds the manus prints are fairly large, by comparison with the pes prints, and are quite regularly spaced (see Figure 6.15). This consistent spacing of large manus prints implies, of course, that the forefeet had an important major role in supporting the track-maker. By comparison, the manus prints of semibipedal dinosaurs tend to be smaller and more erratic in their placement (see Figure 9.14).

In some trackways the manus print lies directly ahead of the pes print (e.g. Figure 6.15a), probably indicating that the forelimbs were erect and rather pillar-like. In other cases the manus print is placed

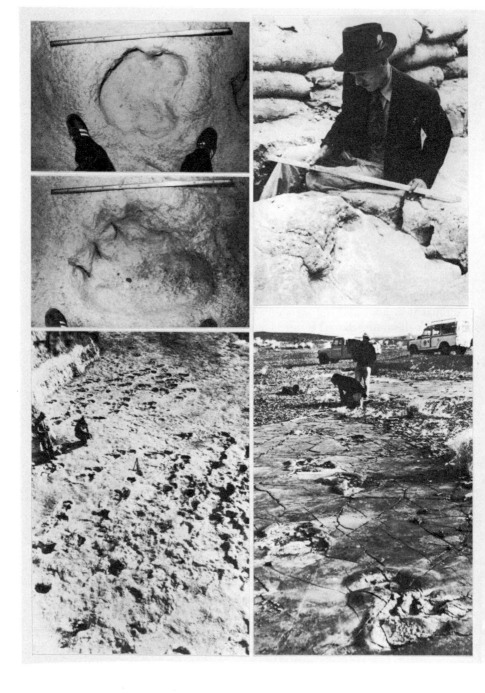

near the outer edge of the pes print (e.g. Figure 6.44a), implying either that the dinosaur swung its forequarters from side to side as it walked along or, more probably, that the forelimbs sprawled out sideways to some extent. Frequently, the manus and pes prints point directly forwards or show negative (outwards) rotation; this is quite unlike the situation in the tracks of bipeds, where the footprints usually show positive (inwards) rotation. This difference in footprint orientation might indicate that quadrupeds tended to swing the foot directly forwards, rather than in the outwards-curving arc described by the foot of bipeds. Consequently, each foot would be lifted from the ground for the shortest possible time, which would be consistent with the fact that most quadrupedal dinosaurs were very heavy animals. In any case there was no need for a quadrupedal dinosaur to swing its foot outwards, or to roll its body to the opposite side for support, as there would have been at least two, and possibly three, other feet on the ground at the same time. Perhaps the closest match to the pattern of limb movements in quadrupedal dinosaurs might be seen today in elephants.

Quadrupedal dinosaurs favoured a walking gait, as did bipedal dinosaurs, but there is no evidence that they used a trotting or running gait. This does not necessarily mean that these animals never ran, but they may have done so only on rare occasions. Figure 9.2 shows that bipedal dinosaurs walked with relatively long strides (mean SL/h 1.29) whereas quadrupeds walked with relatively short strides (mean SL/h 0.95). This difference implies that quadrupedal dinosaurs used a slow walking gait, a conclusion that is supported by some independent observations.

Plate 12 Examples of sauropod tracks. Top left and Centre left: *Brontopodus birdi*, manus and pes prints of a sauropod, from the Lower Cretaceous of Texas, USA; scale bar is 1 m long, and in both cases the front margin of the print is to left. Manus print shows typical 'horseshoe' shape, without obvious indications of individual digits. Top right: Roland T. Bird, the discoverer of sauropod tracks, measuring a hindfoot print in the bed of the Paluxy River, Texas, in 1940. Bottom left: The Davenport Ranch site in Texas, USA, with tracks made by a herd of at least 23 Cretaceous sauropods. The animals travelled from the lower left to the top right. Bottom right: Superbly preserved trackway of a solitary sauropod, in the Jurassic of Niger. This trackway was followed for more than 96 m. (Photograph courtesy of CNRS.)

First, there is the simple observation that most quadrupedal dinosaurs were big animals. With body weights estimated at several tonnes, and often more, it is barely conceivable that the largest quadrupedal dinosaurs could have lengthened their stride by introducing an unsupported interval. These animals must have been restricted to the slower gaits by virtue of their immense weight. The limitations imposed by great body mass were well summarized by J.D. Currey (1977: 156):

> . . . the fact remains that one cannot design very large quad-rupeds without resorting to graviportal legs, and graviportal legs would make fast locomotion very expensive in muscle and in stresses at the joint, and in general would render it imprudent.

In reality, the word 'imprudent' means potentially disastrous, because the stresses generated during locomotion can fracture the limb bones of even small and medium-sized animals; as Currey has noted (1977: 155):

> Fracture is by no means a rare event in wild populations [of mammals]. For instance Buikstra (1975) shows that 40 % of the adults of a complete population of macaques (*Macaca mulatta*) showed healed fractures of the long bones or clavicles. Presumably there were many individuals whose fractures did not heal and who therefore died. Racehorses often break their legs while galloping, even without stumbling, and many ballet dancers near retiring age show limbs with a number of healed fatigue fractures (Schneider *et al.* 1974).

The risk of shattering the limb bones would have been even greater in big quadrupedal dinosaurs attempting to use any gait faster than a slow walk (see Rothschild 1988).

The fact that quadrupedal dinosaurs had forelimbs and hindlimbs of unequal length might also have prohibited their using any gait faster than a walk. Most of these dinosaurs had forelimbs shorter than their hindlimbs, and sometimes markedly so, as in *Stegosaurus* (Figure 9.11). In a walking quadrupedal dinosaur the forelimbs and hindlimbs, though of different sizes, would have taken strides that were equal in number and equal in length. So, at any given speed, relative stride length would have been greater for the (short) forelimb than for the (long) hindlimb. The magnitude of this difference in relative stride length was governed by the ratio of shoulder height (s) to hip height

Figure 9.11 The ornithischian dinosaur *Stegosaurus* was probably restricted to a slow walking gait because its long hindlegs would tend to outstride its short forelegs. Note that the forefeet are nearly as big as the hindfeet, as in other habitual quadrupeds.

(*h*) and by the absolute length of the stride (*SL*). In the case of the sauropod dinosaur *Diplodocus* the length of the forelimb is about 69 % of the length of the hindlimb; if the hindlimb reached a relative stride length of 2.0 (absolute stride length of about 6 m), the forelimb would have had a relative stride length of about 2.9. That is, the forelegs of *Diplodocus* would have been trotting while the animal's hindlegs were merely walking. The case of *Stegosaurus* is even more remarkable. In a skeleton of this dinosaur the forelimb is less than half the length of the hindlimb; if *Stegosaurus* took strides 3 m long its hindlimbs would have been walking (*SL/h* 1.5) while its forelimbs would have been running (*SL/s* 3.3).

Such differences between the gaits of forelimbs and hindlimbs would have become more pronounced as quadrupedal dinosaurs accelerated and took increasingly longer strides. At low speeds, and with short strides, quadrupedal dinosaurs would have found it relatively easy to co-ordinate the striding of forelimbs and hindlimbs. But at higher speeds, and with longer strides, it would have become increasingly difficult to maintain such co-ordination. In other words, quadrupedal

Figure 9.12 Restoration of the primitive ceratopsian dinosaur *Protoceratops*, about 2 m long, from the Upper Cretaceous of Mongolia. The animal is shown in a fast-running bipedal gait resembling that used by some existing lizards. (Adapted from Bakker 1968.)

dinosaurs might have been restricted to short strides, and hence to slow gaits, for simple mechanical reasons: if they attempted to accelerate by taking longer strides their hindlegs might have started to outstride their shorter forelegs. (The sauropod *Brachiosaurus* was exceptional in having forelegs longer than hindlegs; the same constraints apply, but in reverse.)

Many fast-moving mammals overcome this difficulty by swinging the scapula backwards and forwards like a pendulum, thus extending the reach of the forelimb. This requires, of course, that the left and right halves of the shoulder skeleton should be separated, because the left and right forelimbs will be moving independently. There is no certain evidence that quadrupedal dinosaurs did likewise: left and right halves of the shoulder skeleton seem to have been firmly joined in the midline, so that the scapula probably remained immobile. Lizards resolve the same problem of limb disparity in still another way. They introduce an unsupported interval for the forelimbs alone, so that their fast running gait includes short episodes of bipedalism and produces very distinctive trackways (Snyder 1952, figs 1–8). In some cases fast-running lizards are able to withdraw their forelimbs from the ground and become fully bipedal, until they start to decelerate and fall back on to all fours. R.T. Bakker has suggested (1968) that small ceratopsians such as *Protoceratops* might have behaved in similar fashion (Figure 9.12), though there is no trackway evidence to substantiate this possibility.

Quadrupedal dinosaurs that were unable to run, or even to trot very fast, might have been easy prey for bipedal predators. Their only defence – aside from behavioural responses such as herding – might have lain in increasing body size, with or without the added protection of horns or bony spikes and plates. Such consequences might possibly explain the rarity of small quadrupedal dinosaurs in general and the almost total lack of quadrupedal predators among the dinosaurs.

Sauropods as knuckle-walkers?

The structure of the sauropod manus is not well known, though in all cases the digits were rather short and there was usually a prominent claw on the pollex (digit I). Nevertheless it is evident that the manus had an unusual posture. It seems to have been steeply digitigrade, or perhaps nearly unguligrade, in contrast to the pes, which was semi-plantigrade like the feet of elephants and mammoths (compare Figures 6.13 and 6.14). W. Langston has described the 'stilt-like' sauropod manus as 'a semitubular column in which the digits, of subequal length, were completely joined and enclosed' (1974: 96). The most perplexing fact is that prints of the forefeet rarely show any trace of the large pollex claw. Instead, the manus prints are horseshoe-shaped or semicircular impressions that lack any definite indications of separate digits (see Figures 6.15, 6.16; Plate 12, p. 278, top left). It was suggested by G. de Beaumont and G. Demathieu (1980) that the pollex claw failed to leave an imprint because sauropods were knuckle-walkers – that they supported themselves on the distal ends of the metacarpals, with the digits curled up behind. However, the sauropod pollex comprised only two phalanges and might not have been sufficiently flexible to allow such knuckle-walking. Moreover, the distal surface of the first phalanx faced dorso-medially, implying that the ungual phalanx might have been retracted by hyperextension. It seems possible that the pollex claw was normally raised clear of the ground, much as it was in the related prosauropods (see Figure 6.20). Whatever its orientation, this claw seems to have played no major role in sauropod locomotion and presumably had some other function. It could scarcely have been an easily manouevred weapon but it might have served as a hook for pulling down vegetation. A rather similar idea was advanced by R.T. Bakker (1968: 19), who suggested that sauropods used their clawed feet to 'tear into the edible portions of tree trunks'.

Figure 9.13 The prosauropod *Plateosaurus*, perhaps equally adept at walking on all fours and on hindlegs alone. In quadrupedal gait the shoulder joint was well below the level of the hip joint, and the prominent claw on the forefoot was retracted (see Figure 6.20).

GAITS OF SEMIBIPEDAL DINOSAURS

Some dinosaurs, such as the ornithopods and prosauropods, were able to switch between bipedal and quadrupedal gaits according to circumstances (Figure 9.13). The shift from quadrupedal to bipedal gait entailed a number of correlated changes in the trackway pattern left by the hindfeet: an increase in stride length and pace angulation, and a concomitant decrease in trackway width (Figure 9.14). The change in stride length implies that these dinosaurs normally used the bipedal gait for relatively rapid progression but came down on all fours when moving more slowly.

In the quadrupedal gait the hindfeet took fairly long strides, leaving a broad trackway with reduced pace angulation and consistent pace length. Such consistency in pace length probably indicates that the hindlimbs supported much of the track-maker's body mass. By comparison, the manus prints sometimes have an erratic distribution, implying that the forelimbs had a relatively minor role in supporting the track-maker. It is also noticeable that the forefeet often left a wider trackway than the hindfeet (e.g. Figures 6.22, 9.14) and that pace length for the forefeet exceeds pace length for the hindfeet. As the forelimbs were shorter than the hindlimbs, this remarkably wide spacing of left and right forefeet implies that the shoulder region was depressed below the level of hips. It is likely that head, neck and shoulders were carried close to the ground and that the elbows were

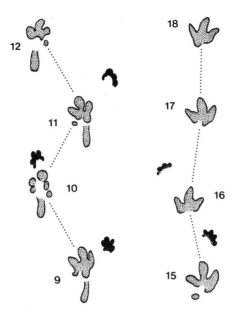

Figure 9.14 *Moyenisauropus natator*, an ornithopod trackway from the Lower Jurassic of Lesotho. The diagram shows two portions of a single trackway, with the pes prints identified by number; the manus prints are indicated by solid shading. Note the characteristic changes in the trackway pattern as the animal switched from a quadrupedal gait (footprints 9–12) to a fully bipedal gait (footprints 16–18): the pes prints lose impressions of the metapodium and of the hallux (digit I), trackway width is reduced (reflected by increased pace angulation), stride length is increased, and positive (inwards) rotation of the pes prints becomes more pronounced. The final stride is 1.08 m long. (Adapted from P. Ellenberger 1974.)

stuck out sideways to some extent. This low-slung and somewhat sprawling posture, which was probably useful in foraging, drinking and feeding, is known to have been adopted by prosauropods and by ornithopods (Thulborn 1989; Heilmann 1927, fig. 108).

In the case of ornithopods the manus prints are considerably narrower than the pes prints, seeming to confirm that most of the track-maker's body weight was supported on the hindlimbs (Figure 9.15). Consequently, the ornithopods might easily have lifted one or both of the forefeet from the ground without difficulty. Prosauropod trackways differ in having manus prints that are nearly as broad as

Figure 9.15 Quadrupedal adult and bipedal juvenile of the semibipedal ornithopod *Iguanodon*. Note that the forefoot is relatively small (by comparison with the hindfoot) and has an erect posture, almost as if walking on finger-tips (see Figure 6.30). The juvenile has to use its fast (bipedal) gait to match the speed of an adult that is travelling in its slow (quadrupedal) gait.

the pes prints (Plate 13, p. 306, top right), thus resembling the tracks of fully quadrupedal dinosaurs such as sauropods and ceratopsians. This may indicate that prosauropods were much less accomplished bipeds than the ornithopods, perhaps spending more of their time on four legs than on two.

The picture is slightly different for theropods, which have sometimes been described as obligate bipeds and may not have resorted to quadrupedal locomotion so frequently as ornithopods. However, the tracks of some early coelurosaurs that moved on all fours (see Figure 6.10b; Plate 11, p. 208, centre) are unusual in showing a narrow and quite regular trackway for the manus prints. Also, the pes prints may be rotated negatively (outwards), as in habitually quadrupedal dinosaurs. These trackway characteristics may indicate that such small theropods were well accustomed to moving on all fours, with their forelimbs straightened into an erect position and the head raised well above the ground (Olsen and Baird 1986, fig. 6.17C).

10

Speeds of dinosaurs

I can go over a territory of country with the velocity of the wind,
while you are an hour in accomplishing a journey of half a
furlong. In a race I could leave you twenty miles behind me. . .
  The Hare and the Tortoise, Aesop's *Fables* (6th century BC)

RELATIVE SPEEDS

Biologists have identified two extremes of locomotor ability among
living land animals. At one extreme there are graviportal forms such
as elephants – big, heavy and slow-moving animals with their limbs
structurally adapted for weight-bearing. At the other extreme there
are cursorial forms, like greyhounds and ostriches, which have a suite
of distinctive adaptations for fast running. An equivalent series of
locomotor adaptations exists in dinosaurs, ranging from the obviously
graviportal sauropods to the ostrich-like ornithomimids.

  In practice, the various sorts of mammals and ground-dwelling birds
may be ranked in order of their running ability, as reflected in the
number and extent of their cursorial adaptations. W.P. Coombs
(1978a) described the most important of these adaptations as follows:

 1. an optimum body weight of about 50 kg, but not over 500 kg or
    below 5 kg;
 2. long limbs relative to other body dimensions;
 3. major limb joints with hinge-like mobility;
 4. a relatively short propodium (upper arm or thigh region);
 5. the two bones of the epipodium (forearm or shin region) reduced
    to a single one, either by fusion (of radius and ulna in forelimb)
    or by loss of one bone (fibula in the hindlimb);
 6. a long and slender metapodium ('palm' region of forelimb or 'sole'
    region of hindlimb);
 7. bones of the metapodium interlocked, fused together or reduced
    to a single one;
 8. manus and pes with pronounced median symmetry;
 9. innermost and outermost digits reduced or lost entirely; and
10. digitigrade to unguligrade stance.

By cataloguing the extent of these cursorial adaptations in living land animals, Coombs (1978a) was able to define four levels of running ability:

Graviportal: with limbs largely or entirely adapted for weight-bearing. Usually without cursorial adaptations (e.g. elephants).

Mediportal: with limbs primarily adapted for weight-bearing, but with some definite indications of cursorial adaptations – such as digitigrade stance, loss of inner and outer digits from the foot, interlocking or fusion of the metapodials (e.g. rhinos and hippos).

Subcursorial: with limbs showing moderate development of most cursorial adaptations. Some subcursorial animals are excellent runners, at least for short distances (e.g. pigs, cats and dogs).

Cursorial: with limbs showing extensive development of cursorial adaptations (e.g. many ungulates, ratites and kangaroos).

'A runner of 'average' ability, if such exists, would be on the borderline between mediportal and subcursorial. Graviportal and mediportal animals are poor to fair runners, [whereas] subcursorial and cursorial animals are good to excellent runners' (Coombs 1978a: 395).

Coombs then proceeded to classify the dinosaurs into these same four levels of running ability (see Figure 10.1). He concluded that bipedal dinosaurs were, on the whole, better at running than quadrupedal dinosaurs. The only possible exceptions were small and lightly built ceratopsians, such as *Protoceratops*: these were probably the speediest of all the quadrupedal dinosaurs and might be classified as advanced subcursorial. The remaining quadrupedal dinosaurs had cursorial adaptations equal or inferior to those of hippos. Sauropods and stegosaurs were unmistakably graviportal, while ankylosaurs and large ceratopsians were ranked as low-grade to middle-grade mediportal. Small bipedal dinosaurs, both ornithopods and theropods, were high-grade subcursorial or cursorial, whereas the bigger bipeds were generally subcursorial, with some of the carnosaurs verging on cursorial. Ornithomimids and long-legged coelurosaurs such as *Coelophysis* were the speediest of all dinosaurs, though they might have been outstripped by living bipeds such as kangaroos and ostriches.

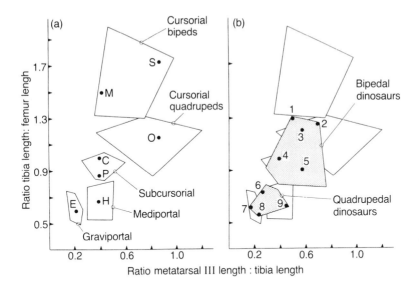

Figure 10.1 A comparison of selected hindlimb proportions in mammals, ground-dwelling birds and dinosaurs. (a) Relative running abilities of mammals and some ground-dwelling birds, related to their hindlimb proportions. Representative animals are indicated by code-letters: S, ostrich *Struthio*; M, kangaroo *Macropus*; O, white-tailed deer *Odocoileus*; C, dog *Canis*; P, pig *Sus*; H, hippo *Hippopotamus*; E, Indian elephant *Elephas*. (b) The same diagram, with dinosaurian data superimposed. Representative dinosaurs indicated by code-numbers: 1, small ornithopod *Heterodontosaurus*; 2, ornithomimid *Dromiceiomimus*; 3, primitive ceratopsian *Protoceratops*; 4, hadrosaur *Lambeosaurus*; 5, carnosaur *Tyrannosaurus*; 6, ankylosaur *Nodosaurus*; 7, stegosaur *Stegosaurus*; 8, sauropod *Camarasaurus*; 9, ceratopsian *Triceratops*. (Adapted from Coombs 1978a.)

It is difficult to measure this overall picture of dinosaur running abilities against that in mammals and ground-dwelling birds. For instance, the ornithomimids or so-called ostrich dinosaurs may have been the best runners among all the dinosaurs. . . but were they faster or slower than ostriches? From anatomical comparisons R.T. Bakker (1972) and D.A. Russell (1972) deduced that these long-legged dinosaurs might have matched or bettered the running speed of the ostrich (possibly 70–80 km/h). Yet, on the other hand, structural differences between ostriches and ornithomimids may indicate that these dinosaurs were not as speedy as ostriches (Coombs 1978a;

Thulborn 1982). At the opposite end of the spectrum, among graviportal animals, there is less controversy: it seems unlikely that sauropods could have surpassed the maximum speeds attained by elephants.

These comparisons give a useful idea of running abilities among dinosaurs in general, though there is no clear indication of the actual speeds attained by these animals.

ABSOLUTE SPEEDS

Several methods are available to determine the absolute speeds of dinosaurs, though some of them are less easily applied than others because they entail lengthy calculations or because they are appropriate only in special circumstances.

ALEXANDER'S METHOD

From observations of diverse living animals, including small mammals, humans, horses, elephants and birds, R.McN. Alexander (1976) determined the following relationship between SL, h and V:

$$SL/h \simeq 2.3(V^2/gh)^{0.3} \tag{10.1}$$

In this equation all linear measurements are in metres, the animal's speed (V) is expressed in metres per second, and g represents the acceleration of free fall. This general relationship may be applied to large and small animals, both bipeds and quadrupeds, at gaits from walk to run, and it does not seem to be seriously affected by variations in the consistency of the substrate. With additional data from fast-moving African ungulates Alexander, Langman and Jayes (1977: 298) refined equation 10.1 to give:

$$SL/h \simeq 1.8(V^2/gh)^{0.39} \tag{10.2}$$

They concluded that equation 10.2 is appropriate for animals that are cantering or galloping whereas equation 10.1 is better applied to animals using a slower gait.

To estimate the speeds of certain dinosaurs, Alexander (1976) transformed equation 10.1 to give:

$$V \simeq 0.25g^{0.5}SL^{1.67}h^{-1.17} \tag{10.3}$$

Table 10.1 Speeds predicted for dinosaurs by R.McN. Alexander (1976)

Track-maker	FL (cm)	h (m)	SL (m)	SL/h	V (m/s)	V* (km/h)
Bipedal dinosaur	53	2.1	3.0	1.4	2.0	7.3
Bipedal dinosaur	50	2.0	3.0	1.5	2.2	7.8
Bipedal dinosaur	24	1.0	2.4	2.5	3.6	12.7
Bipedal dinosaur	27	1.1	1.3	1.2	1.2	4.0
Bipedal dinosaur	28–35	1.1–1.4	1.8–2.5	1.7	2.2	7.9
Bipedal dinosaur	15–20	0.6–0.8	1.5–1.8	2.4	2.9	10.4
Sauropod	76	3.0	2.5	0.8	1.0	3.5
Sauropod	38	1.5	1.6	1.1	1.1	3.8

Note: In all cases height at the hip (*h*) was calculated as 4*FL*; velocity (*V*) was estimated by means of equation 10.3.

* Conversions to km/h have been added here.

He then applied this equation to data from dinosaur trackways, where *SL* could be measured directly and *h* could be estimated from the size of the footprints (Table 10.1). Subsequently, Alexander's method has been applied to hundreds of trackways, thereby providing some important insights into the locomotion and behaviour of dinosaurs (see summary in Table 10.2). The speeds estimated in this way are generally rather low, with few exceeding 10 km/h, though J.O. Farlow (1981) has reported the tracks of a few theropods that might have been running as fast as 40 km/h. This method seems to provide realistic estimates, though it may tend to underestimate moderate speeds and to overestimate high ones (Alexander 1989, fig. 6).

Equation 10.3 is appropriate for dinosaurs that used a walking gait, with the ratio *SL/h* (relative stride length) less than 2.0, but it may not be applicable to the tracks of dinosaurs that were running, with *SL/h* greater than 2.9. For running dinosaurs, equation 10.2 may be modified as follows (after Thulborn and Wade 1984):

$$V \simeq [gh(SL/1.8h)^{2.56}]^{0.5} \qquad (10.4)$$

The speeds of those few dinosaurs that used a trotting gait, with *SL/h* between 2.0 and 2.9, may be calculated as the mean of two estimates, using equations 10.3 and 10.4 (Thulborn 1984: 245).

The foregoing methods are easily applied to dinosaur tracks,

Table 10.2 Speed estimates derived from dinosaur trackways by means of Alexander's (1976) method: representative summary

Dinosaurs/ichnotaxa	N	h (m)	V (km/h)	Source
Bipedal dinosaurs	6	0.6–2.1	4–13	Alexander 1976
Sauropods	2	1.5–3.0	4	Alexander 1976
Ornithomimid	1	1.2	6	Russell and Béland 1976
Theropods ?*Anchisauripus**	2	0.4–0.6	5–8	Tucker and Burchette 1977
Carnosaur	1	2.6	8	
Ornithopods[†]	10	<1.0	16	Thulborn and Wade 1979
Coelurosaurs[†]	10	<1.0	13	
Theropods	15	1.2–1.9	6–43	Farlow 1981
Ornithopod *Gypsichnites*	1	1.2	7	
Theropods *Irenesauripus*[‡]	2	1.5–2.1	5–10	
Theropod *Irenichnites*	1	0.6	10	Kool 1981
Hadrosaur *Amblydactylus*	1	0.5	4	
?Ankylosaur *Tetrapodosaurus*	1	1.4	3	
?Hadrosaur[§]	1	3.4	9	Thulborn 1981
?Theropods *Columbosauripus*	2	1.0–1.2	8	
Theropod	1	1.4	10	
?Theropods	4	1.0–1.2	6–9	
?Theropods	7	0.8–1.3	4–8	
Theropods *Irenesauripus*[¶]	17	0.8–1.5	5–16	
Theropods *Irenesauripus*″	8	1.5–2.3	3–9	Currie 1983
Theropod *Irenesauripus***	1	1.6	8	
Theropod ?*Irenesauripus*[††]	1	1.1	8	
Theropods *Irenichnites*	3	0.6	7–8	
Hadrosaurs *Amblydactylus*	36	0.6–2.9	2–8	
Ornithopods ?*Gypsichnites*	5	1.0–1.2	6–10	
Coelurosaurs *Grallator*	3	0.2–0.8	3–6	
Theropods *Anchisauripus*	3	0.3–1.2	2–11	
Theropod *Ornithomimipus*	1	1.1	7	
Carnosaur *Megalosauropus*	1	1.3	7	
Carnosaur *Tyrannosauropus*	1	3.2	8	Haubold 1984
Ornithopods *Anomoepus*	2	0.2–0.6	2–3	
Ornithopod *Sauropus*	1	0.9	7	
Ornithopods *Iguanodon*	3	1.0–2.6	4–6	
Ornithopod *Sousaichnium*	1	1.8	4	
Ornithischian *Moraesichnium*	1	1.5	6	

Table 10.2 *cont'd.*

Dinosaurs/ichnotaxa	N	h (m)	V (km/h)	Source
Large ornithopods	2	1.8–2.0	10	
Ornithopods	28	1.1–1.8	6–10	
Small ornithopods	2	0.5–0.6	9–10	Lockley *et al.* 1986
Carnosaur	1	1.4	7	
Sauropods[‡‡]	7	1.4–2.1	3–6	

Note: h is estimated as $4FL$ (unless specified otherwise); V is the mean value per trackway.

* Delair and Sarjeant (1985) classified the larger tracks as *Gigandipus*.
† Overall mean speed for 10 trackways; Thulborn and Wade (1984) give revised estimates for these and other trackways at the same site.
‡ Two ichnospecies, *I. mclearni* and *I. acutus*.
§ Russell and Béland (1976) offer a different interpretation.
¶ Ichnospecies *Irenesauripus mclearni*.
" Ichnospecies *Irenesauripus acutus*.
** Identified as *Irenesauripus* cf. *I. acutus*.
†† Identified as *Irenesauripus* cf. *I. mclearni*.
‡‡ With h estimated as four times footprint width; speed estimates published by Lockley *et al.* (1986) were apparently doubled though computational error.

requiring only a preliminary estimate of h. Moreover, these methods are appropriate for dinosaurs of all sizes, both bipeds and quadrupeds, regardless of their gaits and the substrates they traversed. Some other methods, which are described below, are more limited in their application.

Average walking speeds

Bipedal dinosaurs

Analysis of more than 400 trackways has revealed that bipedal dinosaurs, both theropods and ornithopods, walked with mean SL/h about 1.29 (see Figure 9.2). Consequently, it should be possible to estimate the average walking speeds of bipedal dinosaurs by substituting $1.29h$ for SL in equation 10.1. That equation may then be rewritten to solve for V, and simplified to give:

$$V \simeq (1.42h)^{0.5} \tag{10.5}$$

This equation may be used to predict the average walking speed of

Speeds of dinosaurs

Table 10.3 Examples of average walking speeds predicted for dinosaurs

Bipeds	h^* (m)	V (km/h) Equation 10.5	Equation 10.6
Coelurosaur *Compsognathus*	0.21	1.97	2.01
Coelurosaur *Coelophysis*	0.56	3.21	3.04
Ornithomimid *Struthiomimus*	1.39	5.06	4.45
Carnosaur *Megalosaurus*	1.74	5.66	4.89
Carnosaur *Tyrannosaurus*	3.09	7.54	6.22
Small ornithopod *Fabrosaurus*	0.30	2.35	2.34
Small ornithopod *Hypsilophodon*	0.41	2.75	2.66
Iguanodont *Camptosaurus*	1.36	5.00	4.41
Iguanodont *Iguanodon*	1.51	5.27	4.61
Hadrosaur *Edmontosaurus*	2.32	6.53	5.52

Quadrupeds	h (m)	V (km/h) Equation 10.7	Equation 10.8
Prosauropod *Anchisaurus*	0.46	1.68	2.74
Prosauropod *Plateosaurus*	1.42	2.96	3.17
Sauropod *Diplodocus*	2.76	4.13	3.45
Sauropod *Apatosaurus*	3.14	4.40	3.51
Sauropod *Brachiosaurus*	4.45	5.24	3.67
Stegosaur *Stegosaurus*	1.91	3.43	3.29
Ankylosaur *Euoplocephalus*	1.17	2.68	3.09
Ankylosaur *Panoplosaurus*	1.52	3.07	3.20
Ceratopsian *Leptoceratops*	0.69	2.06	2.89
Ceratopsian *Triceratops*	2.23	3.71	3.36

* Measurements of h are from dinosaur skeletons, described in numerous works listed by Thulborn (1982, pp. 252–3).

any bipedal dinosaur that is represented by a skeleton (where h is measured directly) or by a trackway (where h is estimated from the size of the footprints). How realistic are such predictions? In a sample of 175 trackways made by bipedal dinosaurs the regression of estimated speed on estimated h is expressed by the following equation (from Thulborn 1984):

$$V \simeq 0.56h^{0.42} \qquad (10.6)$$

where h is entered in centimetres and V is solved in kilometres per hour. The predictions of equations 10.5 and 10.6 are in fair agreement, though the former might seem to underestimate the speeds of small bipeds ($h < 28$ cm) and to overestimate the speeds of bigger animals. Even so, the differences are probably of minor importance: for dinosaurs up to 228 cm in height at the hip the predictions differ by less than 1 km/h (Table 10.3).

Quadrupedal dinosaurs

In a sample of 63 trackways made by quadrupedal dinosaurs (mostly sauropods) mean SL/h was found to be 0.93 (see also Figure 9.2).[1] Hence, the average walking speeds of quadrupedal dinosaurs might be computed by substituting $0.93h$ for SL/h in equation 10.1. This equation may then be rewritten to solve for V and simplified to give:

$$V \simeq (0.477h)^{0.5} \qquad (10.7)$$

As before, h is entered in metres and V is solved in metres per second. How well, or poorly, do the predictions of this equation match up with the estimates derived from trackways? In the sample of 63 trackways mentioned earlier there is no very obvious correlation between the estimated size (h) of a track-maker and its predicted speed. Nevertheless, the general relationship between these two variables may be expressed as follows:

$$V \simeq 1.675h^{0.129} \qquad (10.8)$$

where h is expressed in centimetres and V is solved in kilometres per hour. Equations 10.7 and 10.8 are in fair agreement insofar as they predict relatively low speeds for quadrupedal dinosaurs, though the former might seem to understimate the speed of smaller quadrupeds ($h < 1.7$ m) and to overestimate the speeds of bigger ones. For quadrupeds between 52 cm and 332 cm in height at the hip the predictions differ by less than 1 km/h (Table 10.3).

In quadrupedal dinosaurs the forelimbs and hindlimbs, though of different sizes, took strides of equal number and length. It has been suspected that the forelimbs regulated the length of stride, and that the striding of the longer hindlimbs was correspondingly suppressed or adjusted (Thulborn 1982: 245). According to this view, SL/s (where s represents height at the shoulder) might have been about 1.29

whereas SL/h was somewhat less, depending on the ratio of s to h. The value of 0.93 for SL/h implies that the ratio s/h would have been about 0.72 in the sample of 63 track-makers (mostly sauropods) that was mentioned previously. In skeletons of the sauropods *Apatosaurus* and *Diplodocus* the ratio s/h appears to be reasonably close to that value, about 0.68.

The speeds predicted for sauropods (Table 10.3) may be fairly realistic because these animals had erect and pillar-like forelimbs. But the speeds predicted for other quadrupedal dinosaurs may be less accurate because these animals might have had forelimbs that sprawled out sideways to some extent. Such out-turning of the elbow would have brought the shoulder closer to the ground, thus reducing the functional height of the forelimb and restricting the length of stride. Unfortunately, there is some uncertainty about the orientation of the forelimbs in quadrupedal dinosaurs other than sauropods. R.A. Thulborn (1982) suspected that a pronounced sprawling of the fore-limbs might have reduced the speed of stegosaurs and ankylosaurs by as much as 10 %, whereas K. Carpenter (1982b) has maintained that these dinosaurs had erect forelimbs. Clearly, there is a need for better information on this point.

There is yet another complication. In all quadrupedal dinosaurs the forwards reach of the forelimb was restricted to some extent by the orientation of the glenoid, or shoulder socket. The glenoid faced down and back, so that the forelimb could not have been swung forwards very far beyond the vertical. Some quadrupedal dinosaurs might have overcome this limitation by having a freely movable shoulder skeleton, like mammals (Bakker 1975b, 1986b), though this idea has been disputed (Bennett and Dalzell 1973; Coombs 1978a,b). Restricted mobility at the shoulder joint could well have limited the speeds attained by some quadrupedal dinosaurs, though, once again, the matter requires further study.

In summary, the predictions of equations 10.7 and 10.8 should be treated with caution: they may overestimate the speeds of quadrupedal dinosaurs with sprawling forelimbs, and they make no allowance for anatomical restrictions on mobility at the shoulder joint.

DEMATHIEU'S METHODS

In 1984, G.R. Demathieu introduced a different method to estimate the speeds of dinosaurs and other extinct animals. In its original form

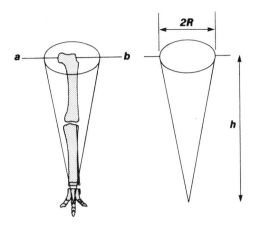

Figure 10.2 Schematic diagram of a dinosaurian hindlimb skeleton likened to a conical pendulum that oscillated round the axis *a–b*. The length of the pendulum would correspond to *h*, the dinosaur's height at the hip. (Adapted from Demathieu 1986.)

this method applied to dinosaur skeletons, though it was later adapted to the study of trackways (Demathieu 1986). The method applies only to animals that used a steady walking gait, and it is explained here in slightly simplified form.

Briefly, Demathieu treated the dinosaur hindlimb as an oscillating pendulum which was assumed to be roughly the shape of an inverted cone (Figure 10.2). By drawing on the principles of dynamics he was then able to estimate the animal's speed as follows:

$$V \simeq (1/2\pi)SLg^{0.5}L^{-0.5} \qquad (10.9)$$

Here *V* is solved in kilometres per hour and *L* represents the length of the simple synchronous pendulum in centimetres (Demathieu 1984: 441). The dimension *L* was calculated as:

$$L = \frac{3R^2 + 2h^2}{5h} \qquad (10.10)$$

where *h* is the height of the cone representing the hindlimb and *R* is the maximum radius of that cone, both in centimetres. In its original application to dinosaur skeletons this method is fairly straightforward: *h* may be measured directly, though it is necessary to estimate *R* and to make an arbitrary assumption about *SL*.

Table 10.4 Steady walking speeds predicted for bipedal dinosaurs*

		Predicted Speeds (km/h)		
Dinosaur	h (m)[†]	Demathieu (1984)	Demathieu[‡] (1986)	Equation 10.6
Hadrosaur *Kritosaurus*	2.35	5.9	5.4	5.5
Iguanodont *Iguanodon*	1.51	4.7	4.5	4.6
Ornithomimid *Struthiomimus*	1.39	4.8	4.2	4.4
Carnosaur *Gorgosaurus*	2.63	6.2	6.1	5.8

* Steady walking speed is assumed to be the same as average walking speed.
[†] Values of h are calculated from osteometric data listed by Galton (1974, table V).
[‡] Estimates derived from table I of Demathieu (1986) on the assumption that SL = 1.29h.

Using these methods Demathieu (1984) estimated steady walking speeds for four bipedal dinosaurs, all represented by skeletons. His estimates are listed in Table 10.4, along with average walking speeds calculated by means of equation 10.6 and by means of a predictive table published by Demathieu in 1986. The three sets of estimates are in good agreement, despite having been obtained by different methods.

In making his original predictions Demathieu (1984) assumed the angle of gait (2θ) to be 40°, though he suspected that this was probably too great for some of the dinosaurs listed in Table 10.4. He was also obliged to estimate the dimension R, and this presented some difficulty because the base of the inverted cone representing the hindlimb (Figure 10.2) was probably elliptical rather than circular. For the hadrosaur *Kritosaurus*, with h of 2.35 m, R was assumed to be 30 cm. It was later suggested (Demathieu 1986) that R was about one-sixth of gleno-acetabular distance. This rule of thumb is useful when dealing with tracks of quadrupeds but is not applicable to those of bipeds, where gleno-acetabular distance is unknown.

In 1986, Demathieu modified his original methods, making them more easily applicable to trackways, and also introduced a new equation to estimate the speed of a track-maker, as follows:

$$V = 0.036(SL/T) \qquad (10.11)$$

Here, SL is entered in centimetres and V is solved in kilometres per hour. The quantity T is the period of oscillation of the pendulum representing the hindlimb, calculated as:

$$T = 0.128451 \sqrt{h} \, (1 + \theta^2/16) \tag{10.12}$$

In many cases, the factor $(1 + \theta^2/16)$ may be disregarded, because it has little effect on the result (Demathieu 1986: 329). Aside from this Demathieu simplified matters by compiling a table to predict the angle of gait and the walking speed at particular values of h and SL (1986: 331).

Demathieu's methods (1984, 1986) take into account numerous factors and seem to offer the advantage of mathematical precision. However, such advantage is probably outweighed by practical difficulties – notably the need to make preliminary estimates and assumptions. For example, it is necessary to estimate h, or an equivalent dimension, and to make assumptions about θ and R. Some of these complications may be avoided by interpolating values of h and SL into Demathieu's (1986) table of predicted speeds. But, even then, Demathieu conceded that some of those predictions might need to be increased by as much as three times, depending on assumptions about a track-maker's frequency of striding. In view of these technicalities it would seem easier to use equation 10.6, which gives broadly similar results.

MAXIMUM SPEEDS

R.T. Bakker (1975a) used data from 17 quadrupedal mammals, ranging in size from jack-rabbit to elephant, to obtain the following relationship between maximum speed and body size:

$$V_{max} \simeq (4.132RHL) - 1.4 \tag{10.13}$$

Here V_{max} represents maximum speed, in kilometres per hour, while RHL, or relative hindlimb length, is a complex measure of body size defined as the sum of the lengths of femur, tibia, tarsus and longest metatarsal (in centimetres) divided by the cube root of body mass (in kilograms). Bakker pointed out that this equation would not apply to exceptionally long-legged animals, such as the giraffe, nor to those, such as the cheetah, which extend stride length by flexing and straightening the spinal column. Subsequently, W.P. Coombs (1978a) used equation 10.13 to predict the theoretical maximum speeds of various dinosaurs – though he did so with extreme caution. Coombs was careful to note that inaccurate measurements of hindlimb bones could seriously affect estimates of *RHL* and, hence, of maximum speed, and that improbably high speeds would be predicted for long-legged dinosaurs such as *Stegosaurus*. An additional element of uncertainty might be introduced when attempting to estimate the body mass of a dinosaur.[2]

The relationship of maximum speed to body mass was also investigated by T. Garland (1983b), who gathered data for 106 species of mammals ranging from a small insectivore (16 g) to an African elephant (6 t). Garland found that the regression of maximum running speed on body mass could be expressed as follows:

$$V_{max} = 23.6W^{0.165} \tag{10.14}$$

where W is in kilograms and V_{max} is solved in kilometres per hour. However, Garland pointed out that the largest mammals are not the fastest, as is implied by such an equation. Instead, there is probably an optimum body size for running ability, as Coombs (1978a) had noted in his review of dinosaur locomotion. In fact, Garland found that his data for mammals were best expressed by means of a polynomial regression equation of the following form:

$$\log_{10}V_{max} \simeq 1.47832 + 0.25892(W_{\log 10}) - 0.06237(W_{\log 10})^2 \tag{10.15}$$

This predicts a maximum running speed of 56 km/h for a mammal at the optimum body size of 119 kg; animals bigger or smaller than this are predicted to achieve lower maximum speeds. If Garland's equation is applied to dinosaurs it yields results that are generally lower than those obtained with Bakker's method (Table 10.5). These differences are an excellent illustration of the great difficulties that arise in trying to ascertain the maximum speeds of *any* animals, living or extinct.

A different approach to estimating maximum speeds was adopted by R.A. Thulborn (1982), who pointed out that equations 10.3 and 10.4 could just as well be applied to dinosaur skeletons as to trackways. In applying these equations to skeletons, *h* is measured directly and *SL* is represented by a selected hypothetical figure. By substituting 2*h* for *SL* in equation 10.3 it is possible to predict the speed of a dinosaur as it shifted from a walking gait (with $SL/h < 2.0$) to a trotting gait (with SL/h between 2.0 and 2.9). That equation may then be simplified to give:

$$V \simeq (6.12h)^{0.5} \tag{10.16}$$

Table 10.5 Examples of maximum speeds predicted for dinosaurs

Dinosaur	W (kg)	h (cm)*	Bakker (1975a)	Garland (1983b)	Thulborn (1982)
Sauropod *Brachiosaurus*	78 258	441	43	18	18
Sauropod *Apatosaurus*	32 418	314	41	24	–
Sauropod *Apatosaurus*	27 869	314	43	25	12
Sauropod *Diplodocus*	10 562	276	52	32	12
Ceratopsian *Triceratops*	8 478	223	45	34	26
Carnosaur *Tyrannosaurus*	6 895	309	68	36	23
Iguanodont *Iguanodon*	4 514	151	37	39	–
Hadrosaur *Corythosaurus*	3 820	246	65	40	20
Ceratopsian *Styracosaurus*	3 686	155	41	41	22
Carnosaur *Gorgosaurus*	3 500	252	69	41	21
Ankylosaur *Panoplosaurus*	3 472	151	41	41	8
Hadrosaur *Anatosaurus*	3 071	232	66	42	20
Hadrosaur *Hadrosaurus*	2 740	233	69	43	20
Hadrosaur *Parasaurolophus*	2 650	229	69	43	20
Carnosaur *Gorgosaurus*	2 400	224	70	44	20
Carnosaur *Antrodemus*	2 087	187	61	45	18
Ankylosaur *Euoplocephalus*	1 900	116	38	45	7
Stegosaur *Stegosaurus*	1 777	179	61	46	7
Iguanodont *Iguanodon*	590	150	74	52	16
Iguanodont *Camptosaurus*	383	136	78	54	15
Ceratopsian *Protoceratops*	177	65	47	56	28
Ornithomimid *Dromiceiomimus*	154	144	111	56	51
Small ornithopod *Dryosaurus*	120	93	78	56	43
Coelurosaur *Deinonychus*	68	87	88	55	42
Coelurosaur *Ornitholestes*	18	48	75	51	34
Coelurosaur *Compsognathus*	6	33	74	44	26
Small ornithopod *Abrictosaurus*	4	24	61	40	19
Coelurosaur *Compsognathus*	3	21	59	39	16

Note: Double listing of a dinosaur indicates two specimens (e.g. coelurosaur *Compsognathus*) or two different weight estimates for a single animal (e.g. sauropod *Apatosaurus*).

* Basic data (W and h) were compiled from numerous sources listed by Thulborn (1982).

Similarly, the substitution of 2.9*h* for *SL* in equation 10.4 provides the following equation, which predicts the speed of a dinosaur as it shifted from a trot (with *SL/h* between 2.0 and 2.9) to a run (with *SL/h* > 2.9):

$$V \simeq (33.13h)^{0.5} \qquad (10.17)$$

In both these equations *h* is expressed in metres and *V* is solved in metres per second.

Thulborn then assumed that heavyweight dinosaurs, with body mass estimated at 1 000 kg or more, were probably restricted to a walking gait and were unlikely to surpass the speeds predicted for them at the walk-trot transition. In other words, it was proposed that equation 10.16 would predict maximum limits of speed for graviportal dinosaurs in general. Next, it was supposed that mediportal dinosaurs were able to shift from a walk into a trotting gait, though they were probably incapable of accelerating into a fast running gait. Consequently, the maximum speeds of these dinosaurs would lie somewhere between the predictions of equations 10.16 and 10.17. Finally, there remained the cursorial dinosaurs. Here it was necessary to determine by how much these dinosaurs exceeded their speeds at the trot–run transition (equation 10.17). In the trackways of fast-moving dinosaurs *SL* reaches a maximum of about 5*h* (Thulborn and Wade 1984). Thus, by substituting 5*h* for *SL* in equation 10.4, it is possible to predict maximum speeds for cursorial dinosaurs such as the ornithomimids (see equation 10.19, below). Examples of the maximum speeds calculated by Thulborn (1982) are shown in Table 10.5. In general, they are considerably lower than those predicted by the equations of Bakker (1975a) and Garland (1983b). Bakker, for example, recently expressed his belief (1986b) that *Tyrannosaurus* could easily have overhauled a galloping white rhino, at a speed greater than 64 km/h. By contrast, Thulborn (1982) calculated that the maximum speed of *Tyrannosaurus* was no greater than about 23 km/h.

How well do these predictions agree with estimates of speed derived from dinosaur tracks? Unfortunately, this question cannot be investigated very thoroughly because there are so few trackways that might have been made by dinosaurs running at or near their maximum speed. Most of the world's examples are preserved at a site known as Lark Quarry, in the Cretaceous rocks of western Queensland, Australia (Plate 14, p. 316). Here there are trackways of more than

160 small bipedal dinosaurs, both ornithopods and theropods, that seem to have been startled into a stampede by an approaching carnosaur (see Chapter 11). Most of the stampeding animals were less than 70 cm high at the hip, and it seems likely that they were running at or near their maximum speeds (Thulborn and Wade 1984: 449–50). Among these running dinosaurs the mean value for maximum SL/h per trackway was found to be 3.93. Thus, by substituting $3.93h$ for SL in equation 10.4 it is possible to predict maximum running speed as:

$$V_{max} \simeq (72.23h)^{0.5} \qquad (10.18)$$

where h is in metres and V_{max} is solved in metres per second. The highest estimate of SL/h for any of the Lark Quarry dinosaurs, an ornithopod with h about 29 cm, is 5.03. If small bipedal dinosaurs did achieve SL/h as high as this, their maximum speeds might be estimated as:

$$V_{max} \simeq (135.99h)^{0.5} \qquad (10.19)$$

There is scant evidence of running ability in bigger bipedal dinosaurs. However, three trackways in the Cretaceous of Texas, USA, have been attributed to running theropods with h about 1.5 m (Farlow 1981). For these animals SL/h is estimated between 3.8 and 4.0 (Thulborn and Wade 1984: 454). In *Saltopoides igalensis*, a theropod trackway from the Lower Jurassic of France (de Lapparent and Montenat 1967), h may have been between 70 and 90 cm, with SL/h as high as 4.9 (Thulborn and Wade 1984: 452). In none of these cases does the track-maker seem to have attained SL/h as high as 5.0.

Equations 10.18 and 10.19 may be used to define a range of maximum speeds for small bipedal dinosaurs, with h up to about 70 cm (see Table 10.6). It might also be legitimate to extrapolate these equations to some medium-sized bipedal dinosaurs, with h up to 1.5 m, but it is certainly not appropriate to do so for very large bipeds or quadrupeds. This is because an animal moving with SL/h as high as 3.73 must include an unsupported interval in the stride, and the ability to do this is generally restricted to animals weighing less than about 800 kg. Many of the large bipedal dinosaurs exceeded this weight limit, as did most of the quadrupedal forms. Nevertheless, these equations might be extended to ornithomimids, which appear to have been remarkably lightly built. One example of the ornithomimid *Dromiceiomimus* had h about 1.22 m and is estimated to have weighed

Table 10.6 Maximum running speeds predicted for cursorial dinosaurs

| | | Predicted maximum speeds (km/h) | |
| | | Equation | Equation |
Dinosaur	h (cm)*	10.18	10.19
Ornithomimid *Gallimimus*	194	43	58
Ornithomimid *Dromiceiomimus*	144	37	50
Ornithomimid *Struthiomimus*	139	36	50
Ornithomimid *Dromiceiomimus*	122	34	46
Small ornithopod *Dryosaurus*	94	30	41
Small ornithopod *Dryosaurus*	93	30	40
Coelurosaur *Deinonychus*	87	29	39
Small ornithopod *Parksosaurus*	74	26	36
Ornithomimid *Gallimimus*	57	23	32
Coelurosaur *Coelophysis*	56	23	31
Coelurosaur *Ornitholestes*	48	21	29
Small ornithopod *Hypsilophodon*	41	20	27
Coelurosaur *Coelophysis*	34	18	24
Coelurosaur *Compsognathus*	33	18	24
Small ornithopod *Heterodontosaurus*	32	17	24
Small ornithopod *Fabrosaurus*	30	17	23
Small ornithopod *Hypsilophodon*	28	16	22
Coelurosaur *Podokesaurus*	26	16	21
Small ornithopod *Abrictosaurus*	24	15	21
Coelurosaur *Compsognathus*	21	14	19

Note: Predictions in Table 10.5 are slightly different because they involve assumptions about stride frequency (see Thulborn 1982).

* Basic data (h) were compiled from numerous sources listed by Thulborn (1982).

about 154 kg (Russell and Béland 1976). Equation 12.18 predicts the maximum speed of this animal to be about 9.3 m/s (33.5 km/h). Among the ornithomimids described by H. Osmólska and her colleagues (1972) the largest example of *Gallimimus* had h about 1.94 m; for this dinosaur, equation 10.18 predicts a maximum speed of 11.8 m/s (42.6 km/h). These speed estimates are based on the assumption

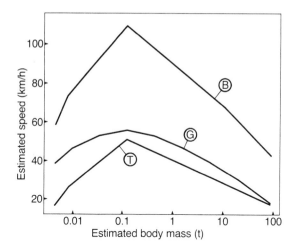

Figure 10.3 Estimated maximum speeds of dinosaurs, related to body mass (plotted on logarithmic scale): B, curve derived from the methods of R.T. Bakker 1975a (equation 10.13 in text); G, curve derived from the findings of T. Garland 1983b (equation 10.15 in text); T, curve based on the findings of R.A. Thulborn 1982. Compare Figure 10.4.

that ornithomimids attained maximum SL/h about 3.93, but it is certainly possible that these dinosaurs were capable of extending stride length beyond that figure, particularly in view of their striking cursorial adaptations (Coombs 1978a). If the example of *Gallimimus* mentioned above attained SL/h as high as 5.03, its maximum speed would have been about 16.2 m/s (from equation 10.19). This speed of nearly 60 km/h, which might conceivably be the highest achieved by any dinosaur, falls rather short of estimates derived from anatomical comparisons (Bakker 1972; Russell 1972) or from the extrapolation of mammalian data (Figures 10.3 and 10.4).

Figure 10.4 compares theoretical predictions of maximum speed with the highest speeds so far estimated from the evidence of dinosaur trackways. In general, the theoretical predictions by Thulborn (1982) are considerably lower than those of Bakker (1975a) and those of equation 10.15 (from Garland 1983b) and are in better agreement with the evidence from dinosaur trackways. Because equations 10.13 (Bakker 1975a) and 10.15 (Garland 1983b) are based on data from living mammals it is not surprising that they should predict the speeds

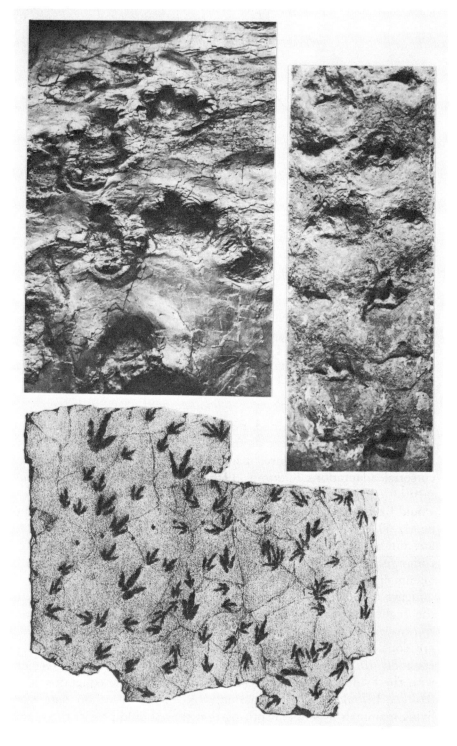

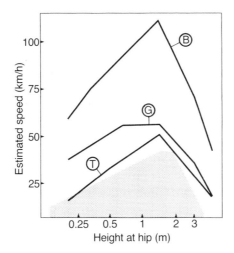

Figure 10.4 Estimated maximum speeds of dinosaurs, related to height at the hip (plotted on logarithmic scale): B, curve derived from the methods of R.T. Bakker 1975a; G, curve derived from the findings of T. Garland 1983b; T, curve based on the findings of R.A. Thulborn 1982. The shaded area encloses all published speed-estimates so far derived from the evidence of dinosaur tracks. The curves differ in shape from those in Figure 10.3 because there is an imperfect correlation between body mass and height at the hip in dinosaurs.

Plate 13 Examples of dinosaur trackways. Top left: Associated trackways of sauropods and a carnosaur at the Upper Jurassic Barkhausen site near Osnabrück, Germany. Several sauropod trackways, *Elephantopoides barkhausensis*, extend from the top to the bottom; large tridactyl prints of a carnosaur, each about 56 cm long, are headed in the opposite direction (slightly to right of centre). Top right: *Navahopus falcipollex*, the trackway of a prosauropod dinosaur, from the Lower Jurassic of Arizona, USA; stride length is about 32 cm. Note the unusually broad trackway and prominent trace of a curved claw on the 'thumb' (e.g. right manus print at extreme top right). Bottom: A magnificent footprint slab from the Hitchcock collection, showing abundant theropod tracks of various sizes; about 2.2 m wide.

of dinosaurs to rival those of mammals. Bakker has argued (1975a,b, 1986a,b) that this is probably true – that dinosaurs did rival or surpass the running abilities of mammals – though this idea has been hotly disputed. Thus, for example, Bakker has continued to defend 'the tremendously heretical concept' of '*Tyrannosaurus* moving at forty-five miles per hour' (1986b: 218). To this idea J.H. Ostrom has retorted that 'the notion of six tons of dinosaur flesh being routinely propelled at forty-five miles an hour is preposterous' (1987: 63). Elsewhere, it has been suggested that dinosaurs were not as speedy as mammals because they were radically different in their locomotor anatomy (Bennett and Dalzell 1973; Coombs 1978a; Thulborn 1982). For instance, dinosaurs had a massive tail, with hindlimb retractor muscles originating behind the thigh, whereas mammals have a reduced tail, with the origins of hindlimb retractor muscles shifted on to the forwardly expanded hip skeleton. These fundamental differences in locomotor anatomy might well imply that dinosaurs also differed from mammals in their running ability. And such scanty evidence as is available from the tracks of running dinosaurs does seem to confirm that dinosaurs were not as speedy as mammals.

NOTES

[1] For the sample of 49 trackways analysed in Figure 9.2 mean SL/h was 0.95 ± 0.29. For the augmented sample of 63 trackways mentioned in the text mean SL/h was 0.93 ± 0.27 and there was a very poor correlation between size (h) and speed (r 0.153, with log-transformed data).

[2] Coombs (1978a) inadvertently omitted the decimal point from the intercept value in Bakker's equation; the same error was later reproduced by Thulborn (1982). It appears from Bakker's original paper (1975a, fig. 21.6) that the correct value is 1.4, as given here in equation 10.13. The speed estimates published by Coombs (1978a, table 2) should be adjusted by adding 13.6 km/h.

11

Assemblages of dinosaur tracks

In the short space of little over two weeks I have laid bare what I believe to be one of the largest single displays of dinosaur tracks ever uncovered, a most spectacular sight to behold, and one that I never expect to match again... Three great sauropod trails 75 to 100 yards long; sections of two others, and carnivore tracks in profusion...

R.T. Bird, Letter of 25 April 1940 (*in* Farlow 1987)

It is common to find tracks of two or more dinosaurs at a single site. Such natural associations may be **monotypic**, comprising footprints of a single type, or **polytypic**, involving several ichnotaxa. In some cases there are only a few trackways, but in others there may be dozens or even hundreds, and occasionally a site is so thoroughly trampled and churned by the feet of dinosaurs that it is impossible to distinguish the individual footprints (e.g. Plate 12, p. 278, bottom left). This dinosaurian trampling of the substrate, which has been termed **dinoturbation** (Dodson *et al.* 1980), sometimes provokes wry comment from palaeontologists. A site trodden by Cretaceous brontosaurs resembled 'a sauropodian barnyard where a wallowing good time was had by all' (R.T. Bird, quoted by Farlow 1987: 3), and S.P. Welles (1971: 30) described a much-trampled area in the Kayenta Formation of Arizona as:

> ... a 'chicken yard' hodge-podge of footprints, few of which can be identified as belonging to a trackway... a single trackway shows about 10 footprints in a straight line heading west – the only animal that seemed to know where he was going!

A natural association of tracks is sometimes termed an **ichnocoenosis** (or 'ichnocoenose'). More precisely, an ichnocoenosis may be defined as a natural assemblage of traces resulting from the activities of an association of living organisms. G. Leonardi (1987: 43) has pointed out that the term is applicable only to the trace fossils preserved at a single horizon. The cognate term **ichnofauna** is sometimes used as a synonym for ichnocoenosis though, strictly speaking, it has a slightly different meaning: this term signifies a fauna whose existence and composition are revealed by the evidence of trace fossils. An ichnofauna is roughly equivalent to Edward Hitchcock's 'Lithichnozoa', signifying 'animals made known by their tracks in stone' (1858: 45), and it may comprise the evidence of trace fossils from several horizons (Leonardi 1987: 43). Ichnocoenoses produced by invertebrate animals have proved to be of great practical value for identifying and defining palaeoenvironments (Frey 1978; Seilacher 1978; Curran 1985); by comparison, the palaeoenvironmental significance of dinosaurian ichnocoenoses remains largely unexplored (Lockley 1986a; Pollard 1988).

RANDOMLY ORIENTED TRACKWAYS

Many sites show dinosaur tracks distributed at random (e.g. Plate 13, p. 306, bottom). Sometimes, the tracks are uniform in morphology, representing a single ichnospecies or perhaps even a single species of dinosaur, but in other instances they are heterogeneous and clearly originated from several different sorts of dinosaurs (Figure 11.1).

Some of these occurrences really do seem to be random accumulations, formed when dinosaurs traversed an area by chance, either individually or in groups. But occasionally there is an obvious reason for the local concentration of footprints. This is the case with a rich assemblage of dinosaur tracks described by P. Ellenberger (1974) from the Lower Jurassic of Lesotho (Figure 11.2). Here it seems that a lake or water-hole was the focus of attention for numerous dinosaurs, which presumably came down to the water to drink, to feed or to investigate the behaviour of other animals. Not surprisingly, the various track-makers show a diversity of behaviours, as deduced from their tracks. Some of the ornithopods, for example, seem to have been walking bipedally, occasionally coming down on to all fours, to rest or to dabble through the mud.

Another interesting occurrence was reported by J.K. Balsley (1980) from the roof of a coal mine in Utah (Figure 11.3). Numerous footprint casts of late Cretaceous ornithopods were preserved on the roof of the coal workings, scattered among tree stumps preserved in their position of growth. It is noteworthy that some footprints 'occur

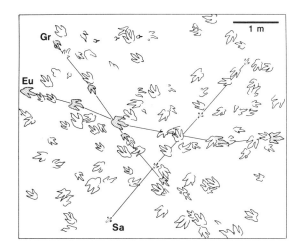

Figure 11.1 Polytypic assemblage of randomly oriented tracks, from the Lower Jurassic of France. The trackways were classified into three ichnogenera, all presumed to represent theropod dinosaurs. Examples of ichnogenera (shaded trackways): Eu, *Eubrontes*; Gr, *Grallator*; Sa, *Saltopoides*. (Adapted from de Lapparent and Montenat 1967.)

around the bases of the trees and are pointed inward as though the animals were browsing on the standing vegetation' (Balsley 1980: 133).

At some sites dinosaur tracks may have accumulated over a period of weeks or even months, so that the substrate resembles a palimpsest, with remnants of old footprints mingled among more recent tracks. The sequence in which the various track-makers traversed a site may be ascertained by searching for three clues. First, there may be some variation in the quality of footprint preservation: the earlier formed prints may be eroded whereas more recently formed prints have sharper outlines. Second, the footprints formed at different times tend to differ in depth because the substrate usually underwent progressive changes in its consistency (Courel and Demathieu 1976; Demathieu 1985; Weems 1987). Initially, the substrate would have been soft and wet, so that track-makers tended to produce deep footprints; thereafter, the substrate would become progressively firmer and drier, so that track-makers left shallower prints (Figure 11.4). The third and obvious clue is simply the overlapping of older footprints by more recent ones (Figure 11.5). These three criteria sometimes permit the

Figure 11.2 Randomly oriented dinosaur tracks preserved round an ancient water-hole, from the Lower Jurassic of Lesotho, southern Africa. Most track-makers were ornithopod dinosaurs, exemplified by the shaded trackway at left (ichnogenus *Moyenisauropus*). Note that this track-maker arrived bipedally, settled down on all fours to dabble in the muddy sediment, and then departed bipedally. Other track-makers included theropods, exemplified by the shaded trackway at right (ichnogenus *Neotrisauropus*). The broad and sinuous trackway at centre was possibly made by a reptile resembling a turtle. (Adapted from P. Ellenberger 1974.)

detailed reconstruction of events at a trackway site (e.g. Figures 11.4, 11.10).

TRACKWAY ASSEMBLAGES WITH PREFERRED ORIENTATION

Some of the most thought-provoking sites are those where dinosaur

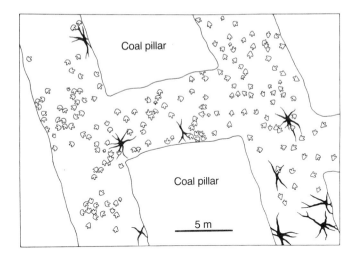

Figure 11.3 Numerous footprints of ornithopod dinosaurs on the roof of a coal mine in the Upper Cretaceous of Utah, USA. Some footprints point directly towards tree-stumps preserved in the position of growth, as if the track-makers had been browsing on the trees. (Adapted from Balsley 1980, Hickey 1980.)

trackways have a parallel or subparallel arrangement. Parallel tracks are more common than one might suspect. After examining the orientation of dinosaur tracks at several well-known sites in the United States, J.H. Ostrom (1972: 298) was prompted to remark: 'It had been my impression that most dinosaur footprint localities preserved random, rather than preferred, trackway orientations. This impression now seems to be false.' Ostrom's verdict has since been substantiated by the discovery of parallel dinosaur tracks at many other sites around the world (e.g. Plate 13, p. 306, top; Plate 14, p. 316).

Even more surprising is the fact that the preferred orientation of dinosaur tracks may remain fairly constant from one horizon to another within a succession of sediments. This was demonstrated by L.C. Godoy and G. Leonardi (1985), who measured the orientation of dinosaur tracks at more than a dozen horizons in the Lower Cretaceous sediments of Brazil, and found that there were consistent directional trends to the NNW or SSE.

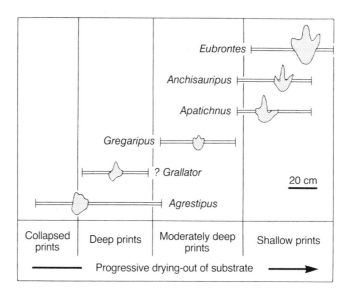

Figure 11.4 Reconstruction of sequence in which various dinosaurs traversed a site in the Upper Triassic of Virginia, USA. The dinosaur tracks were classified into six ichnogenera, four attributed to theropods (*Anchisauripus*, *Apatichnus*, *Eubrontes* and *?Grallator*), one to an ornithopod (*Gregaripus*) and one to a sauropod (*Agrestipus*). A representative right footprint is illustrated for each ichnogenus, at uniform scale. Initially, the substrate was waterlogged, so that the earliest-formed footprints collapsed as the track-makers withdrew their feet. Subsequently, the substrate became firm enough to retain footprints, and as it continued to grow harder and drier the track-makers formed progressively shallower tracks. Note that there is no correlation between the size of a track-maker and the depth of its footprints: some small dinosaurs left deep prints (e.g. *?Grallator*) whereas some bigger and heavier animals produced shallow prints (e.g. *Eubrontes*). (Adapted from Weems 1987.)

Two-directional assemblages

A good example is furnished by a sandstone slab with numerous footprint casts (Figure 11.6), described from the Lower Cretaceous of Germany by U. Lehmann (1978). The footprints are those of large ornithopods, possibly *Iguanodon*, and they represent at least 10 trackways forming a distinctly two-directional traffic.

Such prevailing trends in trackway assemblages certainly testify to dinosaurian comings and goings along preferred routes or, at least, in

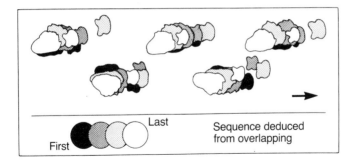

Figure 11.5 Diagram of four nearly coincident sauropod trackways at the Davenport Ranch site, in the Lower Cretaceous of Texas, USA. The sequence in which the four animals traversed the site is ascertained from the overlapping of footprints. (Adapted from Bird 1944, 1985.)

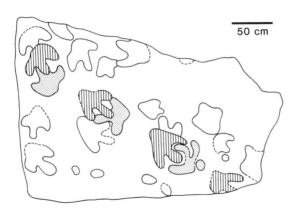

Figure 11.6 Diagram of sandstone slab with 23 footprints attributed to large ornithopod dinosaurs (*?Iguanodon*), from the Lower Cretaceous of Germany. Note that the tracks form a pronounced two-way traffic, as indicated by two representative trackways (differently shaded). (Adapted from Lehmann 1978, Haubold 1984.)

preferred directions. Such directional preferences have been noted in the tracks of existing mammals (e.g. Walther *et al.* 1983, fig. 29), and may sometimes be dictated by physical features of the environment. Thus, at some localities, it is conceivable that dinosaurs were funnelled along a common route by natural barriers such as river-banks, shore-lines, rocky terrain or patches of dense vegetation. For

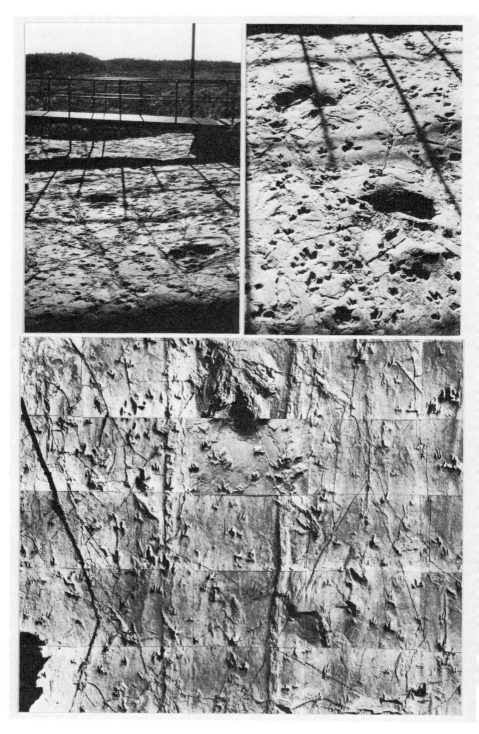

instance, the prevailing direction of the tracks described by Godoy and Leonardi (1985) was found to coincide with the alignment of ripple-marks in the substrate, perhaps indicating that the track-makers had travelled along a shore. A similar relationship between ripple-marks and sauropod tracks was noted by M.G. Lockley (1986a) in the Upper Jurassic Morrison Formation of Colorado (see also Lockley *et al.* 1986). By contrast, G.R. Demathieu (1985) found that reptile trackways in the Middle Triassic of France extended towards and away from shore-lines, rather than alongside them. Demathieu considered that such trackways probably represented direct paths between land and water (see also Gand 1986; Plate 5, p. 68, top right).

There may also be other explanations for two-directional assemblages. For example, the slope of the substrate might have exerted some control on the direction in which animal chose to travel. This is certainly known to be the case with some invertebrates (e.g. traces of cranefly larvae, described by Ekdale and Picard (1985) from the Jurassic of Utah), but is not so clearly demonstrated for vertebrate animals. However, the surface of a modern sand dune, illustrated by E.D. McKee (1982, fig. 11), showed numerous trackways heading straight up or down the line of maximum slope or horizontally along the contours. By contrast only a few track-makers seem to have moved obliquely across the sloping dune face. It is also pertinent to recall that the slope of the substrate might favour the preservation of trackways heading in one particular direction (see Chapter 2). Finally, there is the possibility that parallel trackways resulted from dinosaurs selecting the most direct route across unattractive or potentially

Plate 14 Dinosaur tracks at the Lark Quarry site, mid-Cretaceous of Queensland, Australia. Evidence of a dinosaur stampede about 100 million years ago? Top left: Part of site photographed at sunrise to emphasize surface relief. The first six footprints of a carnosaur's trackway extend from top left to lower right (print number 3 is partly concealed by shadow of raised walkway); stride length is about 3.3 m. Subsequently, more than 160 small dinosaurs, both ornithopods and coelurosaurs, ran in the opposite direction, perhaps in response to the carnosaur's approach. Top right: Closer view of carnosaur footprints 4 and 5, and the numerous smaller dinosaur tracks that surround and overlie them. Bottom: Part of photomosaic showing carnosaur footprints 6 (top centre) and 7 (very shallow print towards lower right), with about 300 footprints of small ornithopods and coelurosaurs. The area shown is about 9 m^2 and represents about one-twentieth of the whole site.

dangerous areas. This effect is seen today in Death Valley, California, where trackways reveal that animals tend to cross the inhospitable valley floor by the shortest practicable route.

It is conceivable that some two-directional assemblages resulted from two groups of dinosaurs traversing a single area (e.g. Ostrom 1972: 296), though the most convincing evidence of gregarious behaviour comes from unidirectional assemblages.

Unidirectional assemblages

Some unidirectional assemblages comprise only a few trackways, so that it is possible to investigate the interactions between one dinosaur and another (e.g. Plate 10, p. 160, bottom left). Elsewhere, dozens or even hundreds of trackways may share a common orientation, testifying to the behaviour of dinosaurs in large groups (e.g. Plate 13, p. 306, top left).

A straightforward example was described by J.B. Delair and P.A. Brown (1974) from the coast of Dorset, in southern England. Here, in the space of a few square metres, the subparallel trackways of three iguanodont dinosaurs were discovered, all headed in a northeasterly direction (Figure 11.7a). From the published diagram of the site it may be estimated that the three track-makers were walking at speeds between 0.6 and 0.8 m/s. The overall correspondence in direction, gait and speed is certainly consistent with the idea that these three dinosaurs were travelling as a group.

Also from the Lower Cretaceous rocks of Dorset comes a more puzzling example (Figure 11.7b). This comprises two nearly coincident trackways, both heading in a southwesterly direction, which were exposed by quarrying in 1961. Unfortunately, there is considerable uncertainty about the identity of the track-makers, which have sometimes been identified as two ornithopods (*Iguanodon?*), sometimes as two theropods (*Megalosaurus?*), and sometimes as one of each (Anonymous 1962, 1964; Charig and Newman 1962; Swaine 1962; Delair 1966; Thulborn 1984). Regardless of these problems, the trackways are unlikely to have been made by two dinosaurs travelling together or in rapid succession. D.B. Norman (1980) pointed out that the left-hand trackway was made by an animal moving on all fours whereas the right-hand track-maker was walking bipedally. More significant is the fact that the left-hand track is much eroded and appears to be considerably older than the right-hand one: its

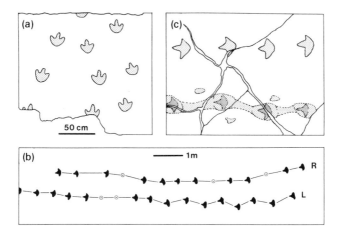

Figure 11.7 Unidirectional dinosaur tracks in the Lower Cretaceous of Dorset, southern England. (a) Three subparallel trackways of ornithopod dinosaurs (?*Iguanodon*). (b) Two nearly coincident trackways; the ringed points indicate damaged or missing footprints. (c) Detail of the same two trackways. Note that the right-hand trackway (above) was made by a biped whereas the left-hand trackway (below) is that of a quadruped and includes small elliptical or bilobed handprints. The pes prints of the left-hand trackway are weathered into basin-like depressions linked together by an erosional channel. These preservational differences imply that the two trackways were made at different times. (Adapted from Delair and Brown 1974 (a), Charig and Newman 1962 (b), Norman 1980, Haubold 1984 (c).)

individual footprints are weathered into almost featureless basins that are connected by a run-off channel that might be mistaken for a trail drag (Figure 11.7c). By contrast, the right-hand trackway comprises sharply defined tridactyl footprints. These differences in preservation imply that the two trackways were formed at different times and that their near-coincidence might be fortuitous. Even so, this explanation is less than satisfactory because a third trackway was discovered about 3 m away, and this also shared the same orientation (see Delair and Lander 1973; Charig 1979: 32). Perhaps these three differently preserved but closely parallel trackways betray the existence of a well-established route that was followed, from time to time, by one or more dinosaurs.

A less perplexing series of parallel tracks was described by P.J. Currie (1983) from the Lower Cretaceous of British Columbia. The 11 trackways were attributed to hadrosaurs, and, of these, Currie (1983: 69) identified at least four that were made by animals walking side by side:

> The four trackways follow the same sinuous curves. . . The trackways are close together at points, but do not intersect. One possible interpretation is that the four animals were walking so close together that when [one of them] . . . changed course suddenly, the courses of the remaining three animals were affected to avoid collision.

From these trackways, and similar ones at a second site, Currie inferred that herds of hadrosaurs were accustomed to travelling on a broad front, with the animals side by side and seldom crossing paths.

A slightly bigger assemblage of unidirectional tracks was described by J.H. Ostrom (1972) from Lower Jurassic sediments at Mt Tom, near Holyoke in Massachusetts (Figure 11.8a). Here, 19 trackways of the ichnogenus *Eubrontes* were found heading in a westerly direction. In addition, three *Eubrontes* trackways were heading to the east, and a few other tracks (*Anchisauripus*, *Grallator*) were aligned in other directions. The numerous *Eubrontes* tracks heading to the west are certainly suggestive of dinosaurs travelling in a group. An even larger and more complicated assemblage, described by G. Leonardi (1984a) from the Upper Cretaceous of Bolivia, included the tracks of about 60 bipedal dinosaurs, all headed in roughly the same direction (Figure 11.8b).

Still more impressive is a series of sauropod trackways at the Davenport Ranch site, in the Lower Cretaceous of Texas (Figure 11.9; Plate 12, p. 278, bottom left). Originally described by R.T. Bird (1944), this assemblage comprised trackways of 23 sauropods, all headed in a single direction. Bird drew up an excellent chart of the trackways, which he attributed to a herd of sauropods on the move, and this idea was recently endorsed by M.G. Lockley (in Farlow 1987: 21), who calculated that the speeds of the track-makers fell into the appropriately narrow range of 1.23 to 1.85 m/s. In 1968, R.T. Bakker suggested that this sauropod herd had a definite social structure, with the smaller individuals in the middle of the group so that their flanks would be protected by bigger animals. J.H. Ostrom (1972: 295) agreed that Bakker's interpretation was entirely plausible, but considered that it was not readily apparent from Bird's (1944) chart of the trackways. Recently, Lockley (1986b) has attempted to unravel the sequence in which the 23 track-makers traversed the site. He pointed out that the

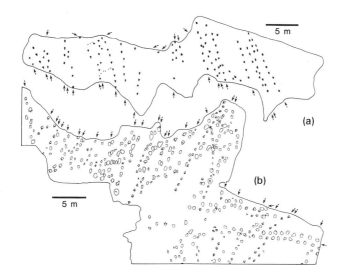

Figure 11.8 Charts of two trackway assemblages, each including a large number of unidirectional tracks. (a) Mt Tom site, in the Lower Jurassic of Massachusetts, USA. Most trackways were assigned to the ichnogenus *Eubrontes*, including 19 that share common orientation towards the west. (b) Part of the Toro Toro site, in the Upper Cretaceous of Bolivia. Many of the numerous tracks heading NNE were attributed to carnosaurs. (Adapted from Ostrom 1972 (a), Leonardi 1984a (b).)

trackways were in two groups, ranged side by side, and suggested that in one of these groups the larger animals had preceded the smaller ones. His conclusion (1986b: 41) is that the Davenport Ranch sauropods

> . . . must have been travelling in a series of groups veering from left to right. Rather than travelling on a broad front with larger forms flanking the group, there must have been much following in line.

According to this interpretation, juvenile sauropods might, literally, have followed in the footsteps of their parents. Elsewhere, in the Upper Jurassic Morrison Formation of Colorado, Lockley and his colleagues have discovered the trackways of sauropods that 'were probably walking in some type of staggered or spearhead formation' (Lockley *et al.* 1986: 1174).

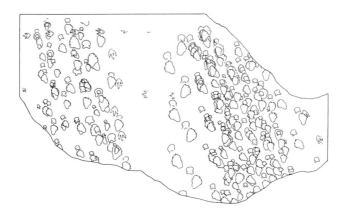

Figure 11.9 Outline chart to show distribution of sauropod trackways at the Davenport Ranch site, in the Lower Cretaceous of Texas, USA. There are trackways of at least 23 sauropods that apparently travelled together as a herd. The area shown is roughly 17 × 10 m. (Adapted from Bird 1944, 1985.)

One of the world's most famous trackway assemblages, currently on display in the American Museum of Natural History, New York, is commonly attributed to a sauropod that was being stalked or pursued by a carnosaur. The footprints of the carnosaur overlap those of the sauropod at several points, and the trackways of both animals veer sideways in almost identical fashion (Plate 10, p. 160, bottom left). J.O. Farlow considered (1987: 22) that these facts are 'difficult to explain if one argues that the two trackways were made independently of each other'. This scenario, with a dinosaurian killer hot on the trail of its victim, has enthralled generations of museum visitors, but in reality the story is not quite so straightforward. In the first place, one of the carnosaur's left footprints is missing from its otherwise perfectly preserved trackway. To account for this missing footprint, R.T. Bird imagined that the carnosaur had fastened on to the sauropod and had been dragged off its feet for an instant, thus performing an involuntary hop. 'This is, perhaps, reading too much into a missing footprint' (Farlow 1987: 22). More significant is the fact that several other trackways, both of sauropods and theropods, were uncovered at the same site. These additional trackways prompted Bird to envisage an entire sauropod herd being harried by a pack of carnosaurs! Unfortunately, this epic interpretation is difficult to substantiate, because

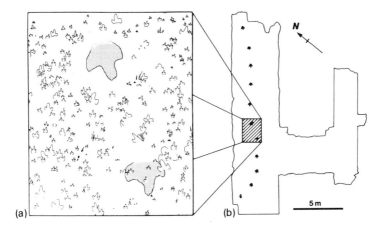

Figure 11.10 Part of a rich trackway assemblage at the Lark Quarry site, in the mid-Cretaceous of Queensland, Australia. The enlarged area shows two footprints of a carnosaur, which approached a water-hole located to the south-west. Subsequently, at least 160 small ornithopods and coelurosaurs, many no bigger than chickens, ran to the north-east, possibly in a stampede triggered by the approach of the carnosaur. Note that footprints of the small dinosaurs are superimposed on those of the carnosaur. (Adapted from Thulborn and Wade 1984.)

some of the theropod tracks form a two-directional pattern and others appear to be oriented randomly. Whatever its true significance, this trackway assemblage certainly involved far more than a single carnosaur trailing a single sauropod.

The hunting activities of a carnosaur have also been invoked to explain an unusual assemblage of tracks at the Lark Quarry site, in western Queensland, Australia (Figure 11.10; Plate 14, p. 316). This site was originally the bed of a broad drainage channel that disgorged into a lake or water-hole to the south-west. At the time most trackways were formed, the area was floored with fine-grained mud of about the ideal consistency for preserving footprints. A carnosaur, about 2.6 m high at the hip, walked down the drainage channel towards the water-hole on a slightly weaving course, veering first to the left and then to the right (see Figure 5.6). The first part of the carnosaur's trackway is noticeably different from the second. Initially, this dinosaur took long strides and its feet plunged right through the mud to rest on firmer sandy sediments beneath. Suddenly, after four strides, it began to take shorter steps and its footprints became shallower. These shallow footprints lack any impression of a 'heel', almost as if the animal were walking more cautiously, on the tips of its toes (see Figure 5.11). After a total of nine strides the carnosaur took an abrupt turn to its right, entering an area now destroyed by erosion. At this point, or shortly thereafter, the carnosaur's approach seems to have triggered a stampede of numerous small ornithopods and coelurosaurs that were gathered around the water-hole to the south-west. At least 160 of these small dinosaurs ran up the drainage channel, to the north-east, leaving thousands of footprints pointing in a single direction. Footprints of both coelurosaurs and ornithopods were found superimposed upon those of the carnosaur.

This reconstruction of events at Lark Quarry has been explained in much greater detail elsewhere (Thulborn and Wade 1984), and only a few remarks are needed here. First, there seems no doubt that the ornithopods and coelurosaurs were running: the mean value for relative stride length (SL/h) was found to be 3.7, in contrast to the mean value of 1.3 for bipedal dinosaurs elsewhere in the world (see Figure 9.2). Next, a group of 160 dinosaurs running in a single direction must surely constitute a stampede or some similar event. And, finally, it cannot be proved beyond doubt that the carnosaur's approach actually initiated the stampede. Nevertheless, it seems plausible that a gathering of ornithopods and coelurosaurs, drinking and foraging round a water-hole, might have been startled by the approach of a big predator. It is risky to read so much significance into the evidence of a single carnosaur trackway, but the only alternative is to admit that the unusual behaviour of the ornithopods and coelurosaurs is inexplicable. Regardless of its interpretation (see also Paul 1988: 36) the Lark Quarry site is unique in one respect: it is the only known site that preserves trackways made by numerous dinosaurs running in a single direction. This is unusual behaviour, by any standards, and is certainly consistent with the scenario of a dinosaur stampede.

Gregarious dinosaurs

The possibility that unidirectional trackway assemblages resulted from dinosaurs travelling in groups was first seriously investigated by J.H. Ostrom in 1972. Ostrom's findings have been substantiated at major trackway sites around the world, and there is now compelling evidence that certain dinosaurs were gregarious (see also Ostrom 1984).

R.E. Weems (1987: 33) recommended that groups of dinosaurs should be termed 'flocks' because this 'is an archosaurian term, whereas the term 'herd' is applied to mammals'. In fact, the term 'flock' is applicable to certain mammals – for example, sheep, goats, camels and lions – while the term 'herd' can be used to denote a group of animals in general (Partridge 1965: 307). W.P. Coombs (1975: 25) considered that

> ... use of the word 'herd' is troublesome because it is usually reserved for ungulate mammals and its application implies mammal-like behavior, interaction, and coordinated activity. Sauropod aggregates might equally well be called 'pods', or 'flocks', or 'troops', or 'packs'. . .

Regardless of these quibbles it seems legitimate to refer to groups of herbivorous dinosaurs (including ornithopods and sauropods) as 'herds' and to groups of theropods as 'packs'.

There are several reports of sites trampled by herds of sauropods (e.g. Bird 1944; Malz 1971; Kaever and de Lapparent 1974; Ishigaki 1986), and at some of these the distribution of trackways seems to indicate that the herds had a definite social structure, with small animals following bigger ones, or with the animals strung out en echelon or in V-formation. Isolated trackways are also fairly common, perhaps indicating that some sauropods were solitary creatures (e.g. Taquet 1972; Dutuit and Ouazzou 1980; Plate 12, p. 278, bottom right).

The evidence is not so clear for prosauropods. It has long been suspected that these dinosaurs did associate in groups (e.g. von Huene 1928), though there are, as yet, no trackway assemblages to support this notion. Authentic prosauropod trackways are uncommon (or have gone unrecognized; see Chapter 6), and seem to have been made by animals travelling alone or, at most, in small groups. However, one rich assemblage of bones has been attributed to a herd of prosauropods that perished in a catastrophic mudflow (Weishampel 1984b).

There is definite evidence of herding behaviour in some of the larger ornithopods, including hadrosaurs. At some sites their tracks are distributed haphazardly, perhaps indicating that the animals associated in loosely organized groups (e.g. Balsley 1980; Lockley *et al.* 1983), but in other cases the trackways are parallel, implying that the animals may have advanced over a broad front. The segregation of large and small hadrosaur tracks in the Cretaceous of Canada led P.J.

Currie (1983: 63) to conclude that 'juveniles were gregarious and stayed together after hatching until they were large enough to join herds of more mature animals'. While the tracks of large ornithopods commonly occur in groups, there also exist trackways made by solitary individuals (e.g. *Sousaichnium*; Plate 11, p. 208, bottom left). Much the same may be said for the smaller ornithopods, which seem to have been gregarious in some instances and solitary in others. Numerous trackways at the Lark Quarry site in Australia (Plate 14, p. 316) indicate that these animals could, and sometimes did, congregate in large numbers, and R.E. Weems (1987) has reported the tracks of an ornithopod herd from the Upper Triassic of Virginia.

Individually, the smaller theropods, or coelurosaurs, were probably incapable of killing larger dinosaurs, but they might perhaps have gathered in packs in order to pull down their prey (see Ostrom 1969a,b on *Deinonychus*). Coelurosaurs that roamed as solitary individuals, or in small groups, were likely to have subsisted on smaller prey, such as insects and lizards, perhaps supplemented with eggs, fruit and carrion. The evidence from Lark Quarry, where tracks of numerous coelurosaurs are intermingled with those of ornithopods, raises another possibility: these two types of dinosaurs might have travelled together as a 'mixed herd', as do various African ungulates today (Krassilov 1980; Thulborn and Wade 1984). In this case one might envisage the coelurosaurs as opportunists, ready to seize insects and other small animals as they were flushed from the vegetation by an ornithopod herd moving through its feeding grounds.

From anatomical and palaeoecological evidence J.O. Farlow (1976) inferred that carnosaurs might have hunted singly, like leopards, or in small packs, like lions or wolves. These conclusions are borne out by the frequent occurrence of carnosaur trackways singly or in small numbers. One would not expect big predators to associate in large groups, yet some carnosaurs may have done so, to judge from G. Leonardi's report (1984a) of about 60 unidirectional trackways in the Upper Cretaceous of Bolivia.

Both the tracks and the skeletons of stegosaurs tend to occur as isolated examples, and much the same is true for ankylosaurs. Such rare and scattered occurrences might indicate that these armoured ornithischians were solitary in their habits, though the evidence is really too inadequate to allow any firm conclusions. Ceratopsian trackways are equally elusive. Nevertheless, it is often assumed that these horned dinosaurs were gregarious creatures, and this is confirmed by the report of bone assemblages resulting from mass mortality of ceratopsian herds (Currie and Dodson 1984). The rarity of tracks made by stegosaurs, ankylosaurs and ceratopsians might also indicate that these animals frequented uplands, plains or densely vegetated environments that were inimical to the preservation of fossil tracks.

Finally, there is the possibility that some dinosaurs were creatures of habit, making regular use of well-defined routes to their preferred feeding grounds and watering-places (Malz 1971). There is certainly evidence that some dinosaurs visited communal rookeries or nesting areas (e.g. Kerourio 1981; Horner 1982, 1984b; Breton *et al.* 1985), and it conceivable that they did so on a regular annual basis. Furthermore, it is now known that dinosaurs were not confined to equable climatic zones, and that some of them existed at high latitudes. Various dinosaurs, including theropods, hadrosaurs and ceratopsians, were able to survive within the Arctic Circle and even at palaeolatitudes between 80° and 85°N (Axelrod 1984; Roehler and Stricker 1984; Spicer and Parrish 1987). And discoveries on the southern coast of Australia reveal that dinosaur communities were established at palaeolatitudes of 60° to 70°S during the Cretaceous (Flannery and Rich 1981). Some dinosaurs might have dwelled permanently in circumpolar regions, perhaps undergoing hibernation during winter, but others may have embarked on seasonal migrations to take advantage of ephemeral food resources. That is, dinosaurs might have moved into higher latitudes during springtime and early summer, to exploit the new season's flush of vegetation, and would return to warmer regions for the winter. Predatory dinosaurs would, presumably, have followed the migrating herds. Annual migrations of dinosaur herds would establish a system of well-defined routes and trails, and some of these might appear in the fossil record as accumulations of parallel trackways.

STRUCTURE OF DINOSAUR COMMUNITIES

A polytypic assemblage of trackways may be regarded as a census of the dinosaurs that inhabited a particular area at a particular time. So, for example, the Lark Quarry assemblage (see Figure 11.10) indicates that carnosaurs, coelurosaurs and ornithopods coexisted in what is now western Queensland during the middle of the Cretaceous. Here,

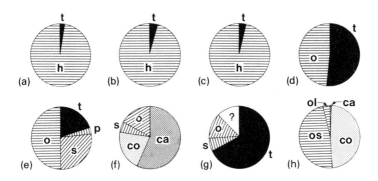

Figure 11.11 Composition of dinosaur communities indicated by skeletal remains (a) and ichnocoenoses (b–h). (a) Skeletal remains in the Morrison Formation (Upper Jurassic) of the USA. (b) Tracks in the Morrison Formation (Upper Jurassic) of the Purgatoire River, Colorado, USA. (c) Tracks in the Mesaverde Formation (Upper Cretaceous) of Colorado, USA. (d) Tracks in the Gething Formation (Lower Cretaceous) of British Columbia, Canada. (e) Tracks in the Bull Run Formation (Upper Triassic) of Virginia, USA. (f) Tracks in the Sousa Formation (Lower Cretaceous) of Paraiba province, Brazil. (g) Combined tracks from 25 localities in the Glen Rose Formation (Lower Cretaceous) of Texas, USA. (h) Tracks in the Winton Formation (mid-Cretaceous) of western Queensland, Australia. Herbivorous dinosaurs are indicated by shading of parallel lines: H, herbivores in general (ornithischians plus sauropods); O, ornithopods; OL, large ornithopods (iguanodonts); OS, small ornithopods; P, prosauropods; S, sauropods. Carnivorous dinosaurs: CA, carnosaurs; CO, coelurosaurs; T, theropods in general. In some track assemblages (b,c) the proportion of carnivores to herbivores matches that in body fossil assemblages (a). Other trackway assssemblages are dominated by carnivore tracks (d,f,g). (Adapted from Lockley 1986a (a–c), and based on data from Currie 1983 (d), Weems 1987 (e), Leonardi 1984a (f), Farlow 1987 (g), Thulborn and Wade 1984 (h).)

as in many other cases, the trackway assemblage provides a valuable insight into the composition of the local dinosaur fauna, even betraying the presence of dinosaurs that are unknown from skeletal remains.

Further, it might be inferred that the relative abundances of ichnotaxa, or footprint types, would correspond to the relative abundance of the various track-makers. Thus, the footprint assemblage at Lark Quarry might indicate that carnosaurs were rare and solitary creatures, that big ornithopods were moderately common, and that the smaller ornithopods and coelurosaurs were both extremely abundant (Figure 11.11h). How reliable are such inferences? In a few cases trackway assemblages do seem to represent a reliable census of the local dinosaur fauna. This has been demonstrated for a trackway assemblage in the Upper Jurassic Morrison Formation of Colorado, where M.G. Lockley and his colleagues (1986) classified the tracks into two groups – those of carnivores (theropods) and those of herbivores (ornithopods and sauropods). The ratio of carnivore tracks to herbivore tracks proved to be almost identical to the ratio of carnivore skeletons to herbivore skeletons in the Morrison Formation (Figure 11.11a,b).

Unfortunately, such instances are rare. More commonly there are grounds for suspecting that a trackway assemblage is biased, so that the diversity and abundance of footprints do not correspond very closely to the diversity and abundance of track-makers. Dinosaurs that were relatively rare might produce disproportionately large numbers of tracks whereas dinosaurs that were common might leave few trackways or none at all. In the case of the Lark Quarry assemblage, for example, it was suspected that coelurosaurs were actually more common than their trackways might indicate (Figure 11.11h): these small and lightweight dinosaurs had broad-spreading feet, and some of them might have traversed the site without leaving any recognizable trackways (Thulborn and Wade 1984: 443). In this instance a preservational bias, favouring bigger and heavier track-makers, might lead one to underestimate the local abundance of coelurosaurs. An observational bias may also be introduced if some tracks are overlooked on account of their poor preservation or unusual size. This, again, is noticeable at the Lark Quarry site: the tracks of the coelurosaurs and ornithopods are eye-catching on account of their abundance and their sharp definition, whereas the much bigger and basin-like footprints of the carnosaur are frequently overlooked by visitors to the site.

The fact that many assemblages are dominated by theropod tracks (Figure 11.11d,f,g) is decidedly puzzling, because one would expect the predatory dinosaurs to have been outnumbered by plant-eaters. The numerical dominance of theropod tracks shows up very clearly in taxonomic lists. For example, H. Haubold's most recent listing (1984) of the world's dinosaur tracks includes 62 ichnotaxa or footprint types made by theropods but only 23 attributed to ornithischians. Several ideas have been advanced to explain this curious imbalance.

First, it is possible that some footprints have been attributed to theropods by mistake. Here it should be recalled that it is sometimes difficult to distinguish the footprints of ornithopods from those of theropods (see Chapter 7). Second, some of the plant-eating dinosaurs may be underrepresented, in terms of fossil tracks, by virtue of their gregarious habits. For instance, J.O. Farlow (1987: 18) has remarked that some of the sauropods seem to have travelled in herds, so that

> . . . one might expect tracksites with sauropod footprints to be uncommon (due to the low probability of a herd's passing through any particular area), but for [the] footprints to be abundant at those sites where they do occur.

By contrast, the bigger theropods are likely to have travelled as lone individuals, or in small groups. Such differences in behaviour might well have resulted in dense concentrations of sauropod tracks at relatively few sites, and in modest numbers of theropod tracks at many sites.

Next, it is likely that the predatory dinosaurs roamed far and wide in search of their prey, while the contemporary plant-eating dinosaurs did not travel so extensively. In other words, the footprints of theropods may be common simply because these animals covered a lot of ground while hunting. A comparable situation exists among mammals, where carnivores travel over four times as far, on a daily basis, as do other mammals of similar size (Garland 1983a). J.O. Farlow has suggested that the same was true in dinosaur communities and that the home ranges of big theropods encompassed hundreds, thousands, or even tens of thousands of square kilometres (Farlow 1987: 19).

Finally, it is possible that theropods frequented environments that were particularly favourable to footprint preservation (Farlow 1987: 18):

> One possible explanation for the abundance of theropod tracks . . . is that environments suitable for footprint preservation may have been good hunting areas, and so frequently patrolled by meat-eating dinosaurs. This may have been not because large numbers of potential prey lived in these coastal settings, but rather because the vegetation . . . may have been less dense . . . making it possible for a hunter to sight migrating game herds further away than in more inland habitats.

In summary, a trackway assemblage may not be an exact and reliable census of a dinosaur community. The various ichnotaxa at any one site do, indeed, testify that various sorts of dinosaurs visited that particular area. But in many instances the relative abundances of those ichnotaxa might have been affected by preservational bias and by the behaviour and habitat preferences of the track-makers.

STRUCTURE OF DINOSAUR POPULATIONS

A monotypic assemblage is sometimes used to reconstruct the composition of a dinosaur population, on the assumption that all the tracks were made by members of a single dinosaur species. Such an assumption is most likely to be valid when all the tracks are preserved together at one site; it becomes more dubious when evidence is gathered from tracks at two or more sites. In any case this assumption should be adopted with some caution because there is no guarantee that ichnotaxa correspond at all closely to dinosaur taxa based on body fossils. It is always possible that different types of dinosaurs could have produced similar footprints.

Leaving aside this problem, the larger tracks in an assemblage might be attributed to adult dinosaurs whereas progressively smaller ones might be regarded as tracks of subadults, adolescents, juveniles, and perhaps even hatchlings. Here, a second assumption has crept in: namely, that a big animal is older than a small one. This seems reasonable enough, though in fact the assumption is valid *only* if rates of growth were constant during the lives of the track-makers. If the growth rate was not constant a small dinosaur might actually be older than a large dinosaur of the same species. To put this another way, some old animals might be stunted whereas some younger ones might have sustained rapid and exuberant growth through particularly favourable conditions. In addition, there may be size variations associated with sexual dimorphism: males might have grown bigger than females, or vice versa. A third and unspoken assumption is that juvenile dinosaurs inhabited the same environments as adults. This certainly seems to be true for some dinosaurs, such as sauropods (see Figure 11.9), and from the evidence preserved at nesting grounds it may be deduced that hadrosaur hatchlings were tended by their parents until they were big enough to fend for themselves (Horner and Makela 1979; Horner 1984b). On the other hand, some trackways indicate that small dinosaurs, probably juveniles, were independent of adults (Coombs 1982) and may even have been segregated from them (Currie 1983). In these latter circumstances one might expect a single site to preserve the tracks of adults or of juveniles, but not necessarily the tracks of both.

Ideally, it should be possible to predict the absolute age of an animal, in years, from its body dimensions (such as *h*) and, hence, from the size of its footprints. This possibility exists for ungulate mammals, where game-keepers can estimate the age of individual animals from footprint dimensions (Bouchner 1982), and it has also been exploited in ecological studies of African elephants (Western *et al.* 1983). Unfortunately, the same approach cannot be extended to dinosaur tracks, because it is so difficult to ascertain the absolute age of any dinosaur – even one that is represented by a perfectly preserved skeleton. Growth bands do occur in dinosaur teeth (Johnston 1979, 1980), though it is far from certain that these were laid down annually (Bolt and de Mar 1980; Boyce 1980; Meinke *et al.* 1980). In any event, dinosaurs are known to have replaced their teeth through life, so that annual banding in a single tooth would represent only a fraction of a dinosaur's total lifespan. However, bone histology does seem to indicate that some dinosaurs were capable of sustaining rapid growth (Reid 1984, 1987). In the humerus of a half-grown sauropod, *Bothriospondylus*, from the Jurassic of Madagascar, A. de Ricqlès (1983) found evidence of about 10 growth rings and estimated that as many as 16 others had been destroyed during bone remodelling. From this evidence de Ricqlès predicted that *Bothriospondylus* would have reached two-thirds of its maximum recorded body length at an age of roughly 43 years. Elsewhere, T.J. Case (1978) extrapolated the growth rates of living reptiles to predict that a European sauropod, *Hypselosaurus*, attained two-thirds of its maximum body length after about 62 years. (Two-thirds of maximum body length was regarded as a significant dimension because this is about the size at which many existing reptiles reach sexual maturity.) These crude estimates give some idea of the time taken for certain dinosaurs to grow to maturity, but they are of no practical value for assessing the absolute age of a trackmaker. Moreover, it is not known whether dinosaurs continued to grow indefinitely throughout their lives, like crocodiles, or whether they ceased growth at a specific size, like birds and mammals.

In short, there are some insuperable problems. The footprints in an ichnospecies may not represent a single species of dinosaur. Even if

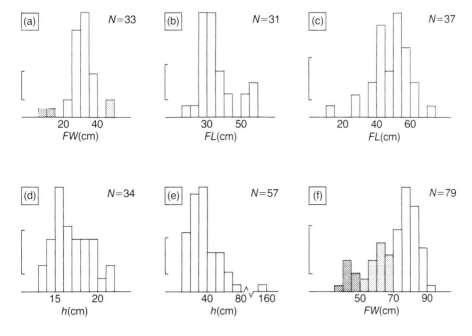

Figure 11.12 Size-frequency distributions for various assemblages of dinosaur tracks. In all cases, the scale bar indicates 10% of sample. Various measures of size are used: footprint width (*FW*), footprint length (*FL*) and estimated height at the hip (*h*). (a) Ornithopod tracks from the Upper Jurassic of Colorado, USA; the largest and smallest tracks (shaded) were attributed to different species. (b) Theropod tracks (ichnogenus *Irenesauripus*) from the Lower Cretaceous of British Columbia, Canada. (c) Ornithopod tracks (ichnogenus *Amblydactylus*) from the Lower Cretaceous of British Columbia, Canada. (d) Coelurosaur tracks (ichnogenus *Skartopus*) from the mid-Cretaceous of Queensland, Australia. (e) Ornithopod tracks (ichnogenus *Wintonopus*) from the mid-Cretaceous of Queensland, Australia. (f) Ornithopod tracks from the Lower Cretaceous of Colorado, USA; small and medium-sized tracks (shaded) were attributed to juveniles. Note that some distributions are skewed to the right (d,e) whereas others are skewed to the left (c,f). Based on data from Lockley *et al.* 1986 (a), Currie 1983 (b,c), Thulborn and Wade 1984 (d,e), Lockley *et al.* 1983 (f).)

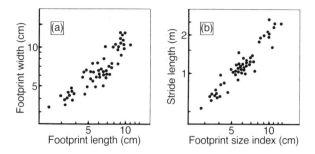

Figure 11.13 Scatter diagrams based on data from small ornithopod trackways (ichnogenus *Wintonopus*) in the mid-Cretaceous of Queensland, Australia. Note that the distribution tends to fall into three size-classes that might represent age-classes. (Adapted from Thulborn and Wade 1984.)

they do, the bigger footprints may not necessarily be those of older animals. And even if the bigger footprints are assumed to be those of older animals it will be impossible to ascertain the absolute ages of the track-makers.

Despite these problems some investigators have identified small footprints as those of juvenile dinosaurs (e.g. Leonardi 1981; Currie 1983; Lockley *et al.* 1983). Such identifications seem most convincing when there is a considerable range in size from the largest ('adult') tracks to the smallest ('juvenile') ones. Occasionally, the tracks of juveniles have been separated from those of adults on an arbitrary basis (e.g. Figure 11.12f), though in some assemblages there appear to be indications of definite size-classes. This is the case with a sample of ornithopod tracks from the Lark Quarry site in Queensland (Figure 11.13). Here it is possible to distinguish three size-classes, though there is, unfortunately, no way of testing the possibility that these might represent cohorts or annual age-classes. Three discrete size-classes are also apparent in a sample of ornithopod tracks from the Upper Jurassic of Colorado (Figure 11.12a), but in this instance it was considered that 'both the very small and very large tracks fall well outside the main cluster, possibly indicating that they were made by different species' (Lockley *et al.* 1986: 1171).

G. Leonardi (1981) was impressed by the rarity of small dinosaur tracks in the Cretaceous of Brazil and inferred that juvenile dinosaurs were correspondingly rare. This conclusion agrees with the notion of stable dinosaur populations, composed mainly of mature animals of extreme longevity. Presumably, the eggs, hatchlings and juveniles would have suffered heavy mortality, and there would have been a low rate of recruitment. However, this particular view of dinosaur populations, which was developed in detail by N.D. Richmond (1965), seems to be contradicted by more recent discoveries revealing that juvenile dinosaurs were surprisingly common in some places. J.R. Horner and R. Makela (1979) found that more than 80 % of dinosaur specimens collected from the Upper Cretaceous Two Medicine Formation of Montana could be identified as juveniles or subadults, and elsewhere there have been several reports of baby dinosaurs or the tracks of juveniles (e.g. Bonaparte and Vince 1979; Coombs 1980a, 1982; Carpenter 1982a; Currie 1983). It now seems clear that some local dinosaur populations were dominated by small animals and that juveniles were far from uncommon.

The numerical importance of small dinosaurs is apparent in a sample of 57 ornithopod trackways preserved at the Lark Quarry site, in Queensland (Figure 11.12e). These tracks provide a size-frequency distribution that is distinctly skewed to the right and rather similar to some of the curves derived by A.J. Boucot (1953) from 'life assemblages' of fossils. Such a size-distribution might be interpreted in any of several ways. First, it might be taken to indicate survivorship in a population of small cursorial ornithopods. According to this interpretation the animals would normally have grown to achieve *h* of about 30 cm. Thereafter, the mortality rate would have reached a peak, with fewer and fewer adults surviving to reach greater and greater size. This interpretation assumes, of course, that the sample of 57 track-makers is truly representative of the population from which it was drawn; it also assumes that there is a strong and consistent correlation between size and age. Alternatively, Figure 11.12e might be interpreted as a survivorship curve drawn from a population of big, and presumably graviportal, ornithopods (with assumptions as before). In this case it would appear that there was a high rate of mortality among juveniles or subadults once these had reached *h* of about 40 cm. The adults, with *h* of 1.5 m or more, would have been comparatively rare and long-lived. A third possibility is that the sample is composed mainly of juvenile ornithopods. This might imply that juveniles were segregated from adults, as may have been the situation among some hadrosaurs (Currie 1983). In these circumstances the steep decline in numbers of track-makers with *h* greater than 30 cm might result from mortality plus the departure of animals to join herds of adults that lived elsewhere. These various possibilities were discussed by R.A. Thulborn and M. Wade (1984), who concluded that the simplest interpretation was probably the first one – that the 57 trackways represent a population of ornithopods which rarely grew to achieve *h* greater than about 70 cm. Ontogenetic studies by G. Callison and H.M. Quimby (1984) have confirmed that some species of dinosaurs, including certain ornithopods, never grew much bigger than a chicken or a turkey.

Other assemblages provide a different picture, with small track-makers outnumbered by large ones (Figure 11.12c,f). Here it seems that population structure may have been closer to the pattern described by N.D. Richmond (1965), with a predominance of big, long-lived individuals and a low rate of recruitment. Such an interpretation would seem appropriate for some of the bigger dinosaurs, such as sauropods and hadrosaurs.

In summary, the subject of demography is fraught with problems, even when biologists investigate populations of living animals. The problems are still greater for palaeontologists who attempt to study populations of extinct animals, and particularly those that are represented by evidence as tenuous as footprints. It may not be legitimate to identify small footprints as those of juvenile dinosaurs, because dinosaurian rates of growth are virtually unknown and may not have been constant. Moreover, some dinosaurs never grew much bigger than chickens, and their footprints might be indistinguishable from those made by the juveniles of bigger dinosaurs. Such intractable problems prohibit any meaningful study of dinosaurian demography based on footprints. For the present, such studies are better founded on dinosaur body fossils, where it might prove possible to ascertain the relationships between absolute age, sexual maturity and body size (e.g. Case 1978; de Ricqlès 1983).

Finally, it is appropriate to clear up one pervasive misconception. The world of dinosaurs is often portrayed as a world of giants. Yet, in reality, many dinosaurs were small creatures, some no bigger than rats or roosters (Plate 10, p. 160, bottom right). This fact is not always obvious in museum galleries, which usually display the biggest and most impressive skeletons, but it is apparent from recent discoveries of skeletons and trackways. In some places small dinosaurs seem to have been extremely common, and these were not necessarily the juveniles of bigger dinosaurs. This simple fact has all too often been ignored, despite its obvious importance for our understanding of dinosaur biology.

References

Abel, O. (1912) *Grundzuge der Paläobiologie der Wirbeltiere*. Schweizerbart, Stuttgart.

Abel, O. (1926) *Amerikafahrt*. Gustav Fischer, Jena.

Abel, O. (1935) *Vorzeitliche Lebensspuren*. Gustav Fischer, Jena.

Ager, D.V. (1963) *Principles of Paleoecology*. McGraw-Hill, New York and London.

Aguirrezabala, L.M. and Viera, L.I. (1980) Icnitas de Dinosaurios en Bretún (Soria). *Munibe*, **32**(3–4): 257–9.

Aguirrezabala, L.M. and Viera, L.I. (1983) Icnitas de Dinosaurios en Santa Cruz de Yangüas (Soria). *Munibe*, **35**(1–2): 1–13.

Alexander, R.McN. (1976) Estimates of speeds of dinosaurs. *Nature, London*, **261**: 129–30.

Alexander, R.McN. (1977) Mechanics and scaling of terrestrial locomotion. In *Scale Effects in Animal Locomotion* (ed. T.J. Pedley). Academic Press, London, pp. 93–110.

Alexander, R.McN. (1981) Factors of safety in the structure of animals. *Science Progress*, **67**: 109–30.

Alexander, R.McN. (1985) Mechanics of posture and gait of some large dinosaurs. *Zoological Journal of the Linnean Society*, **83**: 1–25.

Alexander, R.McN. (1989) Mechanics of fossil vertebrates. *Journal of the Geological Society, London*, **146**: 41–52.

Alexander, R.McN. and Jayes, A.S. (1983) A dynamic similarity hypothesis for the gaits of quadrupedal mammals. *Journal of Zoology, London*, **201**: 135–52.

Alexander, R.McN., Langman, V.A. and Jayes, A.S. (1977) Fast locomotion of some African ungulates. *Journal of Zoology, London*, **183**: 291–300.

Allen, H. (1888) Materials for a memoir on animal locomotion. In *Animal Locomotion; the Muybridge work at the University of Pennsylvania – the method and the result* (by W.D. Marks, H. Allen and F.X. Dercum). Lippincott, Philadelphia, pp. 35–99.

Allen, J.R.L. (1982) *Sedimentary Structures: Their Character and Physical Basis*, Vol. II. Elsevier, Amsterdam.

Allen, P. (1959) The Wealden environment: Anglo-Paris Basin. *Philosophical Transactions of the Royal Society of London*, series B, **242**: 283–346.

Alonso, R.N. (1980) Icnitas de dinosaurios (Ornithopoda, Hadrosauridae) en el Cretácico superior del norte de Argentina. *Acta Geológica Lilloana*, **15**(2): 55–63.

Alonso, R.N. (1987) Valoración icnoavifaunística de ambientes boratíferos. *IV Congreso Latinoamericano de Paleontología, Bolivia, 1987*, **1**: 586–97.

Alonso, R.N. (1989) Late Cretaceous dinosaur trackways in northern Argentina. In *Dinosaur Tracks and Traces* (eds D.D. Gillette and M.G. Lockley). Cambridge University Press, Cambridge, pp. 223–8.

Alonso, R.N. and Marquillas, R.A. (1986) Nueva localidad con huellas de dinosaurios y primer hallazgo de huellas de Aves en la Formación Yacoraite (Maastrichtiano) del Norte Argentino. *Actas, IV Congreso Argentino de Paleontología y Bioestratigrafía, Mendoza, 1986*, **2**: 33–41.

Ambroggi, R. and Lapparent, A.F. de (1954a) Découverte d'empreintes de pas de Reptiles dans le Maestrichtien d'Agadir (Maroc). *Compte Rendu sommaire des Séances de la Société Géologique de France*, sèrie 6, **4**: 50–2.

Ambroggi, R. and Lapparent, A.F. de (1954b) Les empreintes de pas fossiles du Maestrichtien d'Agadir. *Notes et Mémoires du Service des Mines et de la Carte Géologique du Maroc*, **10**: 43–57.

Amstutz, G.C. (1958) Coprolites: a review of the literature and a study of specimens from southern Washington. *Journal of Sedimentary Petrology*, **28**: 498–508.

Anderson, J.F., Hall-Martin, A. and Russell, D.A. (1985) Long-bone circumference and weight in mammals, birds and dinosaurs. *Journal of Zoology, London*, series A, **207**: 53–61.

Anderson, S.M. (1939) Dinosaur tracks in the Lakota Sandstone of the eastern Black Hills, South Dakota. *Journal of Paleontology*, **13**: 361–4.

Andrews, J.E. and Hudson, J.D. (1984) First Jurassic dinosaur footprint from Scotland. *Scottish Journal of Geology*, **20**: 129–34.

Anonymous (1909) Excursions 1908. *Proceedings of the Yorkshire Geological Society*, n.s., **16**: 409–23.

Anonymous (1951) Giant dinosaur's footprints found in Darling Downs colliery. *Queensland Government Mining Journal*, **52**: 582.

Anonymous (1952a) Dinosaur's footprint in Downs colliery. *Queensland Government Mining Journal*, **53**: 107.

Anonymous (1952b) Dinosaur footprints. *Queensland Government Mining Journal*, **53**: 949–50.

Anonymous (1962) The slow march of a Purbeck iguanodon. *New Scientist*, **13**: 186.

Anonymous (1964) A diamond saw for fossil footprints. *New Scientist*, **24**: 530.

Anonymous (1967) Dinosaur State Park. *Bulletin of the Connecticut State Geological and Natural History Survey*, **100**: 3–13.

Anonymous (1982) *Mysteries of the Unexplained*. Readers Digest Association Inc., Pleasantville, New York.

Anonymous (1985) Researchers cold-trail dinosaur with RTV rubber. *Dow Corning Materials News*, January/February: 6–7.

Anonymous (1986a) Footprint from a sparrow-sized dinosaur. *New Scientist*, **109**: 23.

Anonymous (1986b) [*The Making of the Sea of Japan, 100 million years documented by fossils*.] Guide to 4th Special Exhibition, Fukui Prefectural Museum, Fukui City [in Japanese].

Anonymous (1987) Dinosaur tracks in Dorset. *Geology Today*, **3**: 182–3.

Antunes, M.T. (1976) Dinossáurios Eocretácicos de Lagosteiros. *Ciências da Terra (Lisboa)*, **1**: 1–35.

Armstrong, W.G., Halstead, L.B., Reed, F.B. and Wood, L. (1983) Fossil proteins in vertebrate calcified tissues. *Philosophical Transactions of the Royal Society of London*, series B, **301**: 301–43.

Avnimelech, M.A. (1962a) Dinosaur tracks in the Lower Cenomanian of Jerusalem. *Nature, London*, **196**: 264.

Avnimelech, M.A. (1962b) Découverte d'empreintes de pas de Dinosaures dans le Cénomanien inférieur des environs de Jerusalem. *Compte Rendu sommaire des Séances de la Société Géologique de France*, **1962**: 233–5.

Avnimelech, M.A. (1966) Dinosaur tracks in the Judean Hills. *Proceedings of the Israel Academy of Sciences and Humanities, Section of Sciences*, **1**: 1–19.

Axelrod, D.I. (1984) An interpretation of Cretaceous and Tertiary biota in polar regions. *Palaeogeography, Palaeoclimatology, Palaeoecology*, **45**: 105–47.

Axelrod, D.I. (1985) Reply to comments on Cretaceous climatic equability in polar regions. *Palaeogeography, Palaeoclimatology, Palaeoecology*, **49**: 357–9.

Bachofen-Echt, A. (1926) *Iguanodon*-Fährten auf Brioni. *Paläontologische Zeitschrift*, **7**: 172–3.

Baird, D. (1951) Latex molds in paleontology. *The Compass of Sigma Gamma Epsilon*, **28**(4): 339–45.

Baird, D. (1952) Revision of the Pennsylvanian and Permian footprints *Limnopus*, *Allopus* and *Baropus*. *Journal of Paleontology*, **26**: 832–40.

Baird, D. (1954) *Chirotherium lulli*, a pseudosuchian reptile from New Jersey. *Bulletin of the Museum of Comparative Zoology, Harvard University*, **111**: 165–92.

Baird, D. (1957) Triassic reptile footprint faunules from Milford, New Jersey. *Bulletin of the Museum of Comparative Zoology, Harvard University*, **117**: 449–520.

Baird, D. (1963) Fossil footprints or stump holes? *Transactions of the Kansas Academy of Science*, **66**: 397–400.

Baird, D. (1964) Dockum (Late Triassic) reptile footprints from New Mexico. *Journal of Paleontology*, **38**: 118–25.

Baird, D. (1980) A prosauropod dinosaur trackway from the Navajo

Sandstone (Lower Jurassic) of Arizona. In *Aspects of Vertebrate History* (ed. L.L. Jacobs). Museum of Northern Arizona Press, Flagstaff, pp. 219–30.

Baird, D. (1988) Medial Cretaceous carnivorous dinosaur bone and footprints from New Jersey. *Journal of Vertebrate Paleontology*, **8** (suppl.): 8A.

Bakker, R.T. (1968) The superiority of dinosaurs. *Discovery*, **3**(2): 11–22.

Bakker, R.T. (1972) Anatomical and ecological evidence of endothermy in dinosaurs. *Nature, London*, **238**: 81–5.

Bakker, R.T. (1975a) Experimental and fossil evidence for the evolution of tetrapod bioenergetics. In *Perspectives of Biophysical Ecology* (eds D.M. Gates and R.B. Schmerl). Springer Verlag, Berlin, pp. 365–99.

Bakker, R.T. (1975b) Dinosaur renaissance. *Scientific American*, **232**(4): 58–72, 77–8.

Bakker, R.T. (1986a) Dinosaur cruising speed was higher than that of Tertiary mammals. In *First International Symposium on Dinosaur Tracks and Traces, Albuquerque, 1986, Abstracts with Program* (ed. D.D. Gillette). New Mexico Museum of Natural History, Albuquerque, p. 12.

Bakker, R.T. (1986b) *The Dinosaur Heresies: New Theories Unlocking the Mystery of the Dinosaurs and their Extinction*. Morrow, New York.

Ball, L.C. (1933) Fossil footprints. *Queensland Government Mining Journal*, **34**: 384.

Ball, L.C. (1934a) Fossil footprints. Lanefield Colliery, Rosewood District. *Queensland Government Mining Journal*, **35**: 224.

Ball, L.C. (1934b) Fossil footprints. *Queensland Government Mining Journal*, **35**: 297.

Ball, L.C. (1946) Dinosaur footprints. Lanefield Extended Colliery, Rosewood District. *Queensland Government Mining Journal*, **47**: 179.

Ballerstedt, M. (1905) Über Saurierfährten der Wealdenformation Bücke-burgs. *Naturwissenschaftliche Wochenschrift*, **20**: 481–5.

Ballerstedt, M. (1914) Bemerkungen zu den älteren Berichten über Saurier-fährten im Wealdensandstein und Behandlung einer neuen, aus 5 Fussab-drücken bestehenden Spur. *Centralblatt für Mineralogie, Geologie und Paläontologie*, **1914**: 48–64.

Ballerstedt, M. (1922) Zwei grosse, zweizehige Fährten hochbeiniger Bipeden aus dem Wealdensandstein bei Bückeburg. *Zeitschrift der Deutschen Geologischen Gesellschaft, Briefliche Mitteilungen*, **73**: 76–91.

Balsley, J.K. (1980) *Cretaceous Wave-Dominated Delta Systems: Book Cliffs, East-central Utah*. American Association of Petroleum Geologists, Continu-ing Education Department, Tulsa.

Bang, P. and Dahlstrom, P. (1974) *Collins Guide to Animal Tracks and Signs* (translated and adapted by G. Vevers). Collins, London.

Bartholomai, A. (1966) Fossil footprints in Queensland. *Australian Natural History*, **15**(5): 147–50.

Bartholomai, A. and Molnar, R.E. (1981) *Muttaburrasaurus*, a new iguanodontid (Ornithischia: Ornithopoda) dinosaur from the Lower Cretaceous of Queensland. *Memoirs of the Queensland Museum*, **20**: 319–49.

Basan, P.B. (1979) Trace fossil nomenclature: the developing picture. *Palaeogeography, Palaeoclimatology, Palaeoecology*, **28**: 143–67.

Bassett, M.G. and Owens, R.M. (1974) Fossil tracks and trails. *Amgueddfa, Bulletin of the National Museum of Wales*, **1974** (Winter): 1–18.

Bassoullet, J.-P. (1971) Découverte d'empreintes de pas de reptiles dans l'Infralias de la région d'Aïn-Sefra (Atlas saharien – Algérie). *Compte Rendu sommaire des Séances de la Société Géologique de France*, **1971**: 358–9.

Bather, F.A. (1926) U-shaped burrows near Blea Wyke. *Proceedings of the Yorkshire Geological Society*, n.s., **20**: 185–99.

Bather, F.A. (1927) *Arenicoloides*: a suggestion. *Paläontologische Zeitschrift*, **8**: 128–30.

Baud, A. (1978) Pistes de dinosauriens dans les roches vieilles de 200 millions d'années des Alpes valaisannes, un gisement à protéger. *Schweizer Naturschutz*, **1978**(4): 26–7.

Baze, W. (1955) *Just Elephants*. Elek Books, London.

Beasley, H.C. (1896) An attempt to classify the footprints in the New Red Sandstone of this district. *Proceedings of the Liverpool Geological Society*, **7**: 391–409.

Beasley, W.L. (1907) A carnivorous dinosaur: a reconstructed skeleton of a huge saurian. *Scientific American*, **1907** (14 December): 446–7.

Beaumont, G. de (1980) Des Dinosaures dans le Valais. *Musées de Genève*, **202**: 6–13.

Beaumont, G. de and Demathieu, G. (1980) Remarques sur les extrémités antérieures des Sauropodes (Reptiles, Saurischiens). *Compte Rendu des Séances de la Société de Physique et d'Histoire Naturelle de Genève*, n.s., **15**(2): 191–8.

Beckles, S.H. (1851) On supposed casts of footprints in the Wealden. *Quarterly Journal of the Geological Society of London*, **7**: 117.

Beckles, S.H. (1852) On the ornithoidichnites of the Wealden. *Quarterly Journal of the Geological Society of London*, **8**: 396–7.

Beckles, S.H. (1854) On the Ornithoidichnites of the Wealden. *Quarterly Journal of the Geological Society of London*, **10**: 456–64.

Beckles, S.H. (1862) On some natural casts of reptilian footprints in the Wealden beds of the Isle of Wight and Swanage. *Quarterly Journal of the Geological Society of London*, **18**: 443–7.

Bellair, P. and Lapparent, A.F. de (1949) Le Crétacé et les empreintes de pas de dinosauriens d'Amoura (Algérie). *Bulletin de la Société d'Histoire Naturelle de l'Afrique du Nord*, **39**: 168–75.

Bennett, A.F. and Dalzell, B. (1973) Dinosaur physiology: a critique. *Evolution*, **27**: 170–4.

Benton, M.J. and Clark, J.M. (1988) Archosaur phylogeny and the relationships of the Crocodylia. In *The Phylogeny and Classification of Tetrapods*, Vol. 1 (ed. M.J. Benton). Clarendon Press, Oxford, pp. 295–338.

Bernhardi, R. (1834) Thier-Fährten auf Flächen des Bunten Sandsteins bei Hildburghausen. *Neues Jahrbuch für Mineralogie, Geognosie, Geologie und Petrefaktenkunde*, **1834**: 641–2.

Bernier, P. (1984) Les dinosaurs sauteurs. *La Recherche*, **15**: 1438–40.

Bernier, P., Barale, G., Bourseau, J.-P., Buffetaut, E., Demathieu, G., Gaillard, C. and Gall, J.C. (1982) Trace nouvelle de locomotion de chélonien et figures d'émersion associées dans les calcaires lithographiques de Cerin (Kimméridgien supérieur, Ain, France). *Geobios*, **15**: 447–67.

Bernier, P., Barale, G., Bourseau, J.-P., Buffetaut, E., Demathieu, G., Gaillard, C., Gall, J.C. and Wenz, S. (1984) Découverte de pistes de dinosaures sauteurs dans les calcaires lithographiques de Cerin (Kimméridgien supérieur, Ain, France): implications paléoécologiques. *Geobios, Mémoires speciales*, **8**: 177–85.

Bertrand, C.-E. (1903) Les coprolithes de Bernissart. Première partie, les coprolithes qui ont été attribués aux Iguanodons. *Mémoires du Musée Royal d'histoire Naturelle de Belgique*, **1**(4): 1–154.

Beurlen, K. (1950) Neue Fährtenfunde aus der Fränkischen Trias. *Neues Jahrbuch für Geologie und Paläontologie, Monatshefte*, **1950**: 308–20.

Bird, R.T. (1939a) [Untitled letter regarding dinosaur tracks in Colorado.] *Natural History*, **43**: 245.

Bird, R.T. (1939b) Thunder in his footsteps. *Natural History*, **43**: 254–61.

Bird, R.T. (1941) A dinosaur walks into the museum. *Natural History*, **47**: 74–81.

Bird, R.T. (1944) Did *Brontosaurus* ever walk on land? *Natural History*, **53**(2): 60–7.

Bird, R.T. (1953) To capture a dinosaur isn't easy. *Natural History*, **62**: 102–10.

Bird, R.T. (1954) We captured a live brontosaur. *National Geographic Magazine*, **105**(5): 707–22.

Bird, R.T. (1985) *Bones for Barnum Brown: Adventures of a Dinosaur Hunter* (ed. V.T. Schreiber). Texas Christian University Press, Fort Worth.

Biron, P.E. and Dutuit, J.-M. (1981) Figurations sédimentaires et traces d'activité au sol dans le Trias de la formation d'Argana et de l'Ourika (Maroc). *Bulletin du Muséum National d'Histoire Naturelle, Paris*, sèrie 4, section C, **3**: 399–427.

Black, M., Hemingway, J.E. and Wilson, V. (1934) Summer Field Meeting to north-east Yorkshire, August 9th to 20th, 1934. Report by the Directors. *Proceedings of the Geologists' Association*, **45**: 291–306.

Blows, W.T. (1978) *Reptiles on the Rocks*. Isle of Wight Museums Publications, No. 2, Isle of Wight County Council, Newport.

Bock, W. (1952) Triassic reptilian tracks and trends of locomotive evolution. *Journal of Paleontology*, **26**: 395–433.

Bölau, E. (1952) Neue Fossilfünde aus dem Rhät Schonens und ihre paläogeographisch-ökologische Auswertung. *Geologiska Föreningens i Stockholm Förhandlingar*, **74**(1): 44–50.

Bolt, J.R. and Mar, R.E. de (1980) [Untitled comments on growth rings in dinosaur teeth.] *Nature, London*, **288**: 194.

Bonaparte, J.F. (1975) Nuevos materiales de *Lagosuchus talampayensis* Romer (Thecodontia – Pseudosuchia) y su significado en el origen de los Saurischia. *Acta Geológica Lilloana*, **13**: 5–87.

Bonaparte, J.F. (1980) Jurassic tetrapods from South America and dispersal routes. In *Aspects of Vertebrate History* (ed. L.L. Jacobs). Museum of Northern Arizona Press, Flagstaff, pp. 73–97.

Bonaparte, J.F. and Vince, M. (1979) El hallazgo del primer nido de dinosaurios triásicos, (Saurischia, Prosauropoda), Triásico superior de Patagonia, Argentina. *Ameghiniana, Revista de la Asociación Paleontológica Argentina*, **16**: 173–82.

Borgomanero, G. and Leonardi, G. (1981) Um ovo de dinossauro de Aix-en-Provence (Franca) e fragmentos de ovos fosseis de outras procedencias conservados em Curitiba, Parana. *Atas Tercero Simposio regional de Geológia, Sociedade Brasileira de Geológia, Curitiba, 1981*: 213–25.

Bouchner, M. (1982) *A Field Guide in Colour to Animal Tracks and Traces*. Octopus Books, London.

Boucot, A.J. (1953) Life and death assemblages among fossils. *American Journal of Science*, **251**: 25–40, 248.

Bouvé, T.T. (1859) [Untitled comments on earliest discoveries of fossil tracks in the Connecticut Valley.] *Proceedings of the Boston Society of Natural History*, **7**: 49–53.

Boyce, M.S. (1980) [Untitled comments on growth rings in dinosaur teeth.] *Nature, London*, **288**: 194.

Boyd, D.W. (1975) False or misleading traces. In *The Study of Trace Fossils* (ed. R.W. Frey). Springer Verlag, Berlin, pp. 65–83.

Boyd, D.W. and Loope, D.B. (1984) Probable vertebrate origin for certain sole marks in Triassic red beds of Wyoming. *Journal of Paleontology*, **58**: 467–76.

Bradley, W.H. (1957) Physical and ecologic features of the Sagadahoc Bay tidal flat, Georgetown, Maine. *Memoirs of the Geological Society of America*, **67**: 641-82.

Brady, L.F. (1960) Dinosaur tracks from the Navajo and Wingate sandstones. *Plateau*, **32**: 81-2.

Brancas, R., Blaschke, J. and Martinez, J. (1979) Huellas de dinosaurios en Enciso. Análisis, catalogación y localización de las improntas Mesozoicas de los yacimientos de Enciso. *Unidad de Cultura de la Excma, Diputación de Logroño, Publicaciones*, No. 2.

Brand, L. (1979) Field and laboratory studies on the Coconino Sandstone (Permian) vertebrate footprints and their paleoecological implications. *Palaeogeography, Palaeoclimatology, Palaeoecology*, **28**: 25-38.

Branisa, L. (1968) Hallazgo del amonite *Neolobites* en la caliza Miraflores y de huellas de dinosáurios en la Formación El Molino y su significado para la determinación de la edad del Grupo Puca. *Boletín del Instituto Boliviano del Petróleo*, **8**(1): 16-29.

Branson, C.C. (1967) Protest against names for trace fossils. *Oklahoma Geology Notes*, **27**: 151.

Branson, E.B. and Mehl, M.G. (1932) Footprint records from the Paleozoic and Mesozoic of Missouri, Kansas and Wyoming. *Bulletin of the Geological Society of America*, **43**: 383-98.

Breton, G., Fournier, R. and Watte, J.-P. (1985) Le lieu de ponte de dinosaures de Rennes-le-Chateau (Aude): premiers résultats de la campagne de fouilles 1984. In *Les Dinosaures de la Chine à la France*. Muséum d'Histoire Naturelle, Toulouse, pp. 127-40.

Brinkman, D. (1981) The hind limb step cycle of *Iguana* and primitive reptiles. *Journal of Zoology, London*, **181**: 91-103.

Broderick, T.J. (1984) A record of dinosaur footprints from the Chewore Safari Area west of Mana-Angwa. Unpublished report for the Geological Survey of Zimbabwe, Harare, 5pp.

Brodkorb, P. (1978) Catalogue of fossil birds, part 5 (Passeriformes). *Bulletin of the Florida State Museum, Biological Sciences*, **23**: 139-228.

Brodrick, H. (1909) Note on footprint casts from the Inferior Oolite near Whitby, Yorks. *Proceedings of the Liverpool Geological Society*, **10**: 327-35.

Bronner, G. and Demathieu, G. (1977) Premières traces de reptiles archosauriens dans le Trias autochthone des Aiguilles Rouges (Col des Courbeaux-Vieil Émosson, Valais, Suisse). Conséquences paléogéographiques et chronostratigraphiques. *Compte Rendu hébdomadaire des Séances de l'Académie des Sciences, Paris*, sèrie D, **285**: 649-52.

Broom, R. 1911) On the dinosaurs of the Stormberg, South Africa. *Annals of the South African Museum*, **7**: 291-308.

Brown, B. (1908) The *Trachodon* group. *American Museum Journal*, **8**: 51-6.

Brown, B. (1914) *Leptoceratops*, a new genus of Ceratopsia from the Edmonton Cretaceous of Alberta. *Bulletin of the American Museum of Natural History*, **33**: 567–80.

Brown, B. (1916) *Corythosaurus casuarius*: skeleton, musculature and epidermis. *Bulletin of the American Museum of Natural History*, **35**: 709–16.

Brown, B. (1917) A complete skeleton of the horned dinosaur *Monoclonius*, and description of a second skeleton showing skin impressions. *Bulletin of the American Museum of Natural History*, **373**: 281–306.

Brown, B. (1938) The mystery dinosaur. *Natural History*, **41**: 190–202, 235.

Brown, B. (1941) The last dinosaurs. *Natural History*, **48**: 290–5.

Brown, B. and Schlaikjer, E.M. (1942) The skeleton of *Leptoceratops* with the description of a new species. *American Museum Novitates*, No. 1169: 1–15.

Brown, J.C. and Yalden, D.W. (1973) The description of mammals – 2; limbs and locomotion of terrestrial mammals. *Mammal Review*, **3**(4): 107–34.

Bryan, W.H. and Jones, O.A. (1946) The geological history of Queensland; a stratigraphical outline. *Papers of the Geology Department, University of Queensland*, n.s., **2**(12): 1–103.

Buckland, F. (1875) *Log-book of a Fisherman and Zoologist*. Chapman and Hall, London.

Buffetaut, E. (1980) A propos du reste de dinosaurien le plus anciennement décrit: l'interprétation de J.-B. Robinet (1768). *Histoire et Nature*, **14**: 79–84.

Buffetaut, E. (1987) *A Short History of Vertebrate Palaeontology*. Croom Helm, London.

Buffetaut, E. and Ingavat, R. (1985) The Mesozoic vertebrates of Thailand. *Scientific American*, **253**(2): 64–70.

Buffetaut, E., Ingavat, R., Sattayarak, N. and Suteetorn, V. (1985) Les premières empreintes de pas de Dinosaures du Sud-Est asiatique: pistes de Carnosaures du Crétacé inférieur de Thailande. *Compte Rendu de l'Académie des Sciences, Paris*, sèrie 2, **201**: 643–8.

Buick, T.L. (1931) *The Mystery of the Moa, New Zealand's Avian Giant*. Avery, New Plymouth.

Buick, T.L. (1936) *The Discovery of Dinornis: The Story of a Man, a Bone, and a Bird*. Avery, New Plymouth.

Buikstra, J.E. (1975) Healed fractures in *Macaca mulatta*: age, sex and symmetry. *Folia Primatologica*, **23**: 140–8.

Bunker, C.M. (1957) Theropod saurischian footprint discovery in the Wingate (Triassic) Formation. *Journal of Paleontology*, **31**: 973.

Calder, W.A. (1984) *Size, Function, and Life History*. Harvard University Press, Cambridge, Massachusetts.

Callison, G. and Quimby, H.M. (1984) Tiny dinosaurs: are they fully grown? *Journal of Vertebrate Paleontology*, **3**: 200–9.

Carpenter, K. (1982a) Baby dinosaurs from the Late Cretaceous Lance and Hell Creek formations and a description of a new species of theropod. *Contributions to Geology, University of Wyoming*, **20**: 123–34.

Carpenter, K. (1982b) Skeletal and dermal armor reconstruction of *Euoplocephalus tutus* (Ornithischia: Ankylosauridae) from the Late Cretaceous Oldman Formation of Alberta. *Canadian Journal of Earth Sciences*, **19**: 689–97.

Carpenter, K. (1984) Skeletal reconstruction and life restoration of *Sauropelta* (Ankylosauria: Nodosauridae) from the Cretaceous of North America. *Canadian Journal of Earth Sciences*, **21**: 1491–8.

Carrier, D.R. (1987) The evolution of locomotor stamina in tetrapods: circumventing a mechanical constraint. *Paleobiology*, **13**: 326–41.

Casamiquela, R.M. (1964) *Estudios Icnológicos. Problemas y métodos de la Icnologia con aplicación al estudio de pisadas Mesozóicas (Reptilia, Mammalia) de la Patagonia*. Colegio Industrial Pio X, Buenos Aires.

Casamiquela, R.M. (1966) Algunas consideraciones teóricas sobre las andares de los dinosaurios saurisquios. Implicaciones filogenéticas. *Ameghiniana, Revista de la Asociación Paleontológica Argentina*, **4**: 373–85.

Casamiquela, R.M. and Fasola, A. (1968) Sobre pisadas de dinosaurios del Cretácico inferior de Colchagua (Chile). *Publicaciones, Departamento de Geológia, Universidad de Chile, Santiago*, **30**: 1–24.

Casanovas Cladellas, M.L., Perez Lorente, F., Santafé Llopis, J.-V. and Fernandez Ortega, A. (1985) Nuevos datos icnológicos del Cretácico inferior de la Sierra de Cameros (La Rioja, España). *Paleontologia i Evolució*, **19**: 3–18.

Casanovas Cladellas, M.L. and Santafé Llopis, J.-V. (1971) Icnitas de Reptiles Mesozoicos en la Provincia de Logroño. *Acta Geológica Hispánica*, **6**(5): 139–42.

Casanovas Cladellas, M.L. and Santafé Llopis, J.-V. (1974) Dos nuevos yacimientos de Icnitas de Dinosaurios. *Acta Geológica Hispánica*, **9**(3): 88–91.

Casanovas Cladellas, M.L., Santafé Llopis, J.-V. and Garcia, S.L.J. (1984) Las icnitas de 'Los Corrales de Pelejon' en el Cretácico inferior de Galve (Teruel, España). *Paleontologia i Evolució*, **18**: 173–6.

Case, T.J. (1978) Speculations on the growth rate and reproduction of some dinosaurs. *Paleobiology*, **4**: 320–8.

Casier, E. (1960) *Les Iguanodons de Bernissart*. Institut Royal des Sciences Naturelles de Belgique, Brussels.

Caster, K.E. (1939) Were *Micrichnus scotti* Abel and *Artiodactylus sinclairi* Abel of the Newark Series (Triassic) made by vertebrates or limuloids? *American Journal of Science*, **237**: 786–97.

Caster, K.E. (1940) Die Sogenannten 'Wirbeltierespuren' und die *Limulus*

Fährten der solnhofener Plattenkalke. *Paläontologische Zeitschrift*, **22**: 12–29.

Caster, K.E. (1941) Trails of *Limulus* and supposed vertebrates from Solnhofen lithographic limestone. *Pan-American Geologist*, **76**: 241–58.

Caster, K.E. (1944) Limuloid trails from the Upper Triassic (Chinle) of the Petrified Forest National Monument, Arizona. *American Journal of Science*, **242**: 74–84.

Caster, K.E. (1957) Problematica. *Memoirs of the Geological Society of America*, **67**: 1025–32.

Chao, T.K. and Chiang, Y.K. (1974) Microscopic studies on the dinosaurian egg shells from Laiyang, Shantung Province. *Scientia Sinica*, **17**(1): 73–83.

Chardin, P.T. de and Young, C.-C. (1929) On some traces of vertebrate life in the Jurassic and Triassic beds of Shansi and Shensi. *Bulletin of the Geological Society of China*, **8**: 131–3.

Charig, A.J. (1972) The evolution of the archosaur pelvis and hind-limb: an explanation in functional terms. In *Studies in Vertebrate Evolution* (eds K.A. Joysey and T.S. Kemp). Oliver and Boyd, Edinburgh, pp. 121–55.

Charig, A.J. (1979) *A New Look at the Dinosaurs*. W. Heinemann, in association with British Museum (Natural History), London.

Charig, A.J. and Newman, B.H. (1962) Footprints in the Purbeck. *New Scientist*, **14**: 234–5.

Charig, A.J., Attridge, J. and Crompton, A.W. (1965) On the origin of the sauropods and the classification of the Saurischia. *Proceedings of the Linnean Society, London*, **176**: 197–221.

Clemmey, H. (1978) A Proterozoic lacustrine interlude from the Zambian Copperbelt. In *Modern and Ancient Lake Sediments* (International Association of Sedimentologists, Special Publication No. 2) (eds A. Matter and M.E. Tucker). Blackwell, Oxford, pp. 259–78.

Colbert, E.H. (1962) The weights of dinosaurs. *American Museum Novitates*, No. 2076: 1–16.

Colbert, E.H. (1964) Dinosaurs of the Arctic. *Natural History*, **73**(4): 20–3.

Colbert, E.H. (1968) *Men and Dinosaurs*. Evans Brothers, London.

Colbert, E.H. (1981) A primitive ornithischian dinosaur from the Kayenta Formation of Arizona. *Bulletin of the Museum of Northern Arizona*, **53**: 1–61.

Colbert, E.H. (1983) *Dinosaurs, an Illustrated History*. Hammond Incorporated, Maplewood, New Jersey.

Colbert, E.H. and Merrilees, D. (1967) Cretaceous dinosaur footprints from Western Australia. *Journal of the Royal Society of Western Australia*, **50**: 21–5.

Cole, J.R., Godfrey, L.R. and Schafersman, S.D. (1985) Mantracks? The fossils say *No! Creation/Evolution*, **15**: 37–45.

Colliver, F.S. (1956) Triassic footprints in Queensland. *Queensland Naturalist*, **15**: 78–9.

Congleton, J.D. (1988) Early Cretaceous vertebrate fossils from the Koum Basin, northern Cameroon. *Journal of Vertebrate Paleontology*, **8** (suppl.): 12A.

Conrad, K., Lockley, M.G. and Prince, N.K. (1987) Triassic and Jurassic vertebrate-dominated trace fossil assemblages of the Cimarron Valley region: implications for paleoecology and biostratigraphy. *New Mexico Geological Society, Guidebook for 38th Field Conference, 1987*: 127–38.

Coombs, W.P. (1975) Sauropod habits and habitats. *Palaeogeography, Palaeoclimatology, Palaeoecology*, **17**: 1–33.

Coombs, W.P. (1978a) Theoretical aspects of cursorial adaptations in dinosaurs. *Quarterly Review of Biology*, **53**: 393–418.

Coombs, W.P. (1978b) Forelimb muscles of the Ankylosauria (Reptilia, Ornithischia). *Journal of Paleontology*, **52**: 642–57.

Coombs, W.P. (1980a) Juvenile ceratopsians from Mongolia – the smallest known dinosaur specimens. *Nature, London*, **283**: 380–1.

Coombs, W.P. (1980b) Swimming ability of carnivorous dinosaurs. *Science*, **207**: 1198–1200.

Coombs, W.P. (1982) Juvenile specimens of the ornithschian dinosaur *Psittacosaurus*. *Palaeontology*, **25**: 89–107.

Cooper, M.R. (1981) The prosauropod dinosaur *Massospondylus carinatus* Owen from Zimbabwe: its biology, mode of life and phylogenetic significance. *Occasional Papers of the National Museums and Monuments, Zimbabwe*, series B, **6**: 689–840.

Cope, E.D. (1866) On the discovery of the remains of a gigantic dinosaur in the Cretaceous of New Jersey. *Proceedings of the Academy of Natural Sciences, Philadelphia*, **1866**: 275–9.

Cope, E.D. (1867) Account of extinct reptiles which approach birds. *Proceedings of the Academy of Natural Sciences, Philadelphia*, **1867**: 234–5.

Cope, E.D. (1883) On the characters of the skull in the Hadrosauridae. *Proceedings of the Academy of Natural Sciences, Philadelphia*, **1883**: 97–107.

Cott, H.B. (1961) Scientific results of an inquiry into the ecology and economic status of the Nile crocodile (*Crocodilus niloticus*) in Uganda and Northern Rhodesia. *Transactions of the Zoological Society of London*, **29**: 211–356.

Courel, L. and Demathieu, G. (1976) Une ichnofaune reptilienne remarquable dans les grès triasiques de Largentière (Ardèche, France). *Palaeontographica*, Abt. A, **151**: 194–216.

Courel, L. and Demathieu, G. (1984) Les inversions de relief dans les traces fossiles; leur signification. *109ᵉ Congrès National des Sociétés Savantes, Dijon, 1984*, **1**: 373–83.

Courel, L., Demathieu, G. and Buffard, R. (1968) Empreintes de pas de vertébrés et stratigraphie du Trias. *Bulletin de la Société Géologique de France*, sèrie 7, **10**: 275–81.

Courel, L., Demathieu, G. and Gall, J.-C. (1979) Figures sédimentaires et traces d'origine biologique du Trias moyen de la bordure orientale du Massif Central; signification sédimentologique et paléoécologique. *Geobios*, **12**: 379–97.

Cracraft, J. (1971) The functional morphology of the hind limb of the domestic pigeon *Columba livia*. *Bulletin of the American Museum of Natural History*, **144**: 171–268.

Cruickshank, A.R.I. (1986) Archosaur predation on an East African Middle Triassic dicynodont. *Palaeontology*, **29**: 415–22.

Cunningham, J. (1846) On some footmarks of birds and other impressions observed in the New Red Sandstone quarries of Storeton, near Liverpool. *Quarterly Journal of the Geological Society of London*, **2**: 410.

Curran, H.A. ed. (1985) *Biogenic Structures: Their Use in Interpreting Depositional Environments* (Society of Economic Paleontologists and Mineralogists, Special Publication No. 35). Society of Economic Paleontologists and Mineralogists, Tulsa.

Currey, J.D. (1977) Problems of scaling in the skeleton. In *Scale Effects in Animal Locomotion* (ed. T.J. Pedley). Academic Press, London, pp. 153–67.

Currie, P.J. (1981) Bird footprints from the Gething Formation (Aptian, Lower Cretaceous) of northeastern British Columbia, Canada. *Journal of Vertebrate Paleontology*, **1**: 257–64.

Currie, P.J. (1983) Hadrosaur trackways from the Lower Cretaceous of Canada. *Acta Palaeontologica Polonica*, **28**: 63–73.

Currie, P.J. (1989) Dinosaur footprints of western Canada. In *Dinosaur Tracks and Traces* (eds D.D. Gillette and M.G. Lockley). Cambridge University Press, Cambridge, pp. 293–300.

Currie, P.J. and Dodson, P. (1984) Mass death of a herd of ceratopsian dinosaurs. In *Third Symposium on Mesozoic Terrestrial Ecosystems, Tübingen, 1984, Short Papers* (eds W.-E. Reif and F. Westphal). Attempto Verlag, Tübingen, pp. 61–6.

Currie, P.J. and Sarjeant, W.A.S. (1979) Lower Cretaceous dinosaur footprints from the Peace River Canyon, British Columbia, Canada. *Palaeogeography, Palaeoclimatology, Palaeoecology*, **28**: 103–15.

Cushman, J.A. (1904) A new foot-print from the Connecticut Valley. *American Geologist*, **33**: 154–6.

Dagg, A.I. (1973) Gaits in mammals. *Mammal Review*, **3**: 135–54.

Dagg, A.I. and Foster, J.B. (1976) *The Giraffe: Its Biology, Behavior, and Ecology*. Van Nostrand Reinhold, New York.

Dana, S.L. and Dana, J.F. (1845) Analysis of coprolites from the New Red

Sandstone Formation of New England, with remarks by Professor Hitchcock. *American Journal of Science*, **48**: 46–60.

Darby, D.G. and Ojakangas, R.W. (1980) Gastroliths from an Upper Cretaceous plesiosaur. *Journal of Paleontology*, **54**: 548–56.

Dean, D.R. (1969) Hitchcock's dinosaur tracks. *American Quarterly*, **21**: 639–44.

Dean, D.R. (1979) The influence of geology on American literature and thought. In *Two Hundred Years of Geology in America* (ed. C.J. Schneer). University Press of New England, Hanover, New Hampshire, pp. 289–303.

Deane, J. (1844a) On the discovery of fossil footmarks. *American Journal of Science*, **47**: 381–90.

Deane, J. (1844b) Answer to the 'Rejoinder' of Prof. Hitchcock. *American Journal of Science*, **47**: 399–401.

Deane, J. (1861) *Ichnographs from the Sandstone of Connecticut River*. Little, Brown, Boston.

Deck, A. (1865) Notes on the excursion [of Geologists' Association to Hastings]. *Proceedings of the Geologists' Association*, **1**: 248–51.

Deckker, P. de (1988) Biological and sedimentary facies of Australian salt lakes. *Palaeogeography, Palaeoclimatology, Palaeoecology*, **62**: 237–70.

Degenhardt [initials unknown] (1840) [Untitled note on bird track in red sandstone in Mexico.] *Neues Jahrbuch für Mineralogie, Geologie und Paläontologie*, **1840**: 485.

Delair, J.B. (1963) Notes on Purbeck fossil footprints, with descriptions of two hitherto unknown forms from Dorset. *Proceedings of the Dorset Natural History and Archaeological Society*, **84**: 92–100.

Delair, J.B. (1966) New records of dinosaurs and other fossil reptiles from Dorset. *Proceedings of the Dorset Natural History and Archaeological Society*, **87**: 57–66.

Delair, J.B. (1980) Multiple dinosaur trackways from the Isle of Purbeck. *Proceedings of the Dorset Natural History and Archaeological Society*, **102**: 65–7.

Delair, J.B. (1983) Cretaceous dinosaur footprints from the Isle of Wight: a brief history. *Proceedings of the Isle of Wight Natural History and Archaeological Society*, **7**: 609–15.

Delair, J.B. (1989) A history of dinosaur footprint discoveries in the British Wealden. In *Dinosaur Tracks and Traces* (eds D.D. Gillette and M.G. Lockley). Cambridge University Press, Cambridge, pp. 19–25.

Delair, J.B. and Brown, P.A. (1974) Worbarrow Bay footprints. *Proceedings of the Dorset Natural History and Archaeological Society*, **96**: 14–16.

Delair, J.B. and Lander, A.B. (1973) A short history of the discovery of reptilian footprints in the Purbeck Beds of Dorset, with notes on their stratigraphical distribution. *Proceedings of the Dorset Natural History and Archaeological Society*, **94**: 17–20.

Delair, J.B. and Sarjeant, W.A.S. (1975) The earliest discoveries of dinosaurs. *Isis*, **66**: 5–25.

Delair, J.B. and Sarjeant, W.A.S. (1985) History and bibliography of the study of fossil vertebrate footprints in the British Isles: supplement 1973–1983. *Palaeogeography, Palaeoclimatology, Palaeoecology*, **49**: 123–60.

Demathieu, G. (1970) Les empreintes de pas de vertébrés du Trias de la bordure Nord-Est du Massif Central. *Cahiers de Paléontologie, Centre National de la Recherche Scientifique, Paris*, 11: 1–211.

Demathieu, G. (1971) Cinq nouvelles éspèces d'empreintes de Reptiles du Trias de la bordure nord-est du Massif Central. *Compte Rendu de l'Académie des Sciences, Paris*, sèrie D, **272**: 812–14.

Demathieu, G. (1975) Reconstitutions paléoécologiques à partir des données ichnologiques; possibilités et difficultés. *Bulletin de la Société Géologique de France*, sèrie 7, **17**: 896–8.

Demathieu, G. (1984) Utilisation de lois de la mécanique pour l'éstimation de la vitesse de locomotion des vertébrés tétrapodes du passé. *Geobios*, **17**: 439–46.

Demathieu, G.R. (1985) Trace fossil assemblages in Middle Triassic marginal marine deposits, eastern border of the Massif Central, France. In *Biogenic Structures: Their Use in Interpreting Depositional Environments* (ed. H.A. Curran). Society of Economic Paleontologists and Mineralogists, Tulsa, pp. 53–66.

Demathieu, G. (1986) Nouvelles recherches sur la vitesse des vertébrés, auteurs de traces fossiles. *Geobios*, **19**: 327–33.

Demathieu, G. and Gand, G. (1972) Les pistes dinosauroïdes du Trias moyen du plateau d'Antully et leur signification paléozoologique. *Bulletin de la Société d'Histoire Naturelle d'Autun*, **62**: 2–18.

Demathieu, G. and Haubold, H. (1972) Stratigraphische Aussagen der Tetrapodenfährten aus der terrestrischen Trias Europas. *Geologie*, **21**: 802–36.

Demathieu, G. and Omenaca, J.S. de (1976) Estudio del *Rhynchosauroides santanderensis*, n.sp., y otras nuevas huellas de pisadas en el Trias de Santander, con notas sobre el ambiente paleogeografico. *Acta Geológica Hispánica*, **12**: 49–54.

Demathieu, G.R. and Oosterink, H.W. (1983) Die Wirbeltier-Ichnofauna aus dem unteren Muschelkalk von Winterswijk (Die Reptilienfährten aus der Mitteltrias der Niederlande). *Staringia, Nederlandse Geologische Vereniging*, **7**: 1–50.

Demathieu, G.R. and Oosterink, H.W. (1988) New discoveries of ichnofossils from the Middle Triassic of Winterswijk (the Netherlands). *Geologie en Mijnbouw*, **67**: 3–17.

Demathieu, G. and Weidmann, M. (1982) Les empreintes de pas de reptiles dans le Trias du Vieux Émosson (Finhaut, Valais, Suisse). *Eclogae Geologicae Helvetiae*, **75**: 721–57.

Demathieu, G.R. and Wright, R.V.S. (1988) A new approach to the discrimination of chirotherioid ichnospecies by means of multivariate statistics. Triassic eastern border of the French Massif Central. *Geobios*, **21**: 729–39.

Demathieu, G., Ramos, A. and Sopena, A. (1978) Fauna icnologica del Triasico del extremo noroccidental de la Cordillera Iberica (Prov. de Guadalajara). *Estudios Geológicos*, **34**: 175–86.

Desmond, A.J. (1975) *The Hot-Blooded Dinosaurs*. Blond and Briggs, London.

Dietrich, W.O. (1927) Über Fährten ornithopodider Saurier im Oberkirchener Sandstein. *Zeitschrift der Deutschen Geologischen Gesellschaft, Abhandlungen*, **78**: 614–21.

Dijk, D.E. van (1978) Trackways in the Stormberg. *Palaeontologia Africana*, **21**: 113–20.

Dijk, D.E. van, Hobday, D.K. and Tankard, A.J. (1978) Permo-Triassic lacustrine deposits in the Eastern Karoo Basin, Natal, South Africa. In *Modern and Ancient Lake Sediments* (International Association of Sedimentologists, Special Publication No. 2) (eds A. Matter and M.E. Tucker). Blackwell, Oxford, pp. 225–39.

Dingman, R.J. and Galli, C.O. (1965) Geology and ground-water resources of the Pica area, Tarapaca Province, Chile. *Bulletin of the United States Geological Survey*, **1189**: 1–113.

Dodson, P., Behrensmeyer, A.K., Bakker, R.T. and McIntosh, J.S. (1980) Taphonomy and paleoecology of the dinosaur beds of the Jurassic Morrison Formation. *Paleobiology*, **6**: 208–32.

Dollo, L. (1887) Note sur les ligaments ossifiés des dinosauriens de Bernissart. *Archives de Biologie*, **7**: 249–64.

Dollo, L. (1906) Les allures des iguanodons, d'après les empreintes des pieds et de la queue. *Bulletin Biologique de la France et de la Belgique*, **40**: 1–12.

Dubiel, R.F., Blodgett, R.H. and Bown, T.H. (1987) Lungfish burrows in the Upper Triassic Chinle and Dolores Formations, Colorado Plateau. *Journal of Sedimentary Petrology*, **57**: 512–21.

Dughi, R. and Sirugue, F. (1958) Les oeufs de dinosaures du bassin d'Aix-en-Provence – les oeufs du Bégudien. *Compte Rendu hébdomadaire de l'Académie des Sciences, Paris*, **246**: 2386–8.

Dughi, R. and Sirugue, F. (1976) L'extinction des dinosaures à la lumière des gisements d'oeufs du Crétacé terminal du Sud de la France. *Paléobiologie Continentale*, **7**: 1–39.

Dutuit, J.-M. and Ouazzou, A. (1980) Découverte d'une piste de Dinosaure sauropode sur le site d'empreintes de Demnat (Haut-Atlas marocain). *Mémoires de la Société Géologique de France*, n.s., **139**: 95–102.

Duvernoy, G.-L. (1844) Fragments sur les organes genito-urinaires des reptiles et leur produits. *Compte Rendu hébdomadaire des Séances de l'Académie des Sciences, Paris*, **19**: 255–60.

Dzulynski, S. and Kinle, J. (1957) Problematic hieroglyphs of probable organic origin from the Beloveza beds (Western Carpathians). *Annales de la Société Géologique de Pologne (Rocznik Polskiego Towarzystwa Geologicznego)*, **26**: 265–72 [in Polish and English].

Dzulynski, S., Ksiazkiewicz, M. and Kuenen, P.H. (1959) Turbidites in flysch of the Polish Carpathian Mountains. *Bulletin of the Geological Society of America*, **70**: 1089–118.

Eberle, G. (1933) Fährte des Kiebitzregenpfeifers. *Natur und Museum*, **63**: 378–9.

Economos, A.C. (1981) The largest land mammal. *Journal of Theoretical Biology*, **89**: 211–15.

Edmund, A.G. (1960) Tooth replacement phenomena in the lower vertebrates. *Contributions of the Life Sciences Division, Royal Ontario Museum*, **52**: 1–190.

Edwards, M.B., Edwards, R. and Colbert, E.H. (1978) Carnosaurian footprints in the Lower Cretaceous of eastern Spitsbergen. *Journal of Paleontology*, **52**: 940–1.

Efremov, I.A. and Vjushkov, B.P. (1955) [Catalogue of Permian and Triassic terrestrial vertebrates in the territory of the USSR.] *Trudy Paleozoologicheskogo Instituta, Akademiya Nauk SSSR*, **46**: 1–185 [in Russian; for English summary see Olson 1956].

Ekdale, A.A. and Picard, M.D. (1985) Trace fossils in a Jurassic eolianite, Entrada Sandstone, Utah, U.S.A. In *Biogenic Structures: Their Use in Interpreting Depositional Environments* (ed. H.A. Curran). Society of Economic Paleontologists and Mineralogists, Tulsa, pp. 3–12.

Eldredge, N. (1979) Cladism and common sense. In *Phylogenetic Analysis and Paleontology* (eds J. Cracraft and N. Eldredge). Columbia University Press, New York, pp. 165–98.

Ellenberger, F. and Ellenberger, P. (1958) Principaux types de pistes de vertébrés dans les couches du Stormberg au Basutoland (Afrique du Sud). *Compte Rendu sommaire des Séances de la Société Géologique de France*, **1958**: 65–7.

Ellenberger, F. and Ellenberger, P. (1960) Sur une nouvelle dalle à pistes de Vertébrés, découverte au Basutoland (Afrique du Sud). *Compte Rendu sommaire des Séances de la Société Géologique de France*, **1960**: 236–8.

Ellenberger, F. and Fuchs, Y. (1965) Sur la présence de pistes de Vertébrés dans le Lotharingien marin de la région de Severac-le-Chateau (Aveyron). *Compte Rendu sommaire des Séances de la Société Géologique de France*, **1965**: 39–40.

Ellenberger, F. and Ginsburg, L. (1966) Le gisement de Dinosauriens triasiques de Maphutseng (Basutoland) et l'origine des Sauropodes. *Compte Rendu hébdomadaire des Séances de l'Académie des Sciences, Paris*, sèrie D, **262**: 444–7.

Ellenberger, F., Ellenberger, P., Fabre, J. and Mendrez, C. (1963) Deux nouvelles dalles à pistes de Vertébrés fossiles découvertes au Basutoland (Afrique du sud). *Compte Rendu sommaire des Séances de la Société Géologique de France*, **1963**: 315–17.

Ellenberger, F., Ellenberger, P. and Ginsburg, L. (1969) The appearance and evolution of dinosaurs in the Trias and Lias: a comparison between South African Upper Karroo and western Europe based on vertebrate footprints. In *Gondwana Stratigraphy* (IUGS Symposium, Buenos Aires, October 1967). UNESCO, Paris, pp. 333–54.

Ellenberger, F., Ellenberger, P. and Ginsburg, L. (1970) Les dinosaures du Trias et du Lias en France et en Afrique du Sud, d'après les pistes qu'ils ont laissées. *Bulletin de la Société Géologique de France*, sèrie 7, **12**: 151–9.

Ellenberger, P. (1955) Note préliminaire sur les pistes et les restes osseux de vertébrés du Basutoland (Afrique du Sud). *Compte Rendu hébdomadaire des Séances de l'Académie des Sciences, Paris*, **240**: 889–91.

Ellenberger, P. (1970) Les niveaux paléontologiques de première apparition des mammifères primordiaux en Afrique du Sud et leur ichnologie. *Proceedings and Papers of the Second Gondwana Symposium, South Africa, 1970*: 343–70.

Ellenberger, P. (1972) Contribution à la classification des Pistes de Vertébrés du Trias: Les types du Stormberg d'Afrique du Sud (I). *Palaeovertebrata, Mémoire Éxtraordinaire*, Laboratoire de Paléontologie des Vertébrés, Montpellier.

Ellenberger, P. (1974) Contribution à la classification des Pistes de Vertébrés du Trias: Les types du Stormberg d'Afrique du Sud (II). *Palaeovertebrata, Mémoire Éxtraordinaire*, Laboratoire de Paléontologie des Vertébrés, Montpellier.

Emerson, S.B. (1985) Jumping and leaping. In *Functional Vertebrate Morphology* (eds M. Hildebrand, D.M. Bramble, K.F. Liem and D.B. Wake). Belknap Press, Harvard University, Cambridge, Massachusetts, pp. 58–72.

Ennion, E.A.R. and Tinbergen, N. (1967) *Tracks*. Clarendon Press, Oxford.

Ennouchi, E. (1953) A propos des empreintes de Dinosauriens de Demnat

(Est de Marrakech). *Bulletin de la Société des Sciences Naturelles du Maroc*, **32**: 11–16.

Enos, P. (1969) Anatomy of a flysch. *Journal of Sedimentary Petrology*, **39**: 680–723.

Ensom, P.C. (1982) Dorset Geology in 1981. *Proceedings of the Dorset Natural History and Archaeological Society*, **103**: 141.

Ensom, P.C. (1983) Geology in 1982. *Proceedings of the Dorset Natural History and Archaeological Society*, **104**: 201–2.

Ensom, P.C. (1984) Geology in 1983. *Proceedings of the Dorset Natural History and Archaeological Society*, **105**: 165–9.

Erben, H.K. (1970) Ultrastrukturen und Mineralisation rezenter und fossiler Eischalen bei Vögeln und Reptilien. *Biomineralisation Forschungsberichte*, **1**: 1–66.

Erben, H.K. (1972) Ultrastrukturen und Dicke der Wand pathologischer Eischalen. *Abhandlungen, Mathematisch-naturwissenschaftlich Klasse, Akademie der Wissenschaften und der Literatur, Mainz*, **6**: 191–216.

Erben, H.K., Hoefs, J. and Wedepohl, K.H. (1979) Paleobiological and isotopic studies of eggshells from a declining dinosaur species. *Paleobiology*, **5**: 380–414.

Erve, A. van and Mohr, B. (1988) Palynological investigations of the Late Jurassic microflora from the vertebrate locality Guimarota coal mine (Leiria, central Portugal). *Neues Jahrbuch für Geologie und Paläontologie, Monatshefte*, **1988**(4): 246–62.

Faber, F.J. (1958) Fossiele voetstappen in de Muschelkalk van Winterswijk. *Geologie en Mijnbouw*, n.s., **20**: 317–21.

Farlow, J.O. (1976) Speculations about the diet and foraging behavior of large carnivorous dinosaurs. *American Midland Naturalist*, **95**: 186–91.

Farlow, J.O. (1981) Estimates of dinosaur speeds from a new trackway site in Texas. *Nature, London*, **294**: 747–8.

Farlow, J.O. (1985a) Notes. In *Bones for Barnum Brown: Adventures of a Dinosaur Hunter by Roland T. Bird* (ed. V.T. Schreiber). Texas Christian University Press, Fort Worth, pp. 208–19.

Farlow, J.O. (1985b) Speculations about the diet and digestive physiology of herbivorous dinosaurs. *Paleobiology*, **13**: 60–72.

Farlow, J.O. (1987) *A Guide to Lower Cretaceous Dinosaur Footprints and Tracksites of the Paluxy River Valley, Somervell County, Texas*. Baylor University, Texas.

Farlow, J.O., Pittman, J.G. and Hawthorne, J.M. (1989) *Brontopodus birdi*, Lower Cretaceous sauropod footprints from the U.S. Gulf coastal plain. In *Dinosaur Tracks and Traces* (eds D.D. Gillette and M.G. Lockley). Cambridge University Press, Cambridge, pp. 371–94.

Farmer, M.F. (1956) Tracks and trackways of northern Arizona. *Plateau*, **28**(3): 54–66.

Fasola, A. (1966) Hallazgo de huellas de dinosaurios en el Alto Tinguiririca. *Noticiario Mensual, Museo Nacional de Historia Natural, Santiago*, No. 119.

Faul, H. (1951) The naming of fossil footprint 'species'. *Journal of Paleontology*, **25**: 409.

Feduccia, A. (1980) *The Age of Birds*. Harvard University Press, Cambridge, Massachusetts.

Ferrusquia-Villafranca, I., Applegate, S.P. and Espinosa-Arrubarrena, L. (1978) Rocas volcanosedimentarias Mesozoicas y huellas de dinosaurios en la región suroccidental Pacífica de México. *Revista, Instituto de Geología, Universidad Nacional Autónoma de México*, **2**: 150–62.

Flannery, T.F. and Rich, T.H. (1981) Dinosaur digging in Victoria. *Australian Natural History*, **20**: 195–8.

Flynn, L.J., Brillanceau, A., Brunet, M., Coppens, Y., Dejax, J., Duperon-Laudoueneix, M., Ekodeck, G., Flanagan, K.M., Heintz, E., Hell, J., Jacobs, L.L., Pilbeam, D.R., Sen, S. and Djallo, S. (1987) Vertebrate fossils from Cameroon, West Africa. *Journal of Vertebrate Paleontology*, **7**: 469–71.

Frey, R.W. (1973) Concepts in the study of biogenic sedimentary structures. *Journal of Sedimentary Petrology*, **43**: 6–19.

Frey, R.W. (1975) The realm of ichnology, its strengths and limitations. In *The Study of Trace Fossils* (ed. R.W. Frey). Springer-Verlag, Berlin and New York, pp. 13–38.

Frey, R.W. (1978) Behavioral and ecological implications of trace fossils. In *Trace Fossil Concepts* (Society of Economic Paleontologists and Mineralogists, Short Course No. 5) (ed. P.B. Basan). Society of Economic Paleontologists and Mineralogists, Oklahoma City, pp. 43–66.

Frey, R.W. and Pemberton, S.G. (1985) Biogenic structures in outcrops and cores. I. Approaches to Ichnology. *Bulletin of Canadian Petroleum Geology*, **33**: 72–115.

Frey, R.W. and Pemberton, S.G. (1986) Vertebrate lebensspuren in intertidal and supratidal environments, Holocene barrier islands, Georgia. *Senckenbergiana Maritima*, **18**: 45–95.

Frey, R.W. and Pemberton, S.G. (1987) The *Psilonichnus* ichnocoenose, and its relationship to adjacent marine and nonmarine ichnocoenoses along the Georgia coast. *Bulletin of Canadian Petroleum Geology*, **35**: 333–57.

Frith, H.J. and Calaby, J.H. (1969) *Kangaroos*. F.W. Cheshire, Melbourne.

Gabouniya, L.K. (1951) [Dinosaur footprints from the Lower Cretaceous of Georgia.] *Doklady Akademiya Nauk SSSR*, **81**(5): 917–9 [in Russian].

Gabouniya, L.K. (1952) [Dinosaur tracks at Mount Sataplia.] *Priroda*, **1952**(1): 122–3 [in Russian].

Gabouniya, L.K. (1958) [*Dinosaur Tracks.*] Akademiya Nauk SSSR Press, Moscow [in Russian].

Galton, P.M. (1970) The posture of hadrosaurian dinosaurs. *Journal of Paleontology*, **44**: 464–73.

Galton, P.M. (1971a) Manus movements of the coelurosaurian dinosaur *Syntarsus* and the opposability of the theropod hallux [pollex]. *Arnoldia, Miscellaneous Publications of the National Museums of Southern Rhodesia*, 5(15): 1–8.

Galton, P.M. (1971b) The prosauropod dinosaur *Ammosaurus*, the crocodile *Protosuchus*, and their bearing on the age of the Navajo Sandstone of northeastern Arizona. *Journal of Paleontology*, **45**: 781–95.

Galton, P.M. (1973) On the anatomy and relationships of *Efraasia diagnostica* (Huene) n. gen., a prosauropod dinosaur (Reptilia: Saurischia) from the Upper Triassic of Germany. *Paläontologische Zeitschrift*, **47**: 229–55.

Galton, P.M. (1974) The ornithischian dinosaur *Hypsilophodon* from the Wealden of the Isle of Wight. *Bulletin of the British Museum (Natural History), Geology*, **25**: 1–152.

Galton, P.M. (1976) Prosauropod dinosaurs (Reptilia: Saurischia) of North America. *Postilla*, **169**: 1–98.

Galton, P.M. (1977) The ornithopod dinosaur *Dryosaurus* and a Laurasia-Gondwanaland connection in the Upper Jurassic. *Nature, London*, **268**: 230–2.

Galton, P.M. (1982) The postcranial anatomy of the stegosaurian dinosaur *Kentrosaurus* from the Upper Jurassic of Tanzania, East Africa. *Geologica et Palaeontologica*, **15**: 139–60.

Galton, P.M. and Cluver, M.A. (1976) *Anchisaurus capensis* (Broom) and a revision of the Anchisauridae (Reptilia, Saurischia). *Annals of the South African Museum*, **69**: 121–59.

Galton, P.M. and Powell, H.P. (1980) The ornithischian dinosaur *Camptosaurus prestwichii* from the Upper Jurassic of England. *Palaeontology*, **23**: 411–43.

Gand, G. (1976) Étude biométrique de traces de pas laissées par *Gallus gallus*, variété Sussex. *Bulletin de la Société d'Histoire Naturelle du Creusot*, **34**: 19–23.

Gand, G. (1986) Interprétations paléontologique et paléoécologique de quatre niveaux à traces de vertébrés observés dans l'Autunien du Lodévois (Hérault). *Géologie de la France*, **1986**(2): 155–76.

Gand, G., Pellier, F. and Pellier, J.-F. (1976) Sur quelques traces ornithoïdes récoltées dans le Trias moyen de Bourgogne. *Bulletin de la Société d'Histoire Naturelle du Creusot*, **34**: 24–33.

Garcia-Ramos, J.C. and Valenzuela, M. (1977a) Huellas de pisada de vertebrados (Dinosaurios y otros) en el Jurasico Superior de Asturias. *Estudios Geológicos,* **33**: 207–14.

Garcia-Ramos, J.C. and Valenzuela, M. (1977b) Hallazgo de huellas de pisada de vertebrados en el Jurasico de la Costa Asturiana entre Gijon y Ribadesella. *Breviora Geológica Asturica,* **21**(2): 17–21.

Garcia-Ramos, J.C. and Valenzuela, M. (1979) Estudio e interpretacion de la icnofauna (vertebrados e invertebrados) en el Jurasico de la costa asturiana. *Cuadernos Geológicos,* **10**: 23–33.

Garland, T. (1983a) Scaling the ecological cost of transport to body mass in terrestrial mammals. *The American Naturalist,* **121**: 571–87.

Garland, T. (1983b) The relation between maximal running speed and body mass in terrestrial mammals. *Journal of Zoology, London,* **199**: 157–70.

Gaudry, A. (1890) *Les enchaînements du Monde Animal dans les Temps Géologiques. Fossiles Secondaires.* F. Savy, Paris.

Gierliński, G. and Potemska, A. (1987) Lower Jurassic dinosaur footprints from Gliniany Las, northern slope of the Holy Cross Mountains, Poland. *Neues Jahrbuch für Geologie und Paläontologie, Abhandlungen,* **175**: 107–20.

Gillette, D.D. and Lockley, M.G. (eds) (1989) *Dinosaur Tracks and Traces.* Cambridge University Press, Cambridge, 454p.

Gilmore, C.W. (1914) Osteology of the armored Dinosauria in the United States National Museum, with special reference to the genus *Stegosaurus. Bulletin of the United States National Museum,* **89** 1–143.

Gilmore, C.W. (1915) Osteology of *Thescelosaurus,* an orthopodous dinosaur from the Lance Formation of Wyoming. *Proceedings of the United States National Museum,* **49**: 591–616.

Gilmore, C.W. (1924) Collecting fossil footprints in Virginia. *Miscellaneous Collections of the Smithsonian Institution, Washington,* **76**(10): 16–18.

Gilmore, C.W. (1925) Osteology of ornithopodous dinosaurs from the Dinosaur National Monument, Utah. *Memoirs of the Carnegie Museum,* **10**: 385–410.

Gilmore, C.W. (1926) Fossil footprints from the Grand Canyon, Arizona. *Miscellaneous Collections of the Smithsonian Institution, Washington,* **77**(9): 1–48.

Ginsburg, L., Lapparent, A.F. de, Loiret, B. and Taquet, P. (1966) Empreintes de pas de Vertébrés tétrapodes dans les sèries continentales à l'Ouest d'Agadès (République du Niger). *Compte Rendu de l'Académie des Sciences, Paris,* **263**: 28–31.

Glauert, L. (1952) Dinosaur footprints at Broome. *Western Australian Naturalist,* **3**: 82–3.

Godfrey, L.R. and Cole, J.R. (1986) Blunder in their footsteps. *Natural History,* **95**(8): 4–12.

Godoy, L.C. and Leonardi, G. (1985) Direçoēs e comportamento dos dinossauros da localidade de Piau, Sousa, Paraíba (Brasil), Formaçaō Sousa (Cretáceo inferior). *Coletânea de Trabalhos Paleontológicos*, seçaō 2, série Geologia, **27**: 65–73.

Gould, C.N. (1929) Comanchean reptiles from Kansas, Oklahoma, and Texas. *Bulletin of the Geological Society of America*, **40**: 457–62.

Grabbe, H. (1881a) Neue Funde von Saurier-Fährten im Wealdensandstein des Bückeburges. *Verhandlungen des Naturhistorischen Vereins der preussischen Rheinlande und Westfalens, Correspondenzblatt*, **38**: 161–4.

Grabbe, H. (1881b) Die Iguano-Spuren des Bückeburges. *Natur, Halle*, **30**: 424–5.

Granger, W. (1936) The story of the dinosaur eggs. *Natural History*, **38**: 21–5.

Grantham, R.G. (1989) Dinosaur tracks and mega-flutes in the Jurassic of Nova Scotia. In *Dinosaur Tracks and Traces* (eds D.D. Gillette and M.G. Lockley). Cambridge University Press, Cambridge, pp. 281–4.

Gregory, W.K. (1905) The weight of the *Brontosaurus*. *Science*, n.s., **22**: 572.

Gürich, G. (1927) Über Saurier-Fährten aus dem Etjo-Sandstein von Südwestafrika. *Paläontologische Zeitschrift*, **8**: 112–20.

Halstead, L.B. (1970) *Scrotum humanum*, Brookes 1763 – the first named dinosaur. *Journal of Insignificant Research*, **5**: 14–15.

Hamilton, A. (1952) The case of the mysterious 'hand animal'. *Natural History*, **1952**: 296–301, 336.

Hamley, T. (in press) Functions of the tail in bipedal locomotion of lizards, dinosaurs and pterosaurs. *Memoirs of the Queensland Museum*, **28**.

Häntzschel, W., El-Baz, F. and Amstutz, G.C. (1968) Coprolites: an annotated bibliography. *Memoirs of the Geological Society of America*, **108**: 1–132.

Harkness, R. (1850) Notice of a tridactylous footmark from the Bunter Sandstone of Weston Point, Cheshire. *Annals and Magazine of Natural History*, series 2, **6**: 440–2.

Hastings, R.J. (1987) New observations on Paluxy tracks confirm their dinosaurian origin. *Journal of Geological Education*, **35**: 4–15.

Hatcher, J.B. (1903) Osteology of *Haplocanthosaurus* with description of a new species, and remarks on the probable habits of the Sauropoda and the age and origin of the *Atlantosaurus* Beds. *Memoirs of the Carnegie Museum*, **2**: 1–72.

Hatcher, J.B., Marsh, O.C. and Lull, R.S. (1907) The Ceratopsia. *Monographs of the United States Geological Survey*, **49**: 1–157.

Haubold, H. (1966) Ein Pseudosuchier-Fährtenfauna aus dem Buntsandstein Südthüringiens. *Hallesches Jahrbuch für Mitteldeutsche Erdgeschichte*, **8**: 12–48 [in German with English and Russian summaries].

Haubold, H. (1969) Die Evolution der Archosaurier in der Trias aus der Sicht ihrer Fährten. *Hercynia, neue Folge*, **6**: 90–106.

Haubold, H. (1971) Ichnia Amphibiorum et Reptiliorum fossilium. In *Handbuch der Paläoherpetologie*, Teil 18 (ed. O. Kuhn). Gustav Fischer Verlag, Stuttgart, pp. 1–124.

Haubold, H. (1973) Die Tetrapodenfährten aus dem Perm Europas. *Freiberger Forschungshefte, reihe C*, **285**: 5–55.

Haubold, H. (1984) *Saurierfährten* (2nd edition). A. Ziemsen Verlag, Wittenberg Lutherstadt.

Haubold, H. (1986) Archosaur footprints at the terrestrial Triassic-Jurassic transition. In *The Beginning of the Age of Dinosaurs* (ed. K. Padian). Cambridge University Press, Cambridge, pp. 189–201.

Haubold, H. and Sarjeant, W.A.S. (1974) Fossil vertebrate footprints and the stratigraphical correlation of the Keele and Enville Beds of the Birmingham region. *Proceedings of the Birmingham Natural History Society*, **22**: 257–68.

Hay, O.P. (1902) Bibliography and catalogue of the fossil Vertebrata of North America. *Bulletin of the United States Geological Survey*, **179**: 1–868.

Hay, O.P. (1929–30) Second bibliography and catalogue of the fossil Vertebrata of North America (2 vols). *Publications of the Carnegie Institution, Washington*, **390**: 1–916 (Vol. 1, 1929), 1–1074 (Vol. 2, 1930).

Heilmann, G. (1927) *The Origin of Birds*. Appleton, New York [reprinted 1972 by Dover Publications, New York].

Heintz, N. (1962) Geological excursion to Svalbard in connection with the XXI International Geological Congress in Norden 1960. Årbok, Norsk Polarinstitutt, **1960**: 98–106.

Heintz, N. (1963) Casting dinosaur footprints at Spitsbergen. *Curator*, **6**: 217–25.

Heinz, R. (1932) Die Saurierfährten bei Otjihaenamaparero im Hereroland und das Alter des Etjo-Sandsteins in Deutsch-Südwestafrika. *Zeitschrift der Deutschen Geologischen Gesellschaft*, **84**: 569–70.

Heller, F. (1952) Reptilfährten-Funde aus dem Ansbacher Sandstein des Mittleren Keupers von Franken. *Geologische Blätter für Nordost-Bayern und angrezende Gebiete*, **2**: 129–41.

Hellings, R.W., Adams, P.J., Anderson, J.D., Keesey, M.S., Lau, E.L., Standish, E.M., Canuto, V.M. and Goldman, I. (1983) Experimental test of the variability of G using Viking Lander ranging data. *Physical Review Letters*, **51**: 1609–12.

Hendricks, A. (1981) Die Saurierfährte von Münchenhagen bei Rehburg-Loccum (NW-Deutschland). *Abhandlungen aus dem Landesmuseum für Naturkunde zu Münster in Westfalen*, **43**(2): 1–22.

Herrin, T., Gillette, J.L., Campbell, J. and Gillette, D.B. (1986) Dinosaur tracks in Paradise. In *First International Symposium on Dinosaur Tracks and Traces, Albuquerque, 1986, Abstracts with Program* (ed. D.D. Gillette). New Mexico Museum of Natural History, Albuquerque, p. 15.

Hickey, L.J. (1980) Paleobotany. *Geotimes*, **1980**(2): 39–40.

Hildebrand, M. (1965) Symmetrical gaits of horses. *Science*, **150**: 701–8.

Hildebrand, M. (1984) Rotations of the leg segments of three fast-running cursors and an elephant. *Journal of Mammalogy*, **65**: 718–20.

Hill, C.R. (1976) Coprolite of *Ptilophyllum* cuticles from the Middle Jurassic of North Yorkshire. *Bulletin of the British Museum (Natural History), Geology* **27**: 289–94.

Hill, D., Playford, G. and Woods, J.T. (1965) *Triassic Fossils of Queensland.* Queensland Palaeontographical Society, Brisbane.

Hill, D., Playford, G. and Woods, J.T. (1966) *Jurassic Fossils of Queensland.* Queensland Palaeontographical Society, Brisbane.

Hill, H. (1895) On the occurrence of Moa-footprints in the bed of Manawatu River, near Palmerston North. *Transactions and Proceedings of the New Zealand Institute*, **27**: 476–7.

Hitchcock, C.H. (1866) Description of a new reptilian bird from the Trias of Massachusetts. *Annals of the Lyceum of Natural History, New York*, **8**: 301–2.

Hitchcock, C.H. (1889) Recent progress in ichnology. *Proceedings of the Boston Society of Natural History*, **24**: 117–27.

Hitchcock, C.H. (1927) The Hitchcock Lecture upon Ichnology, and the Dartmouth College ichnological collection (edited by N.M. Grier). *American Midland Naturalist*, **10**: 161–97.

Hitchcock, E. (1836) Ornithichnology. Description of the footmarks of birds (Ornithoidichnites) on New Red Sandstone in Massachusetts. *American Journal of Science*, **29**: 307–40.

Hitchcock, E. (1841) *Final Report on the Geology of Massachusetts* (2 vols). J.H. Butler, Northampton, Massachusetts.

Hitchcock, E. (1843) Description of five new species of fossil footmarks, from the red sandstone of the valley of the Connecticut River. *Transactions of the Association of American Geologists and Naturalists*, **1843**: 254–64.

Hitchcock, E. (1844a) Report on Ichnolithology, or fossil footmarks, with a description of several new species, and the coprolites of birds, from the valley of Connecticut River, and of a supposed footmark from the valley of Hudson River. *American Journal of Science*, **47**: 292–322.

Hitchcock, E. (1844b) Rejoinder to the preceding article of Dr. Deane. *American Journal of Science*, **47**: 390–9.

Hitchcock, E. (1848) An attempt to discriminate and describe the animals that made the fossil footmarks of the United States, and especially of New

England. *Memoirs of the American Academy of Arts and Sciences*, series 2, **3**: 129–256.

Hitchcock, E. (1858) *Ichnology of New England. A Report on the Sandstone of the Connecticut Valley, Especially its Fossil Footmarks*. W. White, Boston [reprinted 1974 by Arno Press, New York].

Hitchcock, E. (1865) *Supplement to the Ichnology of New England. A Report to the Government of Massachusetts in 1863* (ed. C.H. Hitchcock). Wright and Potter, Boston.

Hopson, J.A. (1977) Relative brain size and behavior in archosaurian reptiles. *Annual Review of Ecology and Systematics*, **8**: 429–48.

Hopson, J.A. (1979) Paleoneurology. In *Biology of the Reptilia*, Vol. 9 (eds C. Gans, R.G. Northcutt and P.S. Ulinksi). Academic Press, London, pp. 39–146.

Horner, J.R. (1982) Evidence of colonial nesting and 'site fidelity' among ornithischian dinosaurs. *Nature, London*, **297**: 675–6.

Horner, J.R. (1984a) A 'segmented' epidermal tail frill in a species of hadrosaurian dinosaur. *Journal of Paleontology*, **58**: 270–1.

Horner, J.R. (1984b) The nesting behavior of dinosaurs. *Scientific American*, **250**(4): 130–7.

Horner, J.R. and Makela, R. (1979) Nest of juveniles provides evidence of family structure among dinosaurs. *Nature, London*, **282**: 296–8.

Horner, J.R. and Weishampel, D.B. (1988) A comparative embryological study of two ornithischian dinosaurs. *Nature, London*, **332**: 256–7.

Houston, S.H. (1933) Fossil footprints in Comanchean limestone beds, Bandera County, Texas. *Journal of Geology*, **41**: 650–3.

Howard, J.D., Mayou, T.V. and Heard, R.W. (1977) Biogenic sedimentary structures formed by rays. *Journal of Sedimentary Petrology*, **47**: 339–46.

Hoyt, D.F. and Taylor, C.R. (1981) Gait and the energetics of locomotion in horses. *Nature, London*, **292**: 239–40.

Huene, F. von (1925) Ausgedehnte Karroo-Komplexe mit Fossilführung im nordöstlichen Südwestafrika. *Centralblatt für Mineralogie, Geologie und Paläontologie*, Abt. B, **1925**: 151–6.

Huene, F. von (1928) Lebensbild des Sauriervorkommens im obersten Keuper von Trossingen. *Palaeobiologica*, **1**: 103–16.

Huene, F. von (1931a) Verschiedene mesozoische Wirbeltierreste aus Südamerika. *Neues Jahrbuch für Mineralogie, Geologie und Paläontologie, Abhandlungen*, **66**: 181–98.

Huene, F. von (1931b) Die fossilen Fährten im Rhät von Ischigualasto in Nordwest-Argentinien. *Palaeobiologica*, **4**: 99–112.

Huene, F. von (1932) Die fossile Reptil-Ordnung Saurischia, ihre Entwicklung und Geschichte. *Monographien zur Geologie und Paläontologie*, **4**: 1–361.

Huene, F. von (1933) Die südamerikanische Gondwana-Fauna. *Forschungen und Fortschritte,* **9**: 129–30.

Huene, F. von (1941) Eine Fährtenplatte aus dem Stubensandstein des mittleren Keuper der Tübinger Gegend. *Zentralblatt für Mineralogie, Geologie und Paläontologie,* Abt. B, **5**: 138–41.

Huene, F. von (1942) Die Tetrapoden-Fährten im toskanischen Verrucano und ihre Bedeutung. *Neues Jahrbuch für Mineralogie, Geologie und Paläontologie, Beilage-Bände (Abhandlungen),* Abt. B, **86**: 1–34.

Huene, F. von (1950) Die Entstehung der Ornithischia schon früh in der Trias. *Neues Jahrbuch für Mineralogie, Geologie und Paläontologie, Monatshefte,* **1950**: 53–8.

Hughes, T.M. (1884) On some tracks of terrestrial and freshwater animals. *Quarterly Journal of the Geological Society of London,* **40**: 178–86.

Hunt, C.B. (1975) *Death Valley: Geology, Ecology, Archaeology.* University of California Press, Berkeley.

Hunt, A.P., Lucas, S.G. and Kietzke, K.K. (1989) Dinosaur footprints from the Redonda Member of the Chinle Formation (Upper Triassic), east-central New Mexico. In *Dinosaur Tracks and Traces* (eds D.D. Gillette and M.G. Lockley). Cambridge University Press, Cambridge, pp. 277–80.

Huxley, T.H. (1868) On the animals which are most nearly intermediate between birds and reptiles. *Geological Magazine,* **5**: 357–65.

Huxley, T.H. (1870) Further evidence of the affinity between the dinosaurian reptiles and birds. *Quarterly Journal of the Geological Society of London,* **26**: 12–31.

Isham, L.B. (1965) Preparation of drawings for paleontologic publication. In *Handbook of Paleontological Techniques* (eds B. Kummel and D. Raup). W.H. Freeman, San Francisco and London, pp. 459–68.

Ishigaki, S. (1985a) [Dinosaur footprints of the Atlas Mountains (1).] *Nature Study,* **31**(10): 113–6 [in Japanese].

Ishigaki, S. (1985b) [Dinosaur footprints of the Atlas Mountains (2).] *Nature Study,* **31**(12): 136–9 [in Japanese].

Ishigaki, S. (1986) [*The Dinosaurs of Morocco.*] Tsukiji Shokan, Tokyo [in Japanese].

Ishigaki, S. (1989) Footprints of swimming sauropods from Morocco. In *Dinosaur Tracks and Traces* (eds D.D. Gillette and M.G. Lockley). Cambridge University Press, Cambridge, pp. 83–6.

Ishigaki, S. and Fujisaki, T. (1989) Three dimensional representation of *Eubrontes* by the method of Moiré topography. In *Dinosaur Tracks and Traces* (eds D.D. Gillette and M.G. Lockley). Cambridge University Press, Cambridge, pp. 421–5.

Ishigaki, S. and Haubold, H. (1986) Lower Jurassic dinosaur footprints from the central High Atlas, Morocco. In *First International Symposium on Dinosaur Tracks and Traces, Albuquerque, 1986, Abstracts with Program* (ed. D.D. Gillette). New Mexico Museum of Natural History, Albuquerque, p. 16.

Jacobs, L.L., Flanagan, K.M., Brunet, M., Flynn, L.J., Dejax, J. and Hell, J.V. Dinosaur footprints from the Lower Cretaceous of Cameroon, West Africa. In *Dinosaur Tracks and Traces* (eds D.D. Gillette and M.G. Lockley). Cambridge University Press, Cambridge, pp. 349–51.

Jaeger, E. (1939) How to know footprints. *Natural History*, **44**: 226–32.

Jaeger, E. (1948) *Tracks and Trailcraft*. Macmillan, New York.

Jaekel, O. (1914) Über die Wirbeltierfunde in der oberen Trias von Halberstadt. *Paläontologische Zeitschrift*, **1**: 155–215.

Jaekel, O. (1929) Die Spur eines neuen Urvogels (*Protornis bavarica*) und deren Bedeutung für die Urgeschichte der Vögel. *Paläontologische Zeitschrift*, **11**: 201–38.

Jain, S.L. and Sahni, A. (1985) Dinosaurian egg shell fragments from the Lameta Formation at Pisdura, Chandrapur District, Maharashtra. *Geoscience Journal*, **6**: 211–20.

Jenny, J. and Jossen, J.A. (1982) Découverte d'empreintes de pas de Dinosauriens dans le Jurassique inférieur (Pliensbachien) du Haut Atlas central (Maroc). *Compte Rendu hébdomadaire des Séances de l'Académie des Sciences, Paris*, **294**: 223–6.

Jenny, J., Marrec, A. le and Monbaron, M. (1981) Les empreintes de pas de Dinosauriens du Haut Atlas central (Maroc): nouveaux gisements et précisions stratigraphiques. *Geobios*, **14**: 427–31.

Jensen, J.A. (1988) A fourth new sauropod dinosaur from the Upper Jurassic of the Colorado plateau and sauropod bipedalism. *Great Basin Naturalist*, **48**(2): 121–45.

Jepsen, G.L. (1964) Riddles of the terrible lizards. *American Scientist*, **52**: 227–46.

Jerison, H.J. (1969) Brain evolution and dinosaur brains. *American Naturalist*, **103**: 575–88.

Jerison, H.J. (1973) *Evolution of the Brain and Intelligence*. Academic Press, London and New York.

Johnson, K.R. (1986) Paleocene bird and amphibian tracks from the Fort Union Formation, Bighorn Basin, Wyoming. *Contributions to Geology, University of Wyoming*, **24**(1): 1–10.

Johnston, P.A. (1979) Growth rings in dinosaur teeth. *Nature, London*, **278**: 635–6.

Johnston, P.A. (1980) [Untitled comments on growth rings in dinosaur teeth.] *Nature, London*, **288**: 194–5.

Kaever, M. and Lapparent, A.F. de (1974) Les traces de pas de Dinosaures du Jurassique de Barkhausen (Basse Saxe, Allemagne). *Bulletin de la Société Géologique de France*, sèrie 7, **16**: 516–25.

Karaszewski, W. (1966) [Reptile tracks and dragging traces on the Roethian sandstone surface observed in Jarugi near Ostrowiec Swietokrzyski.] *Kwartalnik Geologiczny*, **10**(2): 327–33 [in Polish with English summary].

Karaszewski, W. (1969) [Tracks of Reptilia in the Lower Liassic of the Swietokrzyskie Mountains, Middle Poland.] *Kwartalnik Geologiczny*, **13**(1): 117–19 [in Polish with English summary].

Karaszewski, W. (1975) [Footprints of pentadactyl dinosaurs in the Lower Jurassic of Poland.] *Bulletin de l'Académie Polonaise des Sciences*, sèrie Sciences de la Terre, **23**(2): 133–6 [in Polish with English summary].

Kaup, J.J. (1835) Thier-Fährten von Hildburghausen; *Chirotherium* oder *Chirosaurus*. *Neues Jahrbuch für Mineralogie, Geognosie, Geologie und Petrefaktenkunde*, **1835**: 327–8.

Kerourio, P. (1981) Nouvelles observations sur le mode de nidification et de ponte chez les dinosauriens du Crétacé terminal du Midi de la France. *Compte Rendu sommaire des Séances de la Société Géologique de France*, **1981**: 25–8.

Khomizuri, N.I. (1972) [Dinosaur tracks in Tadzhikistan.] *Priroda*, **1972**(6): 94–5 [in Russian].

Kim, H.M. (1983) Cretaceous dinosaurs from Korea. *Journal of the Geological Society of Korea*, **19**(3): 115–26.

Kitching, J.W. (1979) Preliminary report on a clutch of six dinosaurian eggs from the Upper Triassic Elliot Formation, northern Orange Free State. *Palaeontologia Africana*, **22**: 41–5.

Knowles, P.C. (1980) Dinosaur Tracks. *Geo, Australia's Geographical Magazine*, **1**(2): 64–5.

Kolesnikov, C.M. and Sochava, A.V. (1972) A paleobiochemical study of Cretaceous dinosaur eggshell from the Gobi. *Paleontological Journal*, **6**: 235–45.

Kool, R. (1981) The walking speed of dinosaurs from the Peace River Canyon, British Columbia, Canada. *Canadian Journal of Earth Sciences*, **18**: 823–5.

Krassilov, V.A. (1980) Changes of Mesozoic vegetation and the extinction of dinosaurs. *Palaeogeography, Palaeoclimatology, Palaeoecology*, **34**: 207–224.

Krebs, B. (1965) Die Triasfauna der Tessiner Kalkalpen. XIX. *Ticinosuchus ferox* nov. gen. nov. sp. *Schweizerische Paläontologische Abhandlungen*, **81**: 1–140.

Krebs, B. (1966) Zur Deutung der *Chirotherium*-Fährten. *Natur und Museum*, **96**: 389–96.

Krejci-Graf, K. (1932) Definition der Begriffe Marken, Spuren, Fährten, Bauten, Hieroglyphen und Fucoiden. *Senckenbergiana*, **14**: 19–39.

Kuban, G.J. (1986) A summary of the Taylor site evidence. *Creation/Evolution*, **6**(1): 10–18.

Kuban, G.J. (1989a) Elongate dinosaur tracks. In *Dinosaur Tracks and Traces* (eds D.D. Gillette and M.G. Lockley). Cambridge University Press, Cambridge, pp. 57–72.

Kuban, G.J. (1989b) Color distinctions and other curious features of dinosaur tracks near Glen Rose, Texas. In *Dinosaur Tracks and Traces* (eds D.D. Gillette and M.G. Lockley). Cambridge University Press, Cambridge, pp. 427–40.

Kuenen, P.H. (1957) Sole markings of graded graywacke beds. *Journal of Geology*, **65**: 231–258.

Kuhn, O. (1958a) *Die Fährten der vorzeitlichen Amphibien und Reptilien*. Verlagshaus Meisenbach, Bamberg.

Kuhn, O. (1958b) Zweie neue Arten von *Coelurosaurichnus* aus dem Keuper Frankens. *Neues Jahrbuch für Geologie und Paläontologie, Monatshefte*, **1958**: 437–40.

Kuhn, O. (1963) *Ichnia Tetrapodorum* (Fossilium Catalogus I (Animalia), Pars 101). Junk, s'Gravenhage.

Kummel, B. and Raup, D. eds (1965) *Handbook of Paleontological Techniques*. W.H. Freeman, San Francisco and London.

Kurzanov, S.M. (1987) [Avimimidae and the problem of the origin of birds.] *Trudy Sovetsko-Mongol'skaya Paleontologicheskaya Ekspeditsiya*, **31**: 1–92 [in Russian].

Lambe, L.M. (1917) The Cretaceous theropodous dinosaur *Gorgosaurus*. *Memoirs of the Geological Survey of Canada, Department of Mines*, **100**: 1–84.

Langston, W. (1960) A hadrosaurian ichnite. *Natural History Papers of the National Museum of Canada*, **4**: 1–9.

Langston, W. (1974) Non-mammalian Comanchean tetrapods. *Geoscience and Man*, **8**: 77–102.

Langston, W. (1979) Lower Cretaceous dinosaur tracks near Glen Rose, Texas. In *Lower Cretaceous shallow marine environments in the Glen Rose Formation: Dinosaur Tracks and Plants* (eds B.F. Perkins and W. Langston). American Association of Stratigraphic Palynologists, Dallas, pp. 39–55.

Langston, W. (1986) Stacked dinosaur tracks from the Lower Cretaceous of Texas – a caution for ichnologists. In *First International Symposium on Dinosaur Tracks and Traces, Albuquerque, 1986, Abstracts with Program* (ed. D.D. Gillette). New Mexico Museum of Natural History, Albuquerque, p. 18.

Lanham, U. (1973) *The Bone Hunters*. Columbia University Press, New York and London.

Laporte, L.F. and Behrensmeyer, A.K. (1980) Tracks and substrate reworking by terrestrial vertebrates in Quaternary sediments of Kenya. *Journal of Sedimentary Petrology*, **50**: 1337–46.

Lapparent, A.F. de (1945) Empreintes de pas de Dinosauriens du Maroc exposées dans la Galerie de Paléontologie. *Bulletin du Muséum National d'Histoire Naturelle, Paris*, sèrie 2, **17**: 268–71.

Lapparent, A.F. de (1962) Footprints of dinosaurs in the Lower Cretaceous of Vestspitsbergen-Svalbard. *Årbok, Norsk Polarinstitutt*, **1960**: 14–21.

Lapparent, A.F. de (1966) Nouveaux gisements de reptiles mésozoïques en Éspagne. *Notas y Comunicaciones del Instituto Geologico y Minero de España*, **84**: 103–10.

Lapparent, A.F. de and Aguirre, E. (1956) Algunos yacimientos de Dinosaurios en el Cretácico Superior de la Cuenca de Tremp. *Estudios Geológicos*, **12**: 377–82.

Lapparent, A.F. de and Davoudzadeh, M. (1972) Jurassic dinosaur footprints of the Kerman area, central Iran. *Reports of the Geological Survey of Iran*, **26**: 5–22.

Lapparent, A.F. de and Montenat, C. (1967) Les empreintes de pas de reptiles de l'Infralias du Veillon (Vendée). *Mémoires de la Société Géologique de France*, n.s., **107** 1–44.

Lapparent, A.F. de and Nowgol Sadat, M.A.A. (1975) Une trace de pas de Dinosaure dans le Lias de l'Elbourz, en Iran. *Compte Rendu de l'Académie des Sciences, Paris*, sèrie D, **280**: 161–3.

Lapparent, A.F. de and Oulmi, M. (1964) Une empreinte de pas de Dinosaurien dans le Portlandien de Chassiron (île d'Oleron). *Compte Rendu sommaire des Séances de la Société Géologique de France*, **1964**: 232–3.

Lapparent, A.F. de and Stöcklin, J. (1971) Sur le Jurassique et le Crétacé du Band-e-Turkestan (Afghanistan du Nord-Ouest). *Compte Rendu sommaire des Séances de la Société Géologique de France*, **1971**: 387–8.

Lapparent, A.F. de and Zbyszewski, G. (1957) Les dinosauriens du Portugal. *Memórias dos Serviços Geológicos de Portugal*, n.s., **2**: 1–64.

Lapparent, A.F. de, Joncour, M. le, Mathieu, A. and Plus, B. (1965) Découverte en Éspagne d'empreintes de pas de reptiles mésozoïques. *Boletín de la Real Sociedad Española de Historia Natural*, seccion Geológica, **63**: 225–30.

Lapparent, A.F. de, Zbyszewski, G., Almeida, F.M. de and Veiga Ferreira, O. da (1951) Empreintes de pas de Dinosauriens dans le Jurassique du Cap Mondego (Portugal). *Compte Rendu sommaire des Séances de la Société Géologique de France*, **1951**: 251–2.

Larsonneur, C. and de Lapparent, A.F. (1966) Un dinosaurien carnivore, *Halticosaurus*, dans le Rhétien d'Airel (Manche). *Bulletin de la Société*

Linnéenne de Normandie, sèrie 10, **7**: 108–16.

Leakey, M.D. (1979) Footprints in the ashes of time. *National Geographic Magazine*, **155**: 446–57.

Lehmann, U. (1978) Eine Platte mit Fährten von *Iguanodon* aus dem Obernkirchener Sandstein (Wealden). *Mitteilungen aus dem Geologisch-Paläontologischen Institut der Universität Hamburg*, **48**: 101–14.

Leidy, J. (1856) Notice of remains of extinct reptiles and fishes, discovered by Dr. F.V. Hayden in the Bad Lands of the Judith River, Nebraska Territory. *Proceedings of the Academy of Natural Sciences, Philadelphia*, **8**: 72–3.

Leidy, J. (1958) *Hadrosaurus* and its discovery. *Proceedings of the Academy of Natural Sciences, Philadelphia*, **10**: 213–18.

Leonardi, G. (1979a) Um glossário comparado da icnologia de vertebrados em Português e uma história desta ciência no Brasil. *Cadernos Universitários, Universidade Estadual de Ponta Grossa*, **17**: 1–55 [in Portuguese and English, with glossary in seven languages].

Leonardi, G. (1979b) Nota preliminar sobre seis pistas de dinossauros Ornithischia da Bacia do Rio do Peixe (Cretáceo inferior) em Sousa, Paraíba, Brasil. *Anais da Academia Brasileira de Ciências*, **51**: 501–16.

Leonardi, G. (1979c) New archosaurian trackways from the Rio do Peixe Basin, Paraíba, Brazil. *Annali dell'Università di Ferrara*, n.s., sezione 9 (Scienze Geologiche e Paleontologiche), **5**: 239–49.

Leonardi, G. (1980a) On the discovery of an abundant ichno-fauna (vertebrates and invertebrates) in the Botucatu Formation s.s. in Araraquara, Sao Paulo, Brazil. *Anais da Academia Brasileira de Ciências*, **52**: 559–67.

Leonardi, G. (1980b) Ornithischian trackways of the Corda Formation (Jurassic), Goias, Brazil. *Actas I Congreso Latinoamericano de Paleontología, Buenos Aires, 1978*, **1**: 215–22.

Leonardi, G. (1980c) Dez novas pistas de dinossauros (Theropoda Marsh, 1881) na Bacia do Rio do Peixe, Paraíba, Brasil. *Actas I Congreso Latinoamericano de Paleontología, Buenos Aires, 1978*, **1**: 243–8.

Leonardi, G. (1981) Ichnological data on the rarity of young in North East Brazil dinosaurian populations. *Anais da Academia Brasileira de Ciências*, **53**: 345–6.

Leonardi, G. (1984a) Le impronte fossili di Dinosauri. In *Sulle Orme dei Dinosauri* (by J.F. Bonaparte, E.H. Colbert, P.J. Currie, A. de Ricqlès, Z. Kielan-Jaworowska, G. Leonardi, N. Morello and P. Taquet). Erizzo Editrice, Venice, pp. 165–86.

Leonardi, G. (1984b) Rastros de um Mundo Perdido. *Ciência Hoje*, **2**(15): 48–60.

Leonardi, G. (1985a) The oldest tetrapod record known in the world, and other news. *Ichnology Newsletter*, **14**: 15–16.

Leonardi, G. (1985b) Vale dos Dinossauros: uma janela no noite dos tempos. *Revista Brasileira de Tecnológica*, **16**(1): 23–8.

Leonardi, G. ed. (1987) *Glossary and Manual of Tetrapod Footprint Palaeoichnology*. Departamento Nacional da Produçaõ Mineral, Brasilia [in English, with glossary in eight languages]

Leonardi, G. (1989) Inventory and statistics of the South American dinosaurian ichnofauna and its paleobiological significance. In *Dinosaur Tracks and Traces* (eds D.D. Gillette and M.G. Lockley). Cambridge University Press, Cambridge, pp. 165–78.

Leonardi, G. and Godoy, L.C. (1980) Novas pistas de tetrápodes da Formaçaõ Botucatu no Estado de Saõ Paulo. *Anais do XXXI Congreso Brasileiro de Geologia, Santa Catarina, 1980*, **5**: 3080–9.

Leonardi, G. and Sarjeant, W.A.S. (1986) Footprints representing a new Mesozoic fauna from Brazil. *Modern Geology*, **10**: 73–84.

Lessertisseur, J. (1955) Traces fossiles d'activité animale et leur signification paléobiologique. *Mémoires de la Société Géologique de France*, n.s., **74**: 1–150.

Lewis, D.W. and Titheridge, D.G. (1978) Small scale sedimentary structures resulting from foot impressions in dune sands. *Journal of Sedimentary Petrology*, **48**: 835–7.

Lim, S.-K., Yang, S.-Y. and Lockley, M.G. (1989) Large dinosaur footprint assemblages from the Cretaceous Jindong Formation of southern Korea. In *Dinosaur Tracks and Traces* (eds D.D. Gillette and M.G. Lockley). Cambridge University Press, Cambridge, pp. 333–6.

Lingen, G.J. van der and Andrews, P.B. (1969) Hoof-print structures in beach sand. *Journal of Sedimentary Petrology*, **39**: 350–7.

Llompart, C. (1979) Yacimiento de huellas de pisadas de reptil en el Cretácico superior prepirenaico. *Acta Geológica Hispánica*, **14**: 333–6.

Llompart, C., Casanovas, M.L. and Santafé, J.-V. (1984) Un nuevo yacimiento de Icnitas de Dinosaurios en las facies garumnienses de la Conca de Tremp (Lleida, Espana). *Acta Geológica Hispánica*, **19**: 143–7.

Lockley, M.G. (1985) Vanishing tracks along Alameda Parkway; implications for Cretaceous dinosaurian paleobiology from the Dakota Group, Colorado. In *A Field Guide to Environments of Deposition (and Trace Fossils) of Cretaceous Sandstones of the Western Interior* (ed. C.K. Chamberlain). Society of Economic Paleontologists and Mineralogists, Golden, Colorado, pp. 131–42.

Lockley, M.G. (1986a) The paleobiological and paleoenvironmental importance of dinosaur footprints. *Palaios*, **1**: 37–47.

Lockley, M.G. ed. (1986b) *A Guide to Dinosaur Tracksites of the Colorado*

Plateau and American Southwest. Geology Department, University of Colorado, Denver.

Lockley, M.G. (1987) Dinosaur footprints from the Dakota Group of Eastern Colorado. *The Mountain Geologist*, **24**(4): 107–22.

Lockley, M.G. and Jennings, C. (1987) Dinosaur tracksites of western Colorado and eastern Utah. In *Paleontology and Geology of the Dinosaur Triangle* (ed. W.R. Averett). Museum of Western Colorado, Grand Junction, Colorado, pp. 85–90.

Lockley, M.G., Houck, K.J. and Prince, N.K. (1986) North America's largest dinosaur trackway site: implications for Morrison paleoecology. *Bulletin of the Geological Society of America*, **97**: 1163–76.

Lockley, M.G., Young, B.H. and Carpenter, K. (1983) Hadrosaur locomotion and herding behavior: evidence from footprints in the Mesaverde Formation, Grand Mesa coal field, Colorado. *The Mountain Geologist*, **20**: 5–14.

Lucas, A.M. (1979) Anatomia topographica externa. In *Nomina Anatomica Avium* (eds J.J. Baumel, A.S. King, A.M. Lucas, J.E. Breazile and H.E. Evans). Academic Press, London, pp. 7–18.

Lucas, A.M. and Stettenheim, P.R. (1972) *Avian Anatomy, Integument* (United States Department of Agriculture, Handbook 362). United States Government Printing Office, Washington.

Lucas, S.G., Hunt, A.P. and Kietzke, K.K. (1989) Stratigraphy and age of Cretaceous dinosaur footprints in northeastern New Mexico and northwestern Oklahoma. In *Dinosaur Tracks and Traces* (eds D.D. Gillette and M.G. Lockley). Cambridge University Press, Cambridge, pp. 217–21.

Lull, R.S. (1904) Fossil footprints of the Jura-Trias of North America. *Memoirs of the Boston Society of Natural History*, **5**: 461–557.

Lull, R.S. (1915) Triassic life of the Connecticut valley. *Bulletin of the Connecticut State Geological and Natural History Survey*, **24**: 1–285.

Lull, R.S. (1917) The Triassic flora and fauna of the Connecticut Valley. *Bulletin of the United States Geological Survey*, **597**: 105–127.

Lull, R.S. (1921) The Cretaceous armored dinosaur *Nodosaurus textilis* Marsh. *American Journal of Science*, series 5, **1**: 97–126.

Lull, R.S. (1933) A revision of the Ceratopsia or horned dinosaurs. *Memoirs of the Peabody Museum of Natural History*, **3**: 1–135.

Lull, R.S. (1942) Triassic footprints from Argentina. *American Journal of Science*, **240**: 421–5.

Lull, R.S. (1953) Triassic life of the Connecticut valley (revised edition). *Bulletin of the Connecticut State Geological and Natural History Survey*, **81**: 1–331.

Lull, R.S. and Wright, N.E. (1942) Hadrosaurian dinosaurs of North America. *Special Papers of the Geological Society of America*, **40**: 1–242.

Lyell, C. (1851) Anniversary address of the President. *Quarterly Journal of the Geological Society of London*, **7**: xxv-lxxvi.

Lyell, C. (1855) *Manual of Elementary Geology* (5th edition). Murray, London.

MacClary, J.S. (1939) Mysterious Steps in Purgatory. *Natural History*, **43**: 128.

Madeira, J. and Dias, R. (1983) Novas pistas de dinosaurios no Cretácico Inferior. *Comuniçãcoes dos Serviços Geológicos de Portugal*, **69**(1): 147-58.

Madsen, J.H. (1976) *Allosaurus fragilis*: a revised osteology. *Bulletin of the Utah Geological and Mineral Survey*, **109**: 1-163.

Madsen, S. (1986) The rediscovery of dinosaur tracks near Cameron, Arizona. In *First International Symposium on Dinosaur Tracks and Traces, Albuquerque, 1986, Abstracts with Program* (ed. D.D. Gillette). New Mexico Museum of Natural History, Albuquerque, p. 20.

Majer, S. (1923) Felsökréta dinosaurus nyomok a Kosdi Eocén széntelep feküjében. *Földtani Közlony*, **51-52**: 66-75, 113-4 [in Hungarian with German summary].

Maleev, E.A. (1956) [Armoured dinosaurs from the Upper Cretaceous of Mongolia - Family Ankylosauridae.] *Trudy Paleontologicheskogo Instituta, Akademiya Nauk SSSR*, **62**: 51-91 [in Russian].

Maleev, E.A. (1974) [Giant carnosaurs of the family Tyrannosauridae.] *Trudy Sovetsko-Mongol'skaya Paleontologicheskaya Ekspeditsiya*, **1**: 132-91 [in Russian].

Malz, H. (1971) Ein fossiler 'Wildwechsel' im Wiehengebirge. *Natur und Museum*, **101**: 431-6.

Manabe, M., Hasegawa, Y. and Azuma, Y. (1989) Two new dinosaur footprints from the Early Cretaceous Tetori Group of Japan. In *Dinosaur Tracks and Traces* (eds D.D. Gillette and M.G. Lockley). Cambridge University Press, Cambridge, pp. 309-12.

Mantell, G.A. (1847) *Geological Excursions round the Isle of Wight, and Along the Adjacent Coast of Dorsetshire*. H.G. Bohn, London.

Mantell, G.A. (1850) Notice of the remains of the *Dinornis* and other birds, and of fossils and rock-specimens, recently collected by Mr. Walter Mantell in the Middle Island of New Zealand. *Quarterly Journal of the Geological Society of London*, **6**: 319-43.

Mantell, G.A. (1851) *Petrifactions and their Teachings*. H.G. Bohn, London.

Marsh, O.C. (1896) The dinosaurs of North America. *Annual Report of the United States Geological Survey*, **16**: 133-244.

Marsh, O.C. (1899) Footprints of Jurassic dinosaurs. *American Journal of Science*, series 4, **7**: 227-32.

Martin, H.T. (1922a) Discovery of gigantic footprints in the Coal Measures of Kansas. *Science*, n.s., **55**: 99-100.

Martin, H.T. (1922b) Indications of a gigantic amphibian in the Coal Measures of Kansas. *Science Bulletin, University of Kansas*, **13**: 103–14.

Martin, L.D. and Bennett, D.K. (1977) The burrows of the Miocene beaver *Palaeocastor*, western Nebraska, U.S.A. *Palaeogeography, Palaeoclimatology, Palaeoecology*, **22**: 173–93.

Martinsson, A. (1970) Toponomy of trace fossils. In *Trace Fossils* (Geological Journal, Special Issue No. 3) (eds T.P. Crimes and J.C. Harper). Seel House Press, Liverpool, pp. 323–30.

Matley, C.A. (1941) The coprolites of Pijdura, Central Province. *Records of the Geological Survey of India*, **74**(4): 535–47.

Matsukawa, M. and Obata, I. (1985a) Dinosaur footprints and other indentation in the Cretaceous Sebayashi Formation, Sebayashi, Japan. *Bulletin of the National Science Museum, Tokyo*, series C, **11**(1): 9–36.

Matsukawa, M. and Obata, I. (1985b) Discovery of dinosaur footprints from the Lower Cretaceous Sebayashi Formation, Japan. *Proceedings of the Japan Academy*, series B, **61**: 109–12.

McAllister, J. (1989) Dakota Formation tracks from Kansas: implications for the recognition of tetrapod subaqueous traces. In *Dinosaur Tracks and Traces* (eds D.D. Gillette and M.G. Lockley). Cambridge University Press, Cambridge, pp. 343–8.

McGinnis, H.J. (1982) *Carnegie's Dinosaurs*. Carnegie Institute, Pittsburgh.

McKee, E.D. (1944) Tracks that go uphill. *Plateau*, **16**: 61–72.

McKee, E.D. (1945) Small-scale structures in the Coconino Sandstone of northern Arizona. *Journal of Geology*, **53**: 313–25.

McKee, E.D. (1947) Experiments on the development of tracks in fine cross-bedded sand. *Journal of Sedimentary Petrology*, **17**: 23–8.

McKee, E.D. (1982) Sedimentary structures in dunes of the Namib Desert, South West Africa. *Special Papers of the Geological Society of America*, **188**: 1–64.

McLearn, F.H. (1923) Peace River Canyon coal area, B.C. *Summary Reports of the Geological Survey of Canada*, **1922**, part B: 1–46.

McWhae, J.R.H., Playford, P.E., Lindner, A.W., Glenister, B.F. and Balme, B.E. (1958) The stratigraphy of Western Australia. *Journal of the Geological Society of Australia*, **4**(2): 1–161.

Mehl, M.G. (1931) Additions to the vertebrate record of the Dakota Sandstone. *American Journal of Science*, series 5, **21**: 441–52.

Meinke, D.K., Padian, K. and Kappelman, J. (1980) Growth rings in dinosaur teeth. *Nature, London*, **288**: 193–4.

Mensink, H. and Mertmann, D. (1984) Dinosaurier-Fährten (*Gigantosauropus asturiensis* n.g.n.sp.; *Hispanosauropus hauboldi* n.g.n.sp.) im Jura Asturiens bei La Griega und Ribadesella (Spanien). *Neues Jahrbuch für Geologie und*

Paläontologie, Monatshefte, **1984**: 405–15.

Mesle, G. le and Peron, P.-A. (1881) Sur des empreintes de pas d'oiseaux observées par M. le Mesle dans le sud de l'Algérie. *Compte Rendu des Sessions de l'Association Française pour l'Avancement des Sciences,* 9: 528–33.

Miller, W.E., Britt, B.B. and Stadtman, K.L. (1989) Tridactyl trackways from the Moenave Formation of southwestern Utah. In *Dinosaur Tracks and Traces* (eds D.D. Gillette and M.G. Lockley). Cambridge University Press, Cambridge, pp. 209–15.

Milne, D.H. and Schafersman, S.D. (1983) Dinosaur tracks, erosion marks and midnight chisel work (but no human footprints) in the Cretaceous limestone of the Paluxy River bed, Texas. *Journal of Geological Education,* **31**: 111–23.

Milner, H.B. and Bull, A.J. (1925) The geology of the Eastbourne-Hastings coastline. *Proceedings of the Geologists' Association,* **36**: 291–316.

Mohabey, D.M. (1982) On the occurrence of dinosaurian fossil eggs from Intertrappean limestone, Kheda District, Gujarat. *Current Science,* **52**: 1194.

Mohabey, D.M. (1986) Note on dinosaur foot print from Kheda District, Gujarat. *Journal of the Geological Society of India,* **27**: 456–9.

Mohabey, D.M. (1987) Juvenile sauropod dinosaur from Upper Cretaceous Lameta Formation of Panchmahals District, Gujarat, India. *Journal of the Geological Society of India,* **30**: 210–16.

Molnar, R.E. (1982) A catalogue of fossil amphibians and reptiles in Queensland. *Memoirs of the Queensland Museum,* **20**: 613–33.

Molnar, R.E. (1985) Alternatives to *Archaeopteryx*: a survey of proposed early or ancestral birds. In *The Beginnings of Birds* (eds M.K. Hecht, J.H. Ostrom, G. Viohl and P. Wellnhofer). Freunde des Jura-Museums Eichstätt, Willibaldsburg, pp. 209–17.

Molnar, R.E. and Frey, E. (1987) The paravertebral elements of the Australian ankylosaur *Minmi* (Reptilia: Ornithischia, Cretaceous). *Neues Jahrbuch für Geologie und Paläontologie, Abhandlungen,* **175**: 19–37.

Monbaron, M. (1983) Dinosauriens du Haut Atlas Central (Maroc). *Actes de la Société Jurassienne d'Émulation,* **1983**: 203–34.

Monbaron, M., Dejax, J. and Demathieu, G. (1985) Longues pistes de Dinosaures bipedes à Adrar-n-Ouglagal (Maroc) et repartition des faunes de grands Reptiles dans le domaine atlasique au course du Mésozoique. *Bulletin du Muséum National d'Histoire Naturelle, Paris,* série 4, section C, **7**: 229–42.

Monroe, J.S. (1987) Creationism, human footprints, and flood geology. *Journal of Geological Education,* **35**: 93–103.

Moodie, R.L. (1921) Status of our knowledge of Mesozoic pathology. *Bulletin of the Geological Society of America,* **32**: 321–6.

Moodie, R.L. (1923) *Paleopathology: An Introduction to the Study of Ancient Evidence of Disease.* University of Illinois Press, Urbana.

Moodie, R.L. (1926) Studies in paleopathology, XV. Excess callus following fracture of the fore foot in a Cretaceous dinosaur. *Annals of Medical History*, **8**(1): 73-7.

Moodie, R.L. (1928) The histological nature of ossified tendons found in dinosaurs. *American Museum Novitates*, No. 311: 1-15.

Moodie, R.L. (1929a) Dinosaur tendons. *Science*, **70**: 98.

Moodie, R.L. (1929b) Vertebrate footprints from the Red Beds of Texas. *American Journal of Science*, series 5, **17**: 352-68.

Moodie, R.L. (1930a) Dental abscesses in a dinosaur millions of years old, and the oldest yet known. *Pacific Dental Gazette, San Francisco*, **38**: 435-40.

Moodie, R.L. (1930b) Vertebrate footprints from the Red Beds of Texas, II. *Journal of Geology*, **38**: 548-65.

Moraes, L.J. de (1924) Serras e Montanhas do Nord'este. *Publicaço da Inspectoria Federal de Obras contra as Seccas, Rio de Janeiro*, serie 1(D), **58**: 1-122.

Morales, M. (1986) Dinosaur tracks in the Lower Jurassic Kayenta Formation near Tuba City, Arizona. In *A Guide to Dinosaur Tracksites of the Colorado Plateau and American Southwest* (ed. M.G. Lockley). Geology Department, University of Colorado, Denver, pp. 14-16.

Morales, M. and Colbert, E.H. (1986) Stratigraphic occurrences of dinosaur tracks in the Glen Canyon Group of Arizona. In *First International Symposium on Dinosaur Tracks and Traces, Albuquerque, 1986, Abstracts with Program* (ed. D.D. Gillette). New Mexico Museum of Natural History, Albuquerque, p. 21.

Moratalla, J.J., Sanz, J.L. and Jiménez, S. (1988a) Multivariate analysis on Lower Cretaceous dinosaur footprints: discrimination between ornithopods and theropods. *Geobios*, **21**: 395-408.

Moratalla, J.J., Sanz, J.L. and Jiménez, S. (1988b) Nueva evidencia icnologica de dinosaurios en el Cretácico inferior de La Rioja (Espana). *Estudios Geológicos*, **44**: 119-31.

Morris, J.D. (1980) *Tracking those Incredible Dinosaurs and the People Who Knew Them.* CLP Publishers, San Diego.

Morris, W.J. (1970) Hadrosaurian dinosaur bills – morphology and function. *Los Angeles County Museum, Contributions to Science*, **193**: 1-14.

Morrison, R.G.B. (1981) *A Field Guide to the Tracks and Traces of Australian Animals.* Rigby, Adelaide.

Mossman, D.J. and Sarjeant, W.A.S. (1983) The footprints of extinct animals. *Scientific American*, **248**(1): 64-74, 122.

Mudge, B.F. (1866) Discovery of fossil footmarks in the Liassic (?) Formation in Kansas. *American Journal of Science*, series 2, **41**: 174-6.

Mudge, B.F. (1874) Recent discoveries of fossil footprints in Kansas. *Transactions of the Kansas Academy of Sciences*, **2**: 7–9.

Murie, O.J. (1975) *Field Guide to Animal Tracks* (2nd edition). Houghton Mifflin, Boston.

Murray, J. (1919) *John Murray III, 1808–1892: A Brief Memoir*. Murray, London.

Muybridge, E. (1899) *Animals in Motion*. Chapman and Hall, London [reprinted 1957 by Dover Publications, New York].

Namnandorski, O. (1957) [Giant reptile tracks from Mongolia.] *Priroda*, **44**(5): 110–11 [in Russian].

Napier, J. (1972) *Bigfoot, the Yeti and the Sasquatch in Myth and Reality*. Jonathan Cape, London.

Nelson, C.H., Johnson, K.R. and Barber, J.H. (1987) Gray whale and walrus feeding excavation on the Bering Shelf, Alaska. *Journal of Sedimentary Petrology*, **57**: 419–30.

Newman, B.H. (1970) Stance and gait in the flesh-eating dinosaur *Tyrannosaurus*. *Biological Journal of the Linnean Society*, **2**: 119–23.

Nicholls, E.L. and Russell, A.P. (1985) Structure and function of the pectoral girdle and forelimb of *Struthiomimus altus* (Theropoda: Ornithomimidae). *Palaeontology*, **28**: 643–77.

Nielsen, E. (1949) On some trails from the Triassic beds of East Greenland. *Meddelelser om Grønland*, **149**(4): 1–44.

Nopcsa, F. (1903) Neues über *Compsognathus*. *Neues Jahrbuch für Mineralogie, Geologie und Paläontologie*, **16**: 476–94.

Nopcsa, F. von (1926) *Osteologia Reptilium fossilium et recentium* (Fossilium Catalogus I (Animalia), Pars 27). W. Junk, Berlin.

Nopcsa, F. von (1931) *Osteologia Reptilium fossilium et recentium II: Appendix* (Fossilium Catalogus I (Animalia), Pars 50). W. Junk, Berlin.

Norman, D.B. (1980) On the ornithischian dinosaur *Iguanodon bernissartensis* of Bernissart (Belgium). *Mémoires de l'Institut Royal des Sciences Naturelles de Belgique*, **178**: 1–105.

Novikov, V.P. and Radililovsky, V.V. (1984) [New locations of dinosaur footprints in the Shirkent River Basin (Hissar Mountain).] *Doklady Akademii Nauk Tadzhikskoi SSR*, **27**(10): 605–8 [in Russian].

Novikov, V.P. and Sapozhnikova, I.G. (1981) [New facts about the Ravat dinosaur footprint locality.] *Doklady Akademii Nauk Tadzhikskoi SSR*, **24**(4): 257–61 [in Russian].

Olsen, P.E. (1980) A comparison of the vertebrate assemblages from the Newark and Hartford Basins (Early Mesozoic, Newark Supergroup) of eastern North America. In *Aspects of Vertebrate History* (ed. L.L. Jacobs). Museum of Northern Arizona Press, Flagstaff, Arizona, pp. 35–53.

Olsen, P.E. and Baird, D. (1982) Early Jurassic vertebrate assemblages from the McCoy Brook Fm. of the Fundy Group (Newark Supergroup), Nova Scotia, Canada. *Geological Society of America, Abstracts with Program*, **14**(1–2): 70.

Olsen, P.E. and Baird, D. (1986) The ichnogenus *Atreipus* and its significance for Triassic biostratigraphy. In *The Beginning of the Age of Dinosaurs* (ed. K. Padian). Cambridge University Press, Cambridge, pp. 61–87.

Olsen, P.E. and Galton, P.M. (1984) A review of the reptile and amphibian assemblages from the Stormberg of southern Africa, with special emphasis on the footprints and the age of the Stormberg. *Palaeontologia Africana*, **25**: 87–110.

Olsen, P.E., Remington, C.L. and Cornet, B. (1978) Cyclic change in Late Triassic lacustrine communities. *Science*, **201**: 729–32.

Olson, E.C. (1957) Catalogue of localities of Permian and Triassic terrestrial vertebrates of the territories of the U.S.S.R. *Journal of Geology*, **65**: 196–226 [English summary of Efremov and Vjushkov 1955].

Osborn, H.F. (1912) Integument of the iguanodont dinosaur *Trachodon*. *Memoirs of the American Museum of Natural History*, n.s., **1**: 33–54.

Osborn, H.F. (1917) Skeletal adaptations of *Ornitholestes, Struthiomimus, Tyrannosaurus*. *Bulletin of the American Museum of Natural History*, **35**: 733–71.

Osborn, H.F. (1933) Mounted skeleton of *Triceratops elatus*. *American Museum Novitates*, No. 654: 1–14.

Osmólska, H., Roniewicz, E. and Barsbold, R. (1972) A new dinosaur *Gallimimus bullatus*, n. gen., n. sp. (Ornithomimidae) from the Upper Cretaceous of Mongolia. *Palaeontologia Polonica*, **27**: 103–43.

Ostrom, J.H. (1967) Peabody paleontologists assist new dinosaur-track park. *Discovery*, **2**(2): 21–4.

Ostrom, J.H. (1969a) Osteology of *Deinonychus antirrhopus*, an unusual theropod from the Lower Cretaceous of Montana. *Bulletin of the Peabody Museum of Natural History*, **30**: 1–165.

Ostrom, J.H. (1969b) Terrible claw. *Discovery*, **5**(1): 1–9.

Ostrom, J.H. (1970) Stratigraphy and paleontology of the Cloverly Formation (Lower Cretaceous) of the Bighorn Basin area, Wyoming and Montana. *Bulletin of the Peabody Museum of Natural History*, **35**: 1–234.

Ostrom, J.H. (1972) Were some dinosaurs gregarious? *Palaeogeography, Palaeoclimatology, Palaeoecology*, **11**: 287–301.

Ostrom, J.H. (1978) The osteology of *Compsognathus longipes* Wagner. *Zitteliana, Abhandlungen der Bayerischen Staatssammlung für Paläontologie und historische Geologie*, **4**: 73–118.

Ostrom, J.H. (1984) Social and unsocial behavior in dinosaurs. *Bulletin of the*

Field Museum of Natural History, **55**(9): 10–21.

Ostrom, J.H. (1987) Romancing the dinosaurs. *The Sciences*, **1987**(3): 56–63.

Owen, R. (1841) Description of parts of the skeleton and teeth of five species of the genus *Labyrinthodon*, from the New Red Sandstone of Coton End and Cubbington quarries; with remarks on the probable identity of *Cheirotherium* with this genus of extinct batrachians. *Proceedings of the Geological Society of London*, **3**: 389–97.

Owen, R. (1842) Report on British fossil reptiles. Part II. *Report of the British Association for the Advancement of Science, Plymouth, 1841*, **11**: 60–204.

Padian, K. and Olsen, P.E. (1984) The fossil trackway *Pteraichnus*: not pterosaurian, but crocodilian. *Journal of Paleontology*, **58**: 178–84.

Padian, K. and Olsen, P.E. (1989) Ratite footprints and the stance and gait of Mesozoic theropods. In *Dinosaur Tracks and Traces* (eds D.D. Gillette and M.G. Lockley). Cambridge University Press, Cambridge, pp. 231–41.

Pannell, K. (1986) Dinosaur footprints at Oak Hill, Virginia. In *First International Symposium on Dinosaur Tracks and Traces, Albuquerque, 1986, Abstracts with Program* (ed. D.D. Gillette). New Mexico Museum of Natural History, Albuquerque, p. 22.

Parker, L.R. and Rowley, R.L. (1989) Dinosaur footprints from a coal mine in east-central Utah. In *Dinosaur Tracks and Traces* (eds D.D. Gillette and M.G. Lockley). Cambridge University Press, Cambridge, pp. 361–6.

Parrish, J.M. (1986) Locomotor adaptations in the hindlimb and pelvis of the Thecodontia. *Hunteria, Publications in Paleontology of the Societas Paleontographica Coloradensis*, **1**(2): 1–35.

Parrish, J.M. and Lockley, M.G. (1984) Dinosaur trackways from the Triassic of western Colorado. *Geological Society of America, Abstracts with Programs*, **16**: 250.

Partridge, E. (1965) *Usage and Abusage: A Guide to Good English* (6th edition). Hamish Hamilton, London.

Paul, G.S. (1987) Predation in the meat eating dinosaurs. In *Fourth Symposium on Mesozoic Terrestrial Ecosystems, Short Papers* (Occasional Papers of the Tyrrell Museum of Palaeontology, No. 3) (eds. P.J. Currie and E.H. Koster). Tyrrell Museum of Palaeontology, Drumheller, pp. 171–6.

Paul, G.S. (1988) *Predatory Dinosaurs of the World*. Simon and Schuster, New York and London.

Pawlicki, R., Korbel, A. and Kubiak, H. (1966) Cells, collagen fibrils and vessels in dinosaur bone. *Nature, London*, **211**: 655–7.

Peabody, F.E. (1947) Current crescents in the Triassic Moenkopi Formation. *Journal of Sedimentary Petrology*, 17: 73–6.

Peabody, F.E. (1948) Reptile and amphibian trackways from the Lower Triassic Moenkopi Formation of Arizona and Utah. *Bulletin of the*

Department of Geological Sciences, University of California, **27**: 295-468.

Peabody, F.E. (1955) Taxonomy and the footprints of tetrapods. *Journal of Paleontology*, **29**: 915-18.

Peabody, F.E. (1956a) Gilmore's split-toed footprint from the Grand Canyon Hermit Shale Formation. *Plateau*, **29**(2): 41-3.

Peabody, F.E. (1956b) Ichnites from the Triassic Moenkopi Formation of Arizona and Utah. *Journal of Paleontology*, **30**: 731-40.

Peabody, F.E. (1959) Trackways of living and fossil salamanders. *Publications in Zoology, University of California*, **63**(1): 1-72.

Penner, M.M. (1985) The problem of dinosaur extinction; contribution of the study of terminal Cretaceous eggshells from southeast France. *Geobios*, **18**: 665-9.

Pennycuick, C.J. (1975) On the running of the gnu (*Connochaetes taurinus*) and other animals. *Journal of Experimental Biology*, **63**: 775-99.

Perkins, B.F. (1974) Paleoecology of a rudist reef complex in the Comanche Cretaceous Glen Rose Limestone of central Texas. *Geoscience and Man*, **8**: 131-73.

Peterson, W. (1924) Dinosaur tracks in the roofs of coal mines. *Natural History*, **24**: 388-91.

Pittman, J.G. (1984) Geology of the De Queen Formation of Arkansas. *Transactions of the Gulf Coast Association of Geological Societies*, **34**: 201-9.

Pittman, J.G. (1989) Stratigraphy, lithology, depositional environment, and track type of dinosaur track-bearing beds of the Gulf coastal plain. In *Dinosaur Tracks and Traces* (eds D.D. Gillette and M.G. Lockley). Cambridge University Press, Cambridge, pp. 135-53.

Pittman, J.G. and Gillette, D.D. (1989) The Briar site: a new sauropod dinosaur tracksite in Lower Cretaceous beds of Arkansas, USA. In *Dinosaur Tracks and Traces* (eds D.D. Gillette and M.G. Lockley). Cambridge University Press, Cambridge, pp. 313-32.

Plateau, H., Giboulet, G. and Roch, E. (1937) Sur la présence d'empreintes de Dinosauriens dans la région de Demnat (Maroc). *Compte Rendu sommaire des Séances de la Société Géologique de France*, **1937**: 241-2.

Pleijel, C. (1975) Nya dinosauriefotspår från skånes Rät-Lias. *Fauna och Flora, Stockholm*, **70**: 116-20.

Pollard, J.E. (1988) Trace fossils in coal-bearing sequences. *Journal of the Geological Society, London*, **145**: 339-50.

Prentice, J.E. (1962) Some sedimentary structures from a Weald Clay sandstone at Warnham Brickworks, Horsham, Sussex. *Proceedings of the Geologists' Association*, **73**: 171-85.

Prince, N.K. (1983) Late Jurassic dinosaur trackways from SE Colorado. *Geology Department Magazine, University of Colorado at Denver*, **2**: 15-19.

Raath, M.A. (1969) A new coelurosaurian dinosaur from the Forest Sandstone of Rhodesia. *Arnoldia, Miscellaneous Publications of the National Museums of Rhodesia*, **4**(28): 1–25.

Raath, M.A. (1972) First record of dinosaur footprints from Rhodesia. *Arnoldia, Miscellaneous Publications of the National Museums of Rhodesia*, **5**(27): 1–5.

Raath, M.A. (1974) Fossil vertebrate studies in Rhodesia: further evidence of gastroliths in prosauropod dinosaurs. *Arnoldia, Miscellaneous Publications of the National Museums of Rhodesia*, **7**(5): 1–7.

Rautman, C.A. and Dott, R.H. (1977) Dish structures formed by fluid escape in Jurassic shallow marine sandstones. *Journal of Sedimentary Petrology*, **47**: 101–6.

Reiche, P. (1938) An analysis of cross-lamination: the Coconino Sandstone. *Journal of Geology*, **46**: 918.

Reichman, O.J. and Aitchison, S. (1981) Mammal trails on mountain slopes: optimal paths in relation to slope angle and body weight. *American Naturalist*, **117**: 416–20.

Reid, R.E.H. (1984) The histology of dinosaurian bone, and its possible bearing on dinosaurian physiology. In *The Structure, Development and Evolution of Reptiles* (Zoological Society of London, Symposium Series, No. 52) (ed. M.W.J. Ferguson). Academic Press, London, pp. 629–63.

Reid, R.E.H. (1987) Bone and dinosaurian 'endothermy'. *Modern Geology*, **11**: 133–54.

Reineck, H.-E. and Howard, J.D. (1978) Alligatorfährten. *Natur und Museum*, **108**: 10–15.

Renders, E. (1984) The gait of *Hipparion* sp. from fossil footprints in Laetoli, Tanzania. *Nature, London*, **308**: 179–81, 866.

Reynolds, R.E. (1983) Jurassic trackways in the Mescal Range, San Bernardino County, California. *Utah Geological and Mineral Survey, Special Studies*, **60**: 46–8.

Reynolds, R.E. (1989) Dinosaur trackways in the Lower Jurassic Aztec Sandstone of California. In *Dinosaur Tracks and Traces* (eds D.D. Gillette and M.G. Lockley). Cambridge University Press, Cambridge, pp. 284–92.

Richmond, N.D. (1965) Perhaps juvenile dinosaurs were always scarce. *Journal of Paleontology*, **39**: 503–5.

Ricqlès, A.J. de (1976) On bone histology of fossil and living reptiles, with comments on its functional and evolutionary significance. In *Morphology and Biology of Reptiles* (Linnean Society Symposium series, No. 3) (eds A.d'A. Bellairs and C.B. Cox). Academic Press, London, pp. 123–50.

Ricqlès, A.J. de (1980) Tissue structure of dinosaur bone; functional significance and possible relation to dinosaur physiology. In *A Cold Look at the Warm-Blooded Dinosaurs* (American Association for the Advancement of Science, Selected Symposium series, No. 28) (eds D.K. Thomas and E.C. Olson). Westview Press, Washington, pp. 103–39.

Ricqlès, A.J. de (1983) Cyclical growth in the long limb bones of a sauropod dinosaur. *Acta Palaeontologica Polonica*, **28**: 225–32.

Riggs, E.S. (1904) Dinosaur footprints from Arizona. *American Journal of Science*, series 4, **17**: 423–4.

Rioult, M. (1978) *Écosystèmes continentaux Mésozoïques de Normandie*. Laboratoire de Géologie Armoricaine, Université de Caen.

Rixon, A.E. (1976) *Fossil Animal Remains: Their Preparation and Conservation*. Athlone Press, University of London, London.

Roehler, H.W. and Stricker, G.D. (1984) Dinosaur and wood fossils from the Cretaceous Corwin Formation in the National Petroleum Reserve, North Slope of Alaska. *Journal of the Alaska Geological Society*, **4**: 35–41.

Rothschild, B.M. (1988) Stress fracture in a ceratopsian phalanx. *Journal of Paleontology*, **62**: 302–3.

Rozhdestvensky, A.K. (1964) [New data on dinosaur occurrences in Kazakhstan and central Asia.] *Tashkentskii Gosudarstvennyi Universitet im V.I. Lenina, Nauchnye Trudy*, n.s., **234**(20): 227–41 [in Russian].

Russell, D.A. (1970) Tyrannosaurs from the Late Cretaceous of western Canada. *Publications in Palaeontology, National Museum of Natural Sciences, Canada*, **1**: 1–34.

Russell, D.A. (1972) Ostrich dinosaurs from the Late Cretaceous of western Canada. *Canadian Journal of Earth Sciences*, **9**: 375–402.

Russell, D.A. (1980) Reflections of the dinosaurian world. In *Aspects of Vertebrate History* (ed. L.L. Jacobs). Museum of Northern Arizona Press, Flagstaff, Arizona, pp. 257–68.

Russell, D.A. (1981) [Untitled comments on estimated speed of a giant bipedal dinosaur.] *Nature, London*, **292**: 274.

Russell, D.A. and Béland, P. (1976) Running dinosaurs. *Nature, London*, **264**: 486.

Russell, D.A., Béland, P. and McIntosh, J.S. (1980) Paleoecology of the dinosaurs of Tendaguru (Tanzania). *Mémoires de la Société Géologique de France*, n.s., **139**: 169–75.

Sahni, A., Rana, R.S. and Prasad, G.V.R. (1984) SEM studies of thin egg shell fragments from the Intertrappean (Cretaceous-Tertiary transition) of Nagpur and Asifabad, peninsular India. *Journal of the Palaeontological Society of India*, **29**: 26–33.

Sams, R.H. (1982) Newly discovered dinosaur tracks, Comal County, Texas. *Bulletin of the South Texas Geological Society*, **23**(1): 19–23.

Sanderson, I.D. (1974) Sedimentary structures and their environmental significance in the Navajo Sandstone, San Rafael Swell, Utah. *Brigham Young University Geology Studies*, **21**: 215–46.

Santa Luca, A.P. (1980) The postcranial skeleton of *Heterodontosaurus tucki* (Reptilia, Ornithischia) from the Stormberg of South Africa. *Annals of the South African Museum*, **79**: 159–211.

Sanz, J.L., Moratalla, J.J. and Casanovas, M.L. (1985) Traza icnologica de un dinosaurio iguanodontido en el Cretácico inferior de Cornago (La Rioja, España). *Estudios Geológicos*, **41**: 85–91.

Sarjeant, W.A.S. (1967) Fossil footprints from the Middle Triassic of Nottinghamshire and Derbyshire. *Mercian Geologist*, **2**: 327–41.

Sarjeant, W.A.S. (1970) Fossil footprints from the Middle Triassic of Nottinghamshire and the Middle Jurassic of Yorkshire. *Mercian Geologist*, **3**: 269–82.

Sarjeant, W.A.S. (1971) Vertebrate tracks from the Permian of Castle Peak, Texas. *Texas Journal of Science*, **22**: 343–66.

Sarjeant, W.A.S. (1974) A history and bibliography of the study of fossil vertebrate footprints in the British Isles. *Palaeogeography, Palaeoclimatology, Palaeoecology*, **16**: 265–378.

Sarjeant, W.A.S. (1975) Fossil tracks and impressions of vertebrates. In *The Study of Trace Fossils* (ed. R.W. Frey). Springer-Verlag, Berlin and New York, pp. 283–324.

Sarjeant, W.A.S. (1981) In the footsteps of the dinosaurs. *Explorers Journal*, December: 164–71.

Sarjeant, W.A.S. (1985) The Beasley collection of photographs and drawings of fossil footprints and bones, and of fossil and recent sedimentary structures. *The Geological Curator*, **4**(3): 133–63.

Sarjeant, W.A.S. (1987) The study of fossil vertebrate footprints; a short history and selective bibliography. In *Glossary and Manual of Tetrapod Footprint Palaeoichnology* (ed. G. Leonardi). Departamento Nacional da Produçaõ Mineral, Brasilia, pp. 1–19.

Sarjeant, W.A.S. (1988) Fossil vertebrate footprints. *Geology Today*, **4**(4): 125–30.

Sarjeant, W.A.S. and Kennedy, W.J. (1973) Proposal of a code for the nomenclature of trace-fossils. *Canadian Journal of Earth Sciences*, **10**: 460–75.

Sarjeant, W.A.S. and Stringer, P. (1978) Triassic reptile tracks in the Lepreau Formation, southern New Brunswick, Canada. *Canadian Journal of Earth Sciences*, **15**: 594–602.

Saunders, J.B., Inman, V.T. and Eberhart, H.D. (1953) The major determinants in normal and pathological gait. *Journal of Bone and Joint Surgery*, series A, **35**: 543–58.

Saxby, S.M. (1846) On the discovery of footmarks in the Greensand of the

Isle of Wight. *The London, Edinburgh and Dublin Philosophical Magazine*, series 3, **29**: 310–12.

Schmidt, H. (1927) Fährten der ältesten Saurier. *Natur und Museum*, **57**: 517–26.

Schmidt, M. (1911) Zur Deutung zweier Problematika des Buntsandsteins. 2. Fährtenplatten im Buntsandstein und englischen Dogger. *Jahresberichte und Mitteilungen des Oberrheinischen geologischen Vereines*, neue Folge, 1(2): 43–6.

Schmidt-Nielsen, K. (1984) *Scaling: Why is Animal Size So Important?* Cambridge University Press, Cambridge.

Schneider, H.J., King, A.Y., Bronso, J.L. and Miller, E.H. (1974) Stress injuries and developmental changes of lower extremities in ballet dancers. *Radiology*, **113**: 627–32.

Schopf, T.J.M. (1981) Evidence from findings of molecular biology with regard to the rapidity of genomic change: implications for species durations. In *Paleobotany, Paleoecology, and Evolution*, Vol. 1 (ed. K.J. Niklas). Praeger, New York, pp. 135–92.

Seilacher, A. (1953) Studien zur Palichnologie. I. Über die Methoden der Palichnologie. *Neues Jahrbuch für Geologie und Paläontologie, Abhandlungen*, **98**: 87–124.

Seilacher, A. (1964) Sedimentological classification and nomenclature of trace fossils. *Sedimentology*, **3**: 253–6.

Seilacher, A. (1978) Use of trace fossil assemblages for recognizing depositional environments. In *Trace Fossil Concepts* (Society of Economic Paleontologists and Mineralogists, Short Course No. 5) (ed. P.B. Basan). Society of Economic Paleontologists and Mineralogists, Oklahoma City, 167–81.

Seymour, R.S. (1979) Dinosaur eggs: gas conductance through the shell, water loss during incubation and clutch size. *Paleobiology*, **5**: 1–11.

Shrock, R.R. (1948) *Sequence in Layered Rocks*. McGraw-Hill, New York.

Shuler, E.W. (1917) Dinosaur tracks in the Glen Rose Limestone near Glen Rose, Texas. *American Journal of Science*, series 4, **44**: 294–8.

Shuler, E.W. (1935) Dinosaur track mounted in the bandstand at Glen Rose, Texas. *Field and Laboratory*, **4**: 9–13.

Sikes, S.K. (1971) *The Natural History of the African Elephant*. Weidenfeld and Nicolson, London.

Simpson, S. (1956) On the trace-fossil *Chondrites*. *Proceedings of the Geological Society of London*, **1537**: 83.

Simpson, S. (1957) On the trace-fossil *Chondrites*. *Quarterly Journal of the Geological Society of London*, **112**: 155–80.

Simpson, S. (1975) Classification of trace fossils. In *The Study of Trace Fossils* (ed. R.W. Frey). Springer-Verlag, Berlin and New York, pp. 39–54.

Sivers, W. von (1929) Irreführende Spuren von Möwen. *Natur und Museum*, **59**: 62–3.

Smith, R.M.H. (1987) Helical burrow casts of therapsid origin from the Beaufort Group (Permian) of South Africa. *Palaeogeography, Palaeoclimatology, Palaeoecology*, **60**: 155–70.

Snyder, R.C. (1952) Quadrupedal and bipedal locomotion in lizards. *Copeia*, **1952**: 64–70.

Sochava, A.V. (1969) Dinosaur eggs from the Upper Cretaceous of the Gobi Desert. *Paleontological Journal*, **3**: 517–27.

Sochava, A.V. (1970) Microtexture of dinosaur eggshells from the Lower Cretaceous of the northern Gobi. *Doklady Akademii Nauk SSSR*, **192**: 203–5.

Sochava, A.V. (1971) Two types of eggshells in Senonian dinosaurs. *Paleontological Journal*, **5**: 353–61.

Sochava, A.V. (1972) The skeleton of an embryo in a dinosaur egg. *Paleontological Journal*, **6**: 527–30.

Soergel, W. (1925) *Die Fährten der Chirotheria. Eine paläobiologische Studie*. Gustav Fischer, Jena, 92p.

Sollas, W.J. (1879) On some three-toed footprints from the Triassic conglomerate of South Wales. *Quarterly Journal of the Geological Society of London*, **35**: 511–16.

Somm, A. and Schneider, B. (1962) Zwei paläontologische und stratigraphische Beobachtungen in der Obertrias der südwestlichen Engadiner Dolomiten (Graubünden). I. Saurierfährten aus der Obertrias der Quattervalsgruppe. *Ergebnisse der wissenschaftlichen Untersuchungen des Schweizerischen Nationalparks*, neue Folge, **7**(47): 353–61.

Sommerkamp, W. (1960) Saurierfährten im Wiehengebirge. *Kosmos (Stuttgart)*, **56**: 228.

Speakman, F.J. (1954) *Tracks, Trails and Signs*. G. Bell, London.

Spicer, R.A. and Parrish, J.T. (1987) Plant megafossils, vertebrate remains, and paleoclimate of the Kogosukruk Tongue (Late Cretaceous), North Slope, Alaska. *Circulars of the United States Geological Survey*, **998**: 47–8.

Srivastava, S., Mohabey, D.M., Sahni, A. and Pant, S.C. (1986) Upper Cretaceous dinosaur egg clutches from Kheda District (Gujarat, India: their distribution, shell ultrastructure and palaeoecology. *Palaeontographica*, Abt. A, **193**: 219–33.

Staines, H.R.E. (1954) Dinosaur footprints at Mount Morgan. *Queensland Government Mining Journal*, **55**: 483–5.

Staines, H.R.E. and Woods, J.T. (1964) Recent discovery of Triassic dinosaur footprints in Queensland. *Australian Journal of Science*, **27**: 55.

Staniland, L.N. (1953) *The Principles of Line Illustration with Emphasis on the*

Requirements of Biological and Other Scientific Workers. Harvard University Press, Cambridge, Massachusetts.

Stechow, E. (1909) Neue Funde von *Iguanodon*-Fährten. *Centralblatt für Mineralogie, Geologie und Paläontologie,* **1909**: 700–5.

Steinbock, R.T. (1989) Ichnology of the Connecticut Valley: a vignette of American science in the early nineteenth century. In *Dinosaur Tracks and Traces* (eds D.D. Gillette and M.G. Lockley). Cambridge University Press, Cambridge, pp. 27–32.

Sternberg, C.M. (1926) Dinosaur tracks from the Edmonton Formation of Alberta. *Canada, Museum Bulletins, Geological Series,* **44**: 85–7.

Sternberg, C.M. (1927) Horned dinosaur group in the National Museum of Canada. *The Canadian Field-Naturalist,* **41**: 67–73.

Sternberg, C.M. (1932) Dinosaur Tracks from Peace River, British Columbia. *Annual Report of the National Museum of Canada,* **1930**: 59–85.

Sternberg, C.M. (1933a) Carboniferous tracks from Nova Scotia. *Bulletin of the Geological Society of America,* **44**: 951–64.

Sternberg, C.M. (1933b) Prehistoric footprints in Peace River. *Canadian Geographic Journal,* **6**(2): 92–102.

Sternberg, C.M. (1935) Hooded hadrosaurs of the Belly River Series of the Upper Cretaceous: a comparison with description of a new species. *Bulletin of the National Museum of Canada,* **77**: 1–37.

Sternberg, C.M. (1940) Ceratopsidae from Alberta. *Journal of Paleontology,* **14**: 468–80.

Sternberg, C.M. (1970) Comments on dinosaurian preservation in the Cretaceous of Alberta and Wyoming. *Publications in Palaeontology, National Museum of Natural Sciences, Canada,* **4**: 1–9.

Stewart, A.D. (1970) Palaeogravity. In *Palaeogeophysics* (ed. S.K. Runcorn). Academic Press, London and New York, pp. 413–34.

Stokes, W.L. (1964) Fossilized stomach contents of a sauropod dinosur. *Science,* **143**: 576–7.

Stokes, W.L. (1978) Animal tracks in the Navajo-Nugget Sandstone. *Contributions to Geology, University of Wyoming,* **16**(2): 103–7.

Stokes, W.L. (1986) Alleged human footprint from Middle Cambrian strata, Millard County, Utah. *Journal of Geological Education,* **34**: 187–90.

Storer, J.E. (1975) Dinosaur tracks, *Columbosauripus ungulatus* (Saurischia: Coelurosauria), from the Dunvegan Formation (Cenomanian) of Northeastern British Columbia. *Canadian Journal of Earth Sciences,* **12**: 1805–7.

Struckmann, C. (1880) Vorläufige Nachricht über das Vorkommen grosser vogelähnlicher Thierfährten (*Ornithoidichnites*) im Hastingssandsteine von Bad Rehburg bei Hannover. *Neues Jahrbuch für Mineralogie, Geologie und Paläontologie,* **1880**: 125–8.

Strunz, C. (1936) Unsere Donner-Echse (*Diplodocus*) in neuer Haltung. *Natur und Volk*, **66**: 371–9.

Swaine, J. (1962) Iguanodon footprints. *New Scientist*, **13**: 520.

Tagart, E. (1846) On markings in the Hastings sands near Hastings, supposed to be the footprints of Birds. *Quarterly Journal of the Geological Society of London*, **2**: 267.

Tanimoto, M. (1988) Tails of sauropods from the Sichuan Basin, China. *Archosaurian Articulations*, **1**(2): 15–16.

Taquet, P. (1972) A la recherche des dinosaures du Niger. *Le Courrier du Centre National de la Recherche Scientifique, Paris*, **1972**(1): 33–6 and rear cover.

Taquet, P. (1976) Géologie et Paléontologie du gisement de Gadoufaoua (Aptien du Niger). *Cahiers de Paléontologie, Centre National de la Recherche Scientifique, Paris*, **1976** [series not numbered]: 1–191.

Taquet, P. (1977a) Dinosaurs of Niger. *Nigerian Field*, **42**(1): 2–10.

Taquet, P. (1977b) Les découvertes récentes de Dinosaures du Jurassique et du Crétacé en Afrique, au Proche et Moyen-Orient et en Inde. *Mémoires hors de Sèrie, Société Géologique de France*, **8**: 325–30.

Tarsitano, S. (1983) Stance and gait in theropod dinosaurs. *Acta Palaeontologica Polonica*, **28**: 251–64.

Teichert, C. (1934) Umgeprägte Spuren in der Schneewüste. *Natur und Volk*, **64**: 73.

Thaler, L. (1962) Empreintes de pas de dinosaures dans les dolomies du Lias inférieur des Causses (note préliminaire). *Compte Rendu sommaire des Séances de la Société Géologique de France*, **1962**: 190–2.

Thorpe, M.R. (1929) A new Triassic fossil field. *American Journal of Science*, series 5, **18**: 277–300.

Thulborn, R.A. (1972) The post-cranial skeleton of the Triassic ornithischian dinosaur *Fabrosaurus australis*. *Palaeontology*, **15**: 29–60.

Thulborn, R.A. (1980) The ankle joints of archosaurs. *Alcheringa*, **4**: 241–61.

Thulborn, R.A. (1981) Estimated speed of a giant bipedal dinosaur. *Nature, London*, **292**: 273–4.

Thulborn, R.A. (1982) Speeds and gaits of dinosaurs. *Palaeogeography, Palaeoclimatology, Palaeoecology*, **38**: 227–56.

Thulborn, R.A. (1984) Preferred gaits of bipedal dinosaurs. *Alcheringa*, **8**: 243–52.

Thulborn, R.A. (1986). Triassic amphibian and reptile tracks of the Brisbane-Ipswich area. In *A Field Guide to Sediments and Fossils of the Ipswich Basin* (ed. B.G. Fordham). Geological Society of Australia (Queensland Division), Brisbane, pp. 20–4.

Thulborn, R.A. (1989) The gaits of dinosaurs. In *Dinosaur Tracks and Traces* (eds D.D. Gillette and M.G. Lockley). Cambridge University Press, Cambridge, pp. 39–50.

Thulborn, R.A. and Hamley, T.L. (1982) The reptilian relationships of *Archaeopteryx. Australian Journal of Zoology*, **30**: 611–34.

Thulborn, R.A. and Wade, M. (1979) Dinosaur stampede in the Cretaceous of Queensland. *Lethaia*, **12**: 275–9.

Thulborn, R.A. and Wade, M. (1984) Dinosaur trackways in the Winton Formation (mid-Cretaceous) of Queensland. *Memoirs of the Queensland Museum*, **21**: 413–517.

Thulborn, R.A. and Wade, M. (1989) A footprint as a history of movement. In *Dinosaur Tracks and Traces* (eds D.D. Gillette and M.G. Lockley). Cambridge University Press, Cambridge, pp. 51–6.

Tongiorgio, M. (1980) Orme di tetrapodi dei Monti Pisani. In *I Vertebrati Fossili Italiani; Catalogo della Mostra, Verona, 1980* (ed. G. Parisi). Commune di Verona, Verona, pp. 77–84.

Triggs, B. (1984) *Mammal Tracks and Signs: A Fieldguide for Southeastern Australia*. Oxford University Press, Melbourne.

Trusheim, F. (1929) Fünfzehige Vogelfährten. *Natur und Museum*, **59**: 63–4.

Tucker, M.E. and Burchette, T.P. (1977) Triassic dinosaur footprints from South Wales: their context and preservation. *Palaeogeography, Palaeoclimatology, Palaeoecology*, **22**: 195–208.

Tylor, A. (1862) On the footprints of an *Iguanodon*, lately found at Hastings. *Quarterly Journal of the Geological Society of London*, **18**: 247–53.

Valenzuela, M., Garcia-Ramos, J.C. and Suarez de Centi, C. (1986) The Jurassic sedimentation in Asturias (N Spain). *Trabajos de Geológia*, **16**: 121–32.

Vialov, O.S. (1966) [*Traces of the Life Activity of Organisms and their Palaeontological Significance.*] Naukova Dumka Press, Kiev [in Russian].

Vialov, O.S. (1972) The classification of the fossil traces of life. *Proceedings of the 24th International Geological Congress, Montreal, 1972*, section 7 (Paleontology): 639–44.

Vianey-Liaud, M., Jain, S.L. and Sahni, A. (1987) Dinosaur eggshells (Saurischia) from the Late Cretaceous Intertrappean and Lameta Formations (Deccan, India). *Journal of Vertebrate Paleontology*, **7**: 408–24.

Viera, L.I. and Aguirrezabala, L.M. (1982) El Weald de Munilla (La Rioja) y sus Icnitas de Dinosaurios (I). *Munibe*, **34**(4): 245–70.

Viera, L.I. and Torres, J.A. (1979) El Wealdico de la zona de Enciso (Sierra de los Cameros) y su fauna de Grandes Reptiles. *Munibe*, **31**(1–2): 141–57.

Voss-Foucart, M.F. (1968) Paléoproteines des coquilles fossiles d'oeufs de dinosauriens du Crétacé supérieur de Provence. *Comparative Bichemistry and Physiology*, **24**: 31–6.

Wade, M. (1979a) A minute – a hundred million years ago. *Hemisphere*, **23**: 16–21.

Wade, M. (1979b. Tracking dinosaurs; the Winton excavation. *Australian Natural History*, **19**: 286–91.

Wade, M. (1989) The stance of dinosaurs and the Cossack Dancer syndrome. In *Dinosaur Tracks and Traces* (eds D.D. Gillette and M.G. Lockley). Cambridge University Press, Cambridge, pp. 73–82.

Walther, F.R., Mungall, E.C. and Grau, G.A. (1983) *Gazelles and Their Relatives; A Study in Territorial Behavior*. Noyes Publications, Park Ridge, New Jersey.

Walther, J. (1917) Über *Chirotherium*. *Zeitschrift der Deutschen Geologischen Gesellschaft, Monatsberichte*, **69**: 181–4.

Warren, A.A. and Hutchinson, M.N. (1987) The skeleton of a new hornless rhytidosteid (Amphibia, Temnospondyli). *Alcheringa*, **11**: 291–302.

Weaver, J.C. (1983) The improbable endotherm: the energetics of the sauropod dinosaur *Brachiosaurus*. *Paleobiology*, **9**: 173–82.

Weems, R.E. (1987) A Late Triassic footprint fauna from the Culpeper Basin Northern Virginia (U.S.A.). *Transactions of the American Philosophical Society*, **77**(1): 1–79.

Weishampel, D.B. (1984a) Interactions between Mesozoic plants and vertebrates: fructifications and seed predation. *Neues Jahrbuch für Geologie und Paläontologie, Abhandlungen*, **167**: 224–50.

Weishampel, D.B. (1984b) Trossingen: E. Fraas, F. von Huene, R. Seemann, and the 'Schwabische Lindwurm' *Plateosaurus*. In *Third Symposium on Mesozoic Terrestrial Ecosystems, Tübingen, 1984, Short Papers* (eds W.-E. Reif and F. Westphal). Attempto Verlag, Tübingen, pp. 249–53.

Welles, S.P. (1971) Dinosaur footprints from the Kayenta Formation of northern Arizona. *Plateau*, **44**: 27–38.

West, I. and El-Shahat, A. (1984) Dinosaur footprints and early cementation of Purbeck bivalve beds. *Proceedings of the Dorset Natural History and Archaeological Society*, **106**: 169–70.

Western, D., Moss, C. and Georgiadis, N. (1983) Age estimation and population age structure of elephants from footprint dimensions. *Journal of Wildlife Management*, **47**: 1192–7.

Whittington, H.B. and Conway Morris, S. eds (1985) Extraordinary fossil biotas: their ecological and evolutionary significance. *Philosophical Transactions of the Royal Society of London*, series B, **311**: 1–192.

Whyte, M.A. and Romano, M. (1981) A footprint in the sands of time. *Journal of the Geological Society, University of Sheffield*, **7**: 323–30.

Wilfarth, M. (1937) Deutungsversuch der Fährte *Kouphichnium*. *Zentralblatt für Mineralogie, Geologie und Paläontologie*, Abt. B, **1937**: 329–33.

Willard, B. (1935) Chemung tracks and trails from Pennsylvania. *Journal of Paleontology*, **9**: 43–56.

Williams, D.L.G., Seymour, R.S. and Kerourio, P. (1984) Structure of fossil dinosaur eggshell from the Aix Basin, France. *Palaeogeography, Palaeoclimatology, Palaeoecology*, **45**: 23–37.

Williams, W.L. (1872) On the occurrence of footprints of a large bird found at Turanganui, Poverty Bay. *Transactions and Proceedings of the New Zealand Institute*, **4**: 124–7.

Wills, L.J. and Sarjeant, W.A.S. (1970) Fossil vertebrate and invertebrate tracks from boreholes through the Bunter Series (Triassic) of Worcestershire. *Mercian Geologist*, **3**: 399–414.

Wilson, V., Hemingway, J.E. and Black, M. (1934) A synopsis of the Jurassic rocks of Yorkshire. *Proceedings of the Geologists' Association*, **45**: 247–91.

Windsor, D.E. and Dagg, A.I. (1971) The gaits of the Macropodinae (Marsupialia). *Journal of Zoology, London*, **163**: 165–75.

Wing, S.L. and Tiffney, B.H. (1987) The reciprocal interaction of angiosperm evolution and tetrapod herbivory. *Review of Palaeobotany and Palynology*, **50**: 179–210.

Winkler, T.C. (1886) Histoire de l'Ichnologie; étude ichnologique sur les empreintes de pas d'animaux fossiles. *Archives du Musée Teyler*, sèrie 2, **2**: 240–440.

Woodhams, K.E. and Hines, J.S. (1989) Dinosaur footprints from the Lower Cretaceous of East Sussex, England. In *Dinosaur Tracks and Traces* (eds D.D. Gillette and M.G. Lockley). Cambridge University Press, Cambridge, pp. 301–7.

Woodworth, J.B. (1895) Three-toed dinosaur tracks in the Newark Group at Avondale, N.J. *American Journal of Science*, series 3, **50**: 481–2.

Wrather, W.E. (1922) Dinosaur tracks in Hamilton County, Texas. *Journal of Geology*, **30**: 354–60.

Wuest, W. (1934) Reiherspuren mit anschliessenden Trockenrissen. *Natur und Volk*, **64**: 73–4.

Wyckoff, R.W.G. and Davidson, F.D. (1979) Fossil reptilian gelatins. *Comparative Biochemistry and Physiology*, series B, **64**: 229–30.

Yabe, H., Inai, Y. and Shikama, T. (1940a) [Dinosaurian footprints found near Yangshan, Chinchou, Manchoukuo.] *Journal of the Geological Society of Japan*, **47**: 169–70 [in Japanese with English summary].

Yabe, H., Inai, Y. and Shikama, T. (1940b) Discovery of dinosaurian footprints from the Cretaceous (?) of Yangshan, Chinchou. *Proceedings of the Imperial Academy of Japan*, **16**: 560–3.

Yalden, D.W. (1984) What size was *Archaeopteryx*? *Zoological Journal of the Linnean Society*, **82**: 177–88.

Yang, S.Y. (1982) On the dinosaur's footprints from the Upper Cretaceous Gyeongsang Group, Korea. *Journal of the Geological Society of Korea*, **18**: 37–48 [in Korean with English summary].

Yang, S.Y. (1986) [Dinosaur footprints in Korea.] *Earth Monthly*, **8**(3): 138–42 [in Japanese].

Young, C.-C. (1947) On *Lufengosaurus magnus* (sp. nov.) and additional finds of *Lufengosaurus huenei* Young. *Palaeontologia Sinica*, **12**: 1–53.

Young, C.-C. (1960) Fossil footprints in China. *Vertebrata Palasiatica*, **4**: 53–66.

Young, C.-C. (1965) Fossil eggs from Nanhsiung, Kwangtung and Kanchou, Kiangsi. *Vertebrata Palasiatica*, **9**: 141–70 [in Chinese and English].

Young, C.-C. (1966) Two footprints from the Jiaoping coal mine of Tungchuan, Shensi. *Vertebrata Palasiatica*, **10**: 68–72 [in Chinese and English].

Young, C.-C. (1979) Footprints from Luanping, Hebei. *Vertebrata Palasiatica*, **17**: 116–7 [in Chinese and English].

Zakharov, S.A. (1964) [On a Cenomanian dinosaur, whose tracks were found in the Shirkent River Valley.] In [*Palaeontology of Tadzhikistan*] (ed. V.M. Reiman). Akademiya Nauk Tadzhikskoi SSR, Dushanbe, pp. 31–5 [in Russian with English summary].

Zakharov, S.A. and Khakimov, F. (1963) [On footprints of a Cenomanian dinosaur in western Tadzhikistan.] *Doklady Akademii Nauk Tadzhikskoi SSR*, **6**(9): 25–7 [in Russian].

Zhao, Z. (1979) Discovery of the dinosaurian eggs and footprint from Neixiang County, Henan Province. *Vertebrata Palasiatica*, **17**: 304–9 [in Chinese with English summary].

Zhen, S., Li, J., Rao, C. and Hu, S. (1986) Dinosaur footprints of Jinning, Yunnan. *Memoirs of the Beijing Natural History Museum*, **33**: 1–19 [in Chinese with English summary].

Zhen, S., Li, J. and Zhen, B. (1983) Dinosaur footprints of Yuechi, Sichuan. *Memoirs of the Beijing Natural History Museum*, **25**: 1–19 [in Chinese with English summary].

Zhen, S., Li, J., Rao, C., Mateer, N.J. and Lockley, M.G. (1989). A review of dinosaur footprints in China. In *Dinosaur Tracks and Traces* (eds D.D. Gillette and M.G. Lockley). Cambridge University Press, Cambridge, pp. 188–97.

Zhen, S., Zhen, B., Mateer, N.J. and Lucas, S.G. (1985) The Mesozoic reptiles of China. *Bulletin of the Geological Institutions of the University of Uppsala*, n.s., **11**: 133–50.

Zweifel, F.W. (1961) *A Handbook of Biological Illustration*. University of Chicago Press, Chicago.

Index

Names of ichnotaxa are in italics, without comment (e.g. *Anchisauripus*). Names of dinosaurs and other animals are distinguished by comment in parentheses, thus: *Anchisaurus* (prosauropod). Bold type indicates pages with illustrations.